# Contemporary Solutions for Advanced Catalytic Materials with a High Impact on Society

# Contemporary Solutions for Advanced Catalytic Materials with a High Impact on Society

Editors

**Simona M. Coman**
**Madalina Tudorache**
**Elisabeth Egholm Jacobsen**

MDPI • Basel • Beijing • Wuhan • Barcelona • Belgrade • Manchester • Tokyo • Cluj • Tianjin

*Editors*
Simona M. Coman
Organic Chemistry,
Biochemistry and Catalysis
University of Bucharest
Bucharest
Romania

Madalina Tudorache
Organic Chemistry
Biochemistry and Catalysis
University of Bucharest
Bucharest
Romania

Elisabeth Egholm Jacobsen
Department of Chemistry
Norwegian University of
Science and Technology
Trondheim
Norway

*Editorial Office*
MDPI
St. Alban-Anlage 66
4052 Basel, Switzerland

This is a reprint of articles from the Special Issue published online in the open access journal *Catalysts* (ISSN 2073-4344) (available at: www.mdpi.com/journal/catalysts/special_issues/mat_catal).

For citation purposes, cite each article independently as indicated on the article page online and as indicated below:

LastName, A.A.; LastName, B.B.; LastName, C.C. Article Title. *Journal Name* **Year**, *Volume Number*, Page Range.

**ISBN 978-3-0365-5892-9 (Hbk)**
**ISBN 978-3-0365-5891-2 (PDF)**

© 2023 by the authors. Articles in this book are Open Access and distributed under the Creative Commons Attribution (CC BY) license, which allows users to download, copy and build upon published articles, as long as the author and publisher are properly credited, which ensures maximum dissemination and a wider impact of our publications.

The book as a whole is distributed by MDPI under the terms and conditions of the Creative Commons license CC BY-NC-ND.

# Contents

Preface to "Contemporary Solutions for Advanced Catalytic Materials with a High Impact on Society" . . . . . . . . . . . . . . . . . . . . . . . . . . . . . . . . . . . . . . . . . . . . . . . . . . . . . . . vii

Magdi El Fergani, Natalia Candu, Iunia Podolean, Bogdan Cojocaru, Adela Nicolaev and Cristian M. Teodorescu et al.
Catalytic Hydrotreatment of Humins Waste over Bifunctional Pd-Based Zeolite Catalysts
Reprinted from: *Catalysts* **2022**, *12*, 1202, doi:10.3390/catal12101202 . . . . . . . . . . . . . . . . . . . 1

Sabina G. Ion, Octavian D. Pavel, Nicolae Guzo, Madalina Tudorache, Simona M. Coman and Vasile I. Parvulescu et al.
Use of Photocatalytically Active Supramolecular Organic–Inorganic Magnetic Composites as Efficient Route to Remove β-Lactam Antibiotics from Water
Reprinted from: *Catalysts* **2022**, *12*, 1044, doi:10.3390/catal12091044 . . . . . . . . . . . . . . . . . . . 19

Susanne Hansen Trøøyen, Lucas Bocquin, Anna Lifen Tennfjord, Kristoffer Klungseth and Elisabeth Egholm Jacobsen
Green Chemo-Enzymatic Protocols for the Synthesis of Enantiopure β-Blockers (S)-Esmolol and (S)-Penbutolol
Reprinted from: *Catalysts* **2022**, *12*, 980, doi:10.3390/catal12090980 . . . . . . . . . . . . . . . . . . . . 33

Anna Kozioł, Kristofer Gunnar Paso and Stanisław Kuciel
Properties and Recyclability of Abandoned Fishing Net-Based Plastic Debris
Reprinted from: *Catalysts* **2022**, *12*, 948, doi:10.3390/catal12090948 . . . . . . . . . . . . . . . . . . . . 45

Elena E. Toma, Giuseppe Stoian, Bogdan Cojocaru, Vasile I. Parvulescu and Simona M. Coman
ZnO/CQDs Nanocomposites for Visible Light Photodegradation of Organic Pollutants
Reprinted from: *Catalysts* **2022**, *12*, 952, doi:10.3390/catal12090952 . . . . . . . . . . . . . . . . . . . . 61

Chong Chen, Yun Li and Jilin Cao
Methane Hydrate Formation in Hollow ZIF-8 Nanoparticles for Improved Methane Storage Capacity
Reprinted from: *Catalysts* **2022**, *12*, 485, doi:10.3390/catal12050485 . . . . . . . . . . . . . . . . . . . . 79

Serly Jolanda Sekewael, Remi Ayu Pratika, Latifah Hauli, Amalia Kurnia Amin, Maisari Utami and Karna Wijaya
Recent Progress on Sulfated Nanozirconia as a Solid Acid Catalyst in the Hydrocracking Reaction
Reprinted from: *Catalysts* **2022**, *12*, 191, doi:10.3390/catal12020191 . . . . . . . . . . . . . . . . . . . . 93

Honghai Wang, Wenda Yue, Shuling Zhang, Yu Zhang, Chunli Li and Weiyi Su
Modification of Silica Xerogels with Polydopamine for Lipase B from *Candida antarctica* Immobilization
Reprinted from: *Catalysts* **2021**, *11*, 1463, doi:10.3390/catal11121463 . . . . . . . . . . . . . . . . . . . 117

Giulia Roxana Gheorghita, Victoria Ioana Paun, Simona Neagu, Gabriel-Mihai Maria, Madalin Enache and Cristina Purcarea et al.
Cold-Active Lipase-Based Biocatalysts for Silymarin Valorization through Biocatalytic Acylation of Silybin
Reprinted from: *Catalysts* **2021**, *11*, 1390, doi:10.3390/catal11111390 . . . . . . . . . . . . . . . . . . . 133

**Morten Andre Gundersen, Guro Buaas Austli, Sigrid Sløgedal Løvland, Mari Bergan Hansen, Mari Rødseth and Elisabeth Egholm Jacobsen**
Lipase Catalyzed Synthesis of Enantiopure Precursors and Derivatives for -Blockers Practolol, Pindolol and Carteolol
Reprinted from: *Catalysts* **2021**, *11*, 503, doi:10.3390/catal11040503 . . . . . . . . . . . . . . . . . . **147**

# Preface to "Contemporary Solutions for Advanced Catalytic Materials with a High Impact on Society"

"Contemporary solution for advanced materials with high impact on society"is a Special Issue dedicated to a workshop with a similar name (with the acronym CoSolMat) organized on the 11th–15th of October 2021 within the GREENCAM project (project no. 18-COP-0041, October 2019–September 2022, project partners University of Bucharest (UB, Bucharest, Romania), and Norwegian University of Science and Technology (NTNU, Trondheim, Norway)).

The GREENCAM project proposes an alliance between Green Chemistry and Advanced Materials in response to the contemporary issues related to the environmental pollution and eco-friendly industry, which emerged from the general concept of environmental preservation. In this way, GREENCAM offers a proper condition for the development of an extensive study in the area of advanced (e.g. nano-) materials with a critical awareness of the theoretical and practical aspects related to the modern concepts of green (nano)synthesis. GREENCAM advocates for the extensive application of the principles of green chemistry in chemical synthesis, especially for the preparation of these advanced materials as a promising perspective for an eco-friendly industry.

Based on the continuous advancements in the area of advanced materials and, as a consequence, the new challenges to be faced, a focus on this matter is highly desirable. UB, represented by Prof. Simona M. Coman and Assoc. Prof. Madalina Tudorache in collaboration with the NTNU represented by Assoc. Prof. Elisabeth Jacobsen launched the idea for collecting valuable publications dedicated to contemporary solutions for issues related to the preparation/characterization and applications of the advanced materials with a high impact on society. The main aim of this Special Issue is to highlight novel developed strategies designed for promoting Green Advanced Materials (GAM), i.e., new concepts and strategies for designing GAMs together with their most important application for improving the health and security of our present-day society. Teachers and researchers from academic and industrial areas with expertise in the GAMs field have offered their contributions. Furthermore, young researchers are well represented, also demonstrating their interests in this research direction. Therefore, this Special Issue has collected original topics, most of which were shared during the CoSolMat workshop, providing new insights into the GAMs field.

We conclude by hoping that this Special Issue will act in the future as a manifesto for the mandatory implementation of GAMs and Green Chemistry in order to ensure the industrial practices required for a healthy and secure society.

Disclaimer: This publication was realized with the financial support of the EEA Financial Mechanism 2014-2021. Its content (text and photos) does not reflect the official opinion of the Programme Operator, the National Contact Point, or the Financial Mechanism Office. The responsibility for the information and views expressed therein lies entirely with the author(s).

**Simona M. Coman, Madalina Tudorache, and Elisabeth Egholm Jacobsen**
*Editors*

Article

# Catalytic Hydrotreatment of Humins Waste over Bifunctional Pd-Based Zeolite Catalysts

Magdi El Fergani [1], Natalia Candu [1], Iunia Podolean [1], Bogdan Cojocaru [1], Adela Nicolaev [2], Cristian M. Teodorescu [2], Madalina Tudorache [1], Vasile I. Parvulescu [1] and Simona M. Coman [1,*]

[1] Department of Organic Chemistry, Biochemistry and Catalysis, Faculty of Chemistry, University of Bucharest, Bdul Regina Elisabeta 4-12, 030016 Bucharest, Romania
[2] National Institute of Materials Physics, Atomistilor 405b, 077125 Magurele-Ilfov, Romania
* Correspondence: simona.coman@chimie.unibuc.ro

**Abstract:** The catalytic hydrotreatment of humins, the solid byproduct produced from the conversion of C6 sugars (glucose, fructose) to 5-hydroxymethylfurfural (HMF), using supported Pd@zeolite (Beta, Y, and USY) catalysts with different amounts of Pd (i.e., 0.5, 1.0 and 1.5 wt%) was investigated under molecular hydrogen pressure. The highest conversion of humins (52.0%) was obtained on 1.5Pd@USY catalyst while the highest amount of humins oil (27.3%) was obtained in the presence of the 1Pd@Beta zeolite sample, at $P_{H_2}$ = 30 bars and T = 250 °C. The major compounds in the humins oil evidenced by GC-MS are alcohols, organic acids, ethers, and alkyl-phenolics. However, although all these classes of compounds are obtained regardless of the nature of the catalyst used, the composition of the mixture differs from one catalyst to another. Furanic compounds were not identified in the reaction products. A possible explanation may be related to their high reactivity under the reaction conditions, in the presence of the Pd-based catalysts these compounds lead to alkyl phenolics, important intermediates in the petrochemical industry.

**Keywords:** palladium; zeolites; bifunctional catalysts; humins; hydrotreatment; biobased compounds

## 1. Introduction

In the last number of decades, significant attention has been paid to the utilization of renewable biomass platform molecules as promising carbon resources that might partly be used as a substitute for fossil fuels. Furan-derivative compounds, for instance, are used in the production of a wide range of chemicals, including 2,5-dimethylfuran, furanedioic acid, pentanoic acid esters, levulinic acid (LA) and its esters, and γ-valerolactone [1,2]. However, due to their high reactivity, the furan-derivative compounds also tend to polymerize in the acid-catalyzed reactions, resulting in large amounts of undesirable polyfuranic polymer by-products (i.e., humins) which severely limit an efficient utilization of the renewable biomass [3,4]. To overcome this inconvenience, the formation of humins should be suppressed by developing efficient catalytic systems able to convert the biomass in a selective way to the targeted products. However, such an objective is difficult to achieve since the formation of humins is favored from the thermodynamic point of view. Alternatively, the humins formed during such processes should be valorized and several reports indicate their potential in the fabrication of materials such as composites [5], catalysts [6], or functional carbon materials [7].

The chemical structure of humins also recommends them as an important source for the production of chemicals. Thus, several reports already suggested humins as a proper feedstock for the production of important compounds such as hydrogen and synthesis gas [8], alkyl phenolics and oligomers (following a depolymerization and hydrodeoxygenation tandem) [9], easily transportable fuels with a high energy density or low molecular weight compounds (such as intermediates for the production of high added value bulk

chemicals like acetic and formic acids) [10]. In most of these approaches, the catalytic alternative provides important advantages [11–13].

However, in spite of its socio-economic importance, the valorization of humins via depolymerization and subsequent hydrodeoxygenation to liquid hydrocarbons is still in the infancy stage. In this context, Ru/C [9] and Pt/C [14] were reported as highly efficient catalysts for the liquefaction of humins, affording conversions of 60–70% at 400 °C, in the presence of isopropanol (IPA) as a hydrogen donor. Other studies carried out at the same temperature investigated the hydrodeoxygenation of fructose-derived humins in methanol, under $H_2$ (30 bars), taking Ru/C, Rh/C, Pt/C, and Pd/C as catalysts. Among these, the conversion reached 75% on Rh/C [15]. The replacement of C with supports such as $TiO_2$, $ZrO_2$, and $CeO_2$ led only to a small increase in conversion (around 80%) [16]. Very recent reports also indicated the high efficiency of a more complex Ru/W-P-Si-O bifunctional catalyst in the selective production of cyclic and aromatic hydrocarbons from humins under mild reaction conditions [17].

Inspired by this state of the art, the aim of this work was to develop an efficient catalytic system for the humins hydrodeoxygenation to functionalized low molecular mass organic molecules (such as alcohols, ketones, and phenolics) that may serve as a source for valuable bio-based chemicals. With this scope, the investigations focused on hierarchical tailored bifunctional Pd@zeolites (Beta, Y, and USY) with enhanced catalytic performances.

## 2. Results and Discussion

The structure of humins is still incompletely understood. However, for humins prepared from C6 sugars, the characterization studies indicated a core-shell morphology with a furan-rich structure containing ether and (hemi) acetal linkages [18–21]. The physico-chemical characterization of the synthesized humins [19] following a hydrothermal treatment of glucose indicated a chemical structure with a morphology in line with these literature studies [18,20,21]. It corresponded to spherical shell-core interconnected particles with a structure consisting of a furan-rich polymeric network linked by functionalized aliphatic chains with aldehyde and beta-hydroxyacids [19,21] (Figure 1).

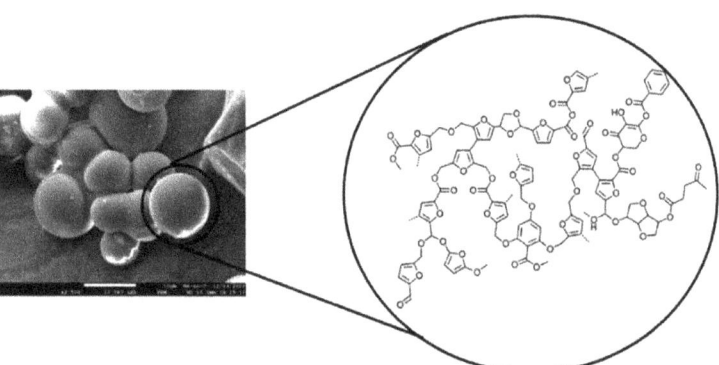

**Figure 1.** The morphology and molecular structure of humins obtained from C6 sugars (Adapted with permission from Ref [19,21], Elsevier, 2022).

### 2.1. Physico-Chemical Characterization of Pd@zeolite Catalysts

Zeolite-supported metal species are usually prepared via a simple impregnation approach [22,23]. However, the resulting catalysts are complex and this complexity is influenced by several factors, such as the zeolite framework, $SiO_2/Al_2O_3$ ratio, the type of exchanged cation, and the strength of the acid sites [24]. This work considered zeolites in the H-form with different $SiO_2/Al_2O_3$ ratios (i.e., 25.0 (H-Beta), 5.2 (H-Y), and 30.0 (H-USY), and different textural properties.

N₂ adsorption-desorption isotherms of liquid nitrogen at −196 °C and the pore size distribution (determined from the BJH method) are shown in Figures 2 and 3, while the textural features of the prepared catalysts and supports (H-Beta, H-Y, and H-USY) are listed in Table 1.

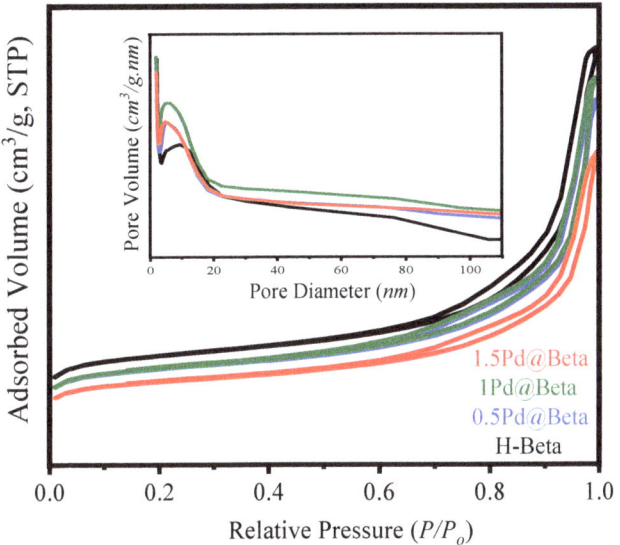

**Figure 2.** N₂ adsorption-desorption isotherms of the H-Beta, 0.5Pd@Beta, 1Pd@Beta, and 1.5Pd@Beta. The pore size distribution is given in the inset.

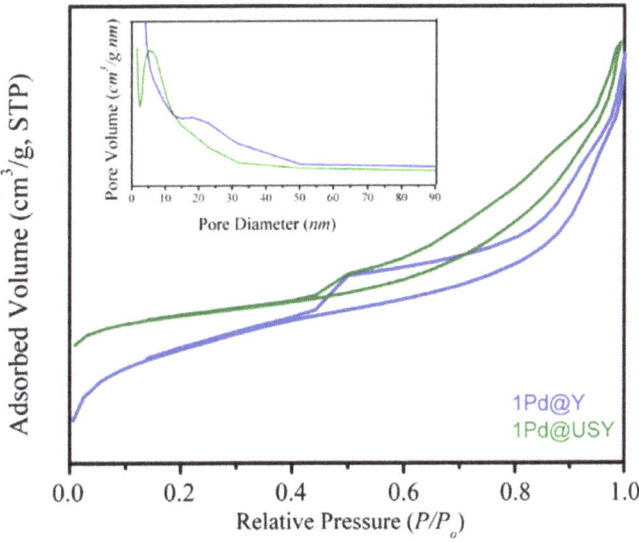

**Figure 3.** N₂ adsorption-desorption isotherms of the 1Pd@Y and 1Pd@USY samples. The pore size distribution is given in the inset.

Table 1. Physico-chemical characterization of the catalytic samples.

| Entry | Catalyst | $S_{BET}$ (m$^2$/g) [a] | $S_{ext}$ (m$^2$/g) [b] | $S_{micro}$ (m$^2$/g) | $V_{total}$ (cm$^3$/g) [c] | $V_{meso}$ (cm$^3$/g) [d] | $V_{micro}$ (cm$^3$/g) [e] |
|---|---|---|---|---|---|---|---|
| 1 | 0.5Pd@Beta | 500 | 177 | 323 | 0.65 | 0.50 | 0.15 |
| 2 | 1Pd@Beta | 504 | 183 | 321 | 0.70 | 0.55 | 0.15 |
| 3 | 1.5Pd@Beta | 433 | 153 | 280 | 0.56 | 0.43 | 0.13 |
| 4 | 1Pd@USY | 668 | 174 | 494 | 0.45 | 0.22 | 0.23 |
| 5 | 1Pd@Y | 352 | 79 | 273 | 0.31 | 0.18 | 0.13 |

[a]—calculated by the BET method; [b]—external surface area calculated using the t-plot method; [c]—the total pore volume determined at a relative pressure (P/P$_0$) of 0.98; [d]—the mesopores volume calculated using the BJH method; [e]—the micropores volume calculated using the t-plot method.

The H-Beta zeolite supported catalysts presented Type IV isotherms, according to the IUPAC classification, and a H4 hysteresis loop at relatively pressures P/P$_0$ of 0.6–1.0 (Figure 2) with a bimodal micro-mesoporosity (Figure 2, inset). In accordance with previous works [25], the mesoporosity of Beta zeolite is the effect of the packing of small zeolite nanocrystals.

The Pd@Y and Pd@USY samples presented a combination of Type I and Type IV isotherms, with the appearance of a micropore filling at low relative pressures (i.e., <0.1) and a hysteresis loop at a relative pressure of 0.45–0.99. This indicates a hierarchical porous system combining micro- and mesoporosity. However, there are differences in the size of the pores, namely, 6.5 nm for 1Pd@Y and 20 nm for 1Pd@USY. The isotherms also showed a sharp rise in adsorbed amount near saturation (P/P$_0$ of 1.0) which is associated with condensation in the inter-particle voids (macropores). The presence of the macropores (cracks and voids) in the two CBV zeolites (i.e., H-Y CBV 600 and H-USY CBV 720) is caused by the steam and acid leaching treatments [26,27].

While the BET surface areas of Beta zeolite-based samples decreased from 680 m$^2$/g for H-Beta, to 504–433 m$^2$/g for Pd@Beta samples (Table 1, entries 1–3), the surface areas of Y- and USY-based catalysts decreased from 780 m$^2$/g to 668 m$^2$/g (1Pd@USY) and from 660 m$^2$/g to 352 m$^2$/g (1Pd@Y), respectively. The $S_{micro}$ decrease paralleled $S_{BET}$ values indicating some micropore blockages due to Pd deposition on the pore mouths during the reduction process [28,29].

The crystalline structure of the zeolite supports is well preserved after its impregnation with the palladium salt (Figures 4–6). In the case of Beta zeolite, the wider full widths at half-maxima of the diffraction peaks at 2θ of 7.79 and 22.53° (indicated with an asterisk in Figure 4) confirm the small crystal size of the employed Beta zeolite [30] and the existence of mesopores as an effect of their packing. The absence of the diffraction lines corresponding to the tetragonal PdO phase (i.e., 2θ angles at 33.6° (002), 42.0° (110), 54.9° (112), 60.3° (103), and 71.7° (211)) and/or metallic Pd (i.e., 2θ angles at 39.8° (111), 46.2° (200), 67.6° (220), and 81.4° (311) (JCPDS no. 46-1043)) [31] also indicates that the size of the PdO and Pd particles is small enough to not be detected by XRD, yielding a considerably high dispersion [29].

The chemical oxidation state of palladium, silicon, and aluminum was determined by XPS for the catalysts in the reduced form. The HR-XPS spectra are given in Figures S1–S4 and the spectral data in Tables S1–S3.

In all the samples, palladium mainly (80–83%) corresponds to a metallic Pd$^0$ state (band located at 335.1–335.2 eV, Figures S1–S4 and Table S1) [32,33]. The difference of 16–19% is associated with PdO clusters (band located at 336.8–337.0 eV) [33,34].

The band corresponding to the Si2p level has been deconvoluted into two components (Figures S1–S4, Table S2) assigned to different ionic bonds [35]. For the Pd@USY catalyst, the main component at 102.5 eV (Table S2) corresponds to silicon in [SiO$_4$$^-$] (i.e., 103.2 eV) in the vicinity of a distorted [AlO$_4$$^-$] component [36] with different coordination numbers [35]. The band at higher binding energy (103.8–105.0 eV) is associated with [SiO$_4$$^-$] components located in the proximity of the defected sites [33].

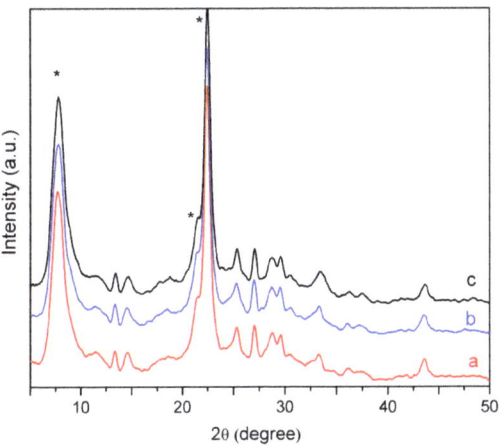

**Figure 4.** XRD patterns for Beta zeolite (**a**), 0.5Pd@Beta (**b**), and 1Pd@Beta (**c**) samples.

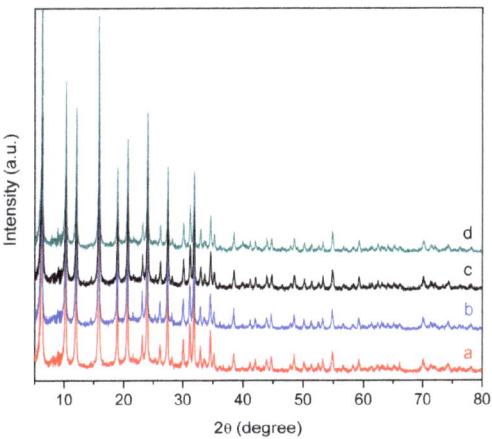

**Figure 5.** XRD patterns for H-Y zeolite (**a**), 0.5Pd@Y (**b**), 1Pd@Y (**c**), and 1.5Pd@Y (**d**) samples.

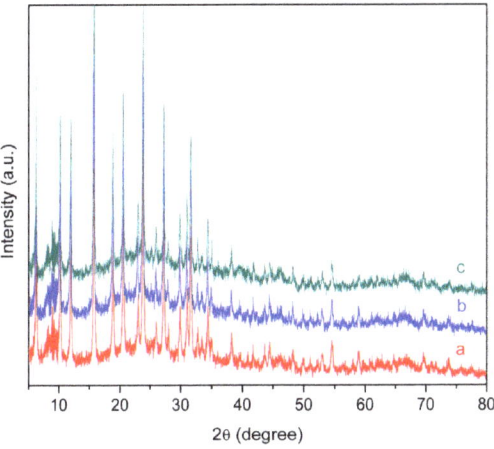

**Figure 6.** XRD patterns for H-USY zeolite (**a**), 0.5Pd@USY (**b**), and 1.5Pd@USY (**c**) samples.

Finally, the band associated with the Al 2p level has also been deconvoluted into two components located at 74.6–74.8 eV that were associated with aluminum in a III-fold state (Figures S1–S4, Table S3) and at 75.5–75.8 eV assigned to Al 2p in distorted [$AlO_4^-$] units for Pd@Beta, and at 73.7–73.8 eV to Al 2p in IV- and VI-fold sites with or without a proton for Pd@Y and Pd@USY [33].

Table 2 compiles the atomic composition determined from the deconvolution normalized by taking the atomic sensitivity factors [37].

Table 2. The main elements and the atomic composition for each sample.

| Sample | Si/Al Bulk | O 1s (%) | Si 2p (%) | Al 2p (%) | Si/Al XPS | Pd 3d5/2 (at%) | $Pd^0/Pd^{n+}$ |
|---|---|---|---|---|---|---|---|
| 1Pd@Beta | 12.5 | 62.07 | 34.9 | 2.95 | 11.8 | 0.08 | 4.8 |
| 1.5Pd@Beta | 12.5 | 62.38 | 34.82 | 2.66 | 13.1 | 0.14 | 4.1 |
| 1Pd@USY | 15.0 | 59.49 | 30.00 | 10.39 | 2.9 | 0.11 | 4.6 |
| 1Pd@Y | 2.6 | 63.58 | 35.11 | 1.23 | 28.5 | 0.07 | 5.1 |

For all the samples, the depletion of the palladium to aluminum ratio may confirm its agglomeration on the zeolite surface, most probably due to a micropore blockage. On the other hand, the reduction of the samples produced an enrichment of the surface in Al for Pd@USY, while for Pd@Y, a depletion of Al was observed. Aluminum enrichment of the surface for Pd@Y is accompanied by an increase in the defects in the [$SiO_4^-$] zeolite framework (Table S2). For Pd@Beta, the surface Si/Al ratio is close to the bulk composition (Table 2).

Figures S5–S7 show the results of thermogravimetric and differential thermal (TG-DTA) analyses for the decomposition of palladium acetate deposited on zeolites, in a nitrogen atmosphere. The first loss of mass, accompanied by an endothermic peak below 100 °C, can be associated with the elimination of water from the zeolite channels while the second mass loss, accompanied by an exothermic effect at 280–330 °C, by the decomposition of palladium acetate [38]. However, as Figure S7 showed, the TG-DTA indicates a different decomposition profile of the Pd precursor. Along with the elimination of water, this is evidence of the other two decomposition steps of the palladium precursor (exothermic peaks at 230–240 °C and 440–460 °C, respectively), thus suggesting a different interaction with the zeolite carrier. As XRD and XPS analysis showed, for all the samples, the reduction step led to highly dispersed metallic $Pd^0$ particles (preponderantly) and PdO clusters located in the depth of the carrier.

$H_2$-TPR allowed establishing the reducibility of the palladium species as an important element for the catalytic hydrotreatment of humins. The $H_2$-TPR profiles of the Pd@zeolites are displayed in Figure 7. In accordance with the literature, the consumed hydrogen corresponds to the reduction of PdO to metallic Pd [39].

As Figure 7 shows, the TPR profiles display at least two maxima at different temperatures, depending on the palladium loading and zeolite nature. The observed differences indicate the presence of different types of palladium species. For the case of Pd@Beta, the loading of palladium was higher, and the temperature corresponding to the $H_2$-TPR peak was higher, i.e., 235 °C (1.5Pd@Beta) > 180 °C (1Pd@Beta) > 160 °C (0.5Pd@Beta). Typically, such a variation indicates the formation of larger PdO nanoparticles for 1.5Pd@Beta and narrower particles for 0.5Pd@Beta, the former more difficult to reduce [40–43]. Indeed, the highest $H_2$ consumption was registered for the 0.5Pd@Beta (2.84 mmol/g) followed by 1Pd@Beta (0.74 mmol/g) and 1.5Pd@Beta (0.06 mmol/g). No $H_2$-TPR peak was observed at a low temperature (<100 °C) indicating a lack of PdO reduction in this range.

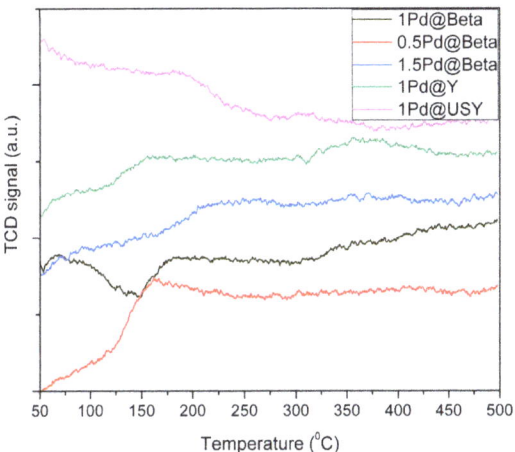

**Figure 7.** H$_2$-TPR profiles for Pd@zeolite samples.

On the other hand, for the same loading of palladium (i.e., 1 wt%) the TPR profiles indicate the presence of peaks at temperatures depending on the Si/Al ratio of the pristine zeolite. Therefore, the higher the content of aluminum in the zeolite framework, the lower the reduction temperature of the PdO particles, i.e., 190 °C (1Pd@USY, Si/Al = 15) > 180 °C (1Pd@Beta, Si/Al = 12.5) > 157 °C (1Pd@Y, Si/Al = 2.6), and the corresponding H$_2$ consumption: 0.91 mmol/g (1Pd@USY) > 0.74 mmol/g (1Pd@Beta) > 0.46 mmol/g (1Pd@Y). An additional peak in the TPR profiles has been determined in the range of 300–400 °C, which can be attributed to the presence of PdO nanoclusters inside the support micropores [40].

The DRIFT spectra collected for both non-calcined and the final catalysts (Figures 8–11) showed specific spectral characteristics (internal vibration of the framework TO$_4$ tetrahedron and vibration related to external linkages between tetrahedral units) of these zeolites [44]. All zeolites display a broad band in the 3000–3400 cm$^{-1}$ region attributed to the O–H stretching of hydrogen-bonded internal silanol groups and hydroxyl stretching of water, and a band at around 1630–1640 cm$^{-1}$ corresponding to the O–H bending mode of water. The bands in the 1017–722 cm$^{-1}$ range are attributed to the symmetric and asymmetric stretching vibrations of the Si–O–T linkages for TO$_4$ (T = Si and/or Al), respectively). Besides these, slight structural differences were detected through the asymmetry of the bands.

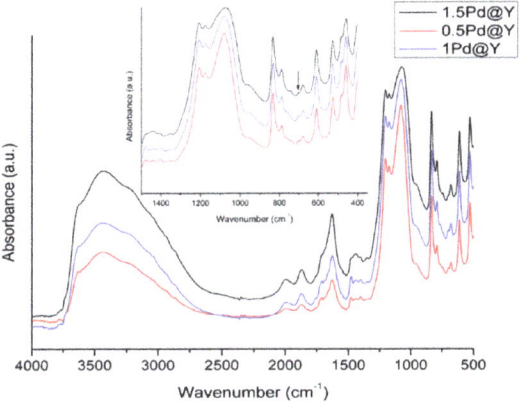

**Figure 8.** IR spectra of the non-calcined 0.5Pd@Y, 1Pd@Y, and 1.5Pd@Y samples and an inset in the 1500–400 cm$^{-1}$ range.

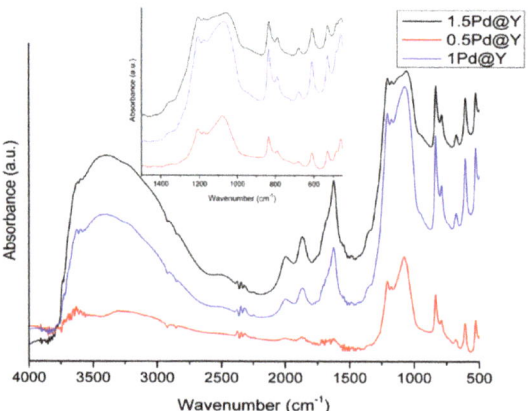

**Figure 9.** DRIFT spectra of the 0.5Pd@Y, 1Pd@Y, and 1.5Pd@Y and an inset in the 1500–400 cm$^{-1}$ range.

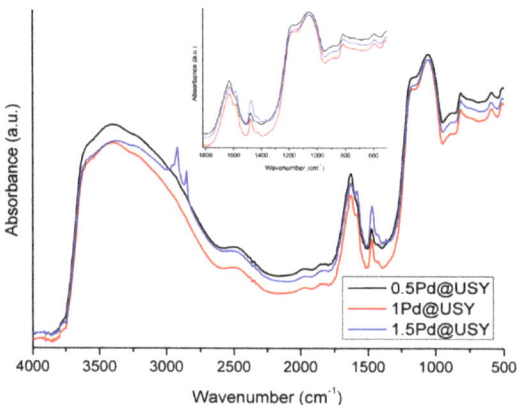

**Figure 10.** DRIFT spectra of the non-calcined 0.5Pd@USY, 1Pd@USY, and 1.5Pd@USY and an inset in the 1800–500 cm$^{-1}$ range.

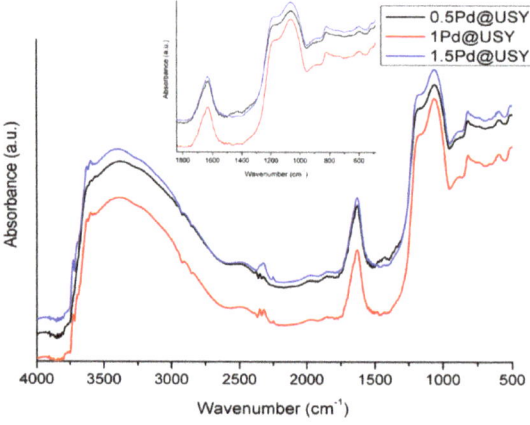

**Figure 11.** DRIFT spectra of the 0.5Pd@USY, 1Pd@USY, and 1.5Pd@USY and an inset in the 1800–500 cm$^{-1}$ range.

The DRIFT spectra for Pd@Y samples (Figures 8 and 9) display bands characteristic of Y zeolite [44]. The bands in the 1380–1530 cm$^{-1}$ range are associated with the $\nu$CH$_3$ deformation. The $\nu$C-O stretching vibrations of the Pd precursor are not visible in the DRIFT spectra of the final catalysts supporting its anchorage (Figure 9). In accordance with Banse and Koel [45], the disappearance of the band at 697 cm$^{-1}$, ascribed to the Pd−O bonds, also validates the formation of the metallic Pd nanoparticles (Figure 9).

H-Y materials typically display an acidic OH band in the vicinity of 3600 cm$^{-1}$ [26]. Therefore, the two bands at 3600 and 3625 cm$^{-1}$ are assigned to Si-O(H)-Al [25] or the OH groups located at the places left by the more easily removable framework aluminum atoms (Figure 9) [46].

For H-USY, the bands located at 1188, 1064, and 824 cm$^{-1}$ (Figures 10 and 11) are characteristic of TO$_4$ tetrahedron units and are attributed to the external asymmetric, internal asymmetric, and external symmetric stretching vibrations of Si–O–T linkages (T = Si and/or Al) [47]. For non-calcined Pd@USY, the bands in the range of 1380–1530 cm$^{-1}$ are associated with the $\nu$CH$_3$ deformation and $\nu$C-O stretching vibrations from the Pd precursor. For the Pd@USY catalysts, these bands vanished due to the effect of the calcination and reduction steps. The broad absorption band in the region 3600–3200 cm$^{-1}$ is assigned to the Si-OH groups interacting with each other through H-bonds. This large absorption band is also indicative of the high surface defectives. Additional bands were also registered at 3720 cm$^{-1}$ and are most probably related to the presence of the Pd(II)-O-H groups (Figure 11) confirming the results of the XPS analysis.

As expected, the NH$_3$-TPD analysis revealed the acidic properties of the Pd@zeolite catalysts (Figure 12 and Table 3). The NH$_3$-TPD profiles (Figure 12) showed at least three maxima in the temperature ranges of 50–150, 150–250, and 250–400 °C, respectively. The peaks in the region of 50–150 °C correspond to the weak adsorption of the ammonia molecules over the surface terminal silanols (Si–OH) [48] while those at 250–400 °C can be assigned to the surface Brønsted acid sites, namely, the bridged hydroxyl group (Si–OH–Al) [49]. The extra-framework Al species indicated by peaks in the 150–250 °C range behave as Lewis acid sites with a weaker strength than the bridged hydroxyl groups [49].

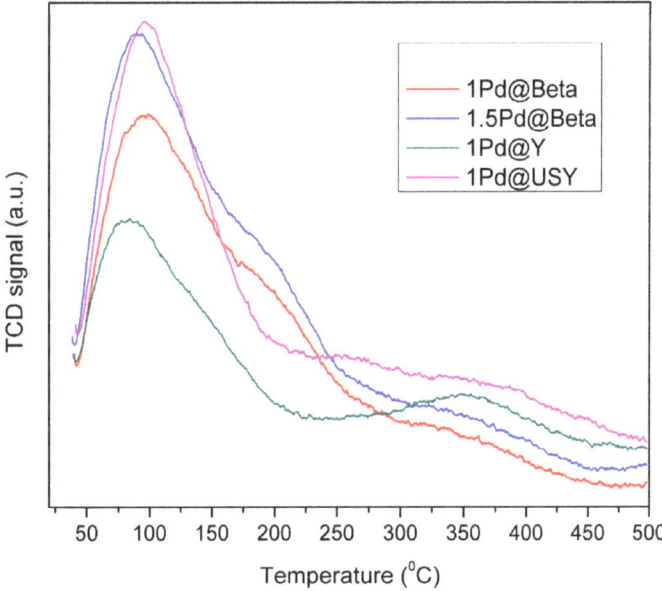

**Figure 12.** Temperature-programmed desorption profiles of ammonia for the investigated Pd@zeolites.

Table 3. Acid site concentration determined through NH3-TPD measurements.

| Catalyst | Acid Site Concentration (µmols NH$_3$/g) | | | | L/B Ratio |
|---|---|---|---|---|---|
| | 50–150 °C | 150–250 °C | 250–400 °C | Total | |
| H-Beta | 66.0 (<300 °C) | | 4.0 (>300 °C) | 70.0 | - |
| 0.5Pd@Beta | 85.6 | 19.9 | 1.01 | 106.51 | 19.7 |
| 1Pd@Beta | 91.51 | 20.08 | 1.81 | 113.40 | 11.1 |
| 1.5Pd@Beta | 117.97 | 10.57 | 3.70 | 132.24 | 2.9 |
| 1 Pd@Y | 54.27 | - | 14.92 | 69.19 | 0 |
| 1Pd@USY | 110.49 | 0.12 (260 °C) | 2.97 | 113.58 | 0.04 |

As Table 3 shows, for the Pd@Beta catalysts, the concentration of the surface Brønsted acid site (250–400 °C) decreased after the palladium deposition to a concentration of 1 wt%, most probably due to the Pd ion exchange which occurs with the replacement of the surface proton from Si–OH–Al. However, increasing the amount of the palladium from 1.0 wt% to 1.5 wt% led to an increase of the Brønsted acidity which, most probably, is related to the presence of the Pd(II)-O-H groups, also in accordance with XPS. In turn, the concentration of the Lewis acid sites (150–250 °C) increased greatly, most probably due to the presence of a high concentration of the well-dispersed PdO species. However, this effect is dependent on the palladium concentration and the nature of the zeolite. In accordance, if Pd@Beta samples registered with the highest concentrations of the Lewis acid sites decreased with the amount of palladium, for Pd@Y, the presence of the Lewis acidity (150–250 °C) is not detectable while for Pd@USY, although in a low concentration, is characterized by enhanced strength.

In summary, the characterization results suggest the presence of both oxide and metallic palladium species both on the outer surface of zeolite crystals (lower degree) and encapsulated within the channels or cavities of zeolites (preponderantly) [23]. During the calcination and reduction processes, the migration of palladium species in the zeolite micropores and their aggregation as larger particles take place. In the case of Beta zeolite, characterized by a large external surface, the higher the amount of palladium, the larger the formed particles and the lower the reduction degree. However, this process is also highly dependent on the zeolite framework [50], namely, a complex system of channels and micropores that can provide strong confinement effects and significantly inhibit particle growth to a particular size region. For a similar amount of palladium (i.e., 1 wt% Pd), the higher Si/Al ratio, the higher the reduction temperature and the higher the H$_2$ consumption, indicating a stronger interaction with the zeolite leading to small particles. In other words, the palladium dispersion is higher.

*2.2. Catalytic Tests*

Recently, Wang et al. [16] reported that in the presence of the supported noble metal catalysts, the hydrotreatment of humins in isopropanol (IPA) as a solvent at 400 °C led to aliphatic and aromatic hydrocarbons and phenolic compounds, alongside alcoholic and ketonic by-products (e.g., acetone and methyl isobutyl ketone (MIBK) most probably formed from the IPA solvent). In the presence of supported noble metals on carbon carriers (i.e., Ru/C, Rh/C, Pt/C, and Pd/C) under 30 bar H$_2$ at 400 °C, the main detected products are aromatic hydrocarbons, phenols, and esters [15].

In this work, the synthesized catalysts were screened for humins hydrotreatment in IPA as a solvent, at 250 °C, 30 bar H$_2$, and reaction times of 6, 12, and 24 h. In most of the reactions, 0.1 g humins and 10 mg catalyst were used (1:10 catalyst to humins mass ratio) but some experiments were also carried out taking 20 mg catalyst (1:5 catalyst to humins mass ratio). Under the reaction conditions, hydrogen can also be generated from the solvent (IPA) via catalytic transfer hydrogenation (CTH) [51,52] providing a larger amount of hydrogen needed in the reaction.

Figures 13 and 14 present the main results obtained in the presence of Pd@Beta and Pd@USY catalysts. After 6 h and with 10 mg of catalyst, the humins conversion was only

25% in the presence of 1Pd@Beta catalyst with a yield to humins oil below 10% (Figure 13). The conversion increased to 34% with the loading of palladium (i.e., 1.5%) but the yield in humins oil remained unchanged. A higher conversion of humins was accompanied by a higher yield in humins oil (20%) only after doubling the amount of catalyst from 10 mg to 20 mg. After 12 h, the conversion of humins increased from 23% (0.5Pd@Beta) to 33.7% (1Pd@Beta) and 43.1% (1.5Pd@Beta), for a catalyst to humins mass ratio of 1:10 (i.e., 10 mg of catalyst). The highest productivity in the humins oil (27.3%) was obtained in the presence of 1Pd@Beta, after 12 h. In the presence of the 1.5Pd@Beta catalyst, a maximum yield in the humins oil (but clearly inferior to that obtained in the presence of 1Pd@Beta catalyst) was obtained after 12 h; a prolonged reaction time of 24 h led to increased conversion of humins but an inferior yield of humins oil, suggesting the gasification of the oil with the increase of the reaction time.

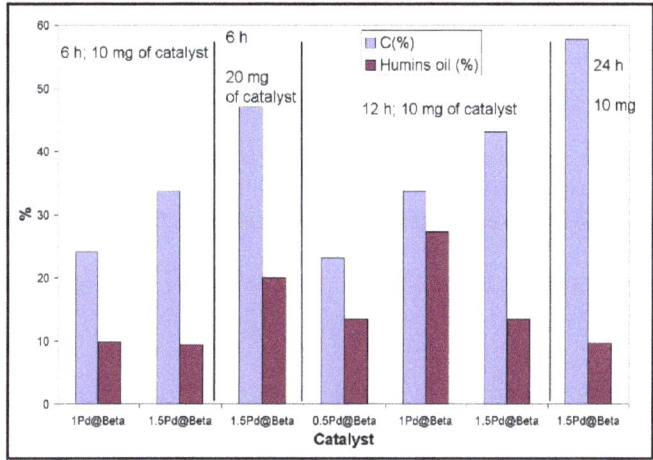

**Figure 13.** Catalytic performances of the Pd@Beta catalysts in the hydrodeoxygenation of humins (reaction conditions: 0.1 g humins, 8 mL IPA, 30 bars, T = 250 °C).

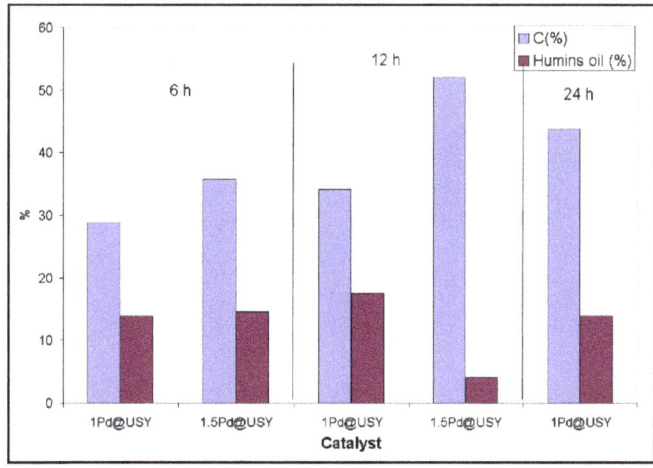

**Figure 14.** Catalytic performances of the Pd@USY catalysts in the hydrodeoxygenation of humins (reaction conditions: 0.1 g humins, 10 mg of catalyst, 8 mL IPA, 30 bars, T = 250 °C).

The nature of zeolite also exerts an influence on the conversion of humins and the yield of humins oil. Therefore, changing the support from Beta to USY zeolite, for the same concentration of palladium (i.e., 1.5%) and similar reaction conditions, led to an increase in the humins conversion from 43.1 to 52.0% (Figures 13 and 14). The textural properties of USY differ from that of Beta thus the catalytic active sites are possibly more easily accessible for rather large molecules for subsequent reactions.

For the series of USY-based-catalysts, the variation in the productivity of the humins oil showed a similar trend, i.e., higher amounts were obtained on the 1Pd@USY catalyst. However, in this case, these were significantly smaller (17.5%) than on 1Pd@Beta.

The lack of a correlation between these two efficiency parameters (i.e., humins conversion and humins oil yield) is, most probably, due to the formation of different quantities of gaseous compounds such as carbon dioxide and light hydrocarbons. Most probably, a higher loading of palladium (i.e., 1.5%) favors not only an enhanced conversion but also advanced depolymerization and hydrodeoxygenation to light hydrocarbons with decreased amounts of humins oil.

The average molecular weight ($M_w$) of untransformed humins fragments is broad, depending on the nature of the catalyst and the reaction conditions (Table 4). The product is a cocktail of molecules with different masses and PD values in the range of 1.5–3.2. It contains both low molecular weight and high molecular weight components that are not GC detectable.

**Table 4.** Molar masses of humins recovered from the hydrodeoxygenation reactions.

| Entry | Catalyst | $M_w$ (Da) | $M_n$ (Da) | PD |
|---|---|---|---|---|
| 1 | 1Pd@Beta | 26,383 | 9600 | 2.7 |
| 2 | 1.5Pd@Beta | 22,835 | 7195 | 3.2 |
| 3 [1] | 1.5Pd@Beta | 17,390 | 5923 | 2.9 |
| 4 [2] | 1.5Pd@Beta | 11,312 | 3902 | 2.9 |
| 5 [3] | 1.5Pd@Beta | 8370 | 4740 | 1.8 |
| 6 | 1Pd@USY | 80,534 | 53,204 | 1.5 |
| 7 | 1.5Pd@USY | 81,631 | 31,419 | 2.6 |

$M_w$—average molecular weight (Da); $M_n$—molecular mass obtained at half the height of the peak (Da); PD—dispersion factor. [1]—12 h, [2]—24 h, [3]—20 mg catalyst.

A typical GC-MS spectrum of the liquid phase is shown in Figure 15. Various products that belong to different organic product classes were observed. Among these products, the major ones are alcohols, hydroxy-acids, ketones, and alkyl-phenolics. An overview of the components identified by GC-MS analysis is given in Table 5. Alcohols are by far the most abundant compound class identified in humins oil, followed by hydroxy-acids and alkyl-phenolics. As humins contains significant amounts of furan units, the presence of furanic in the humins oil was expected. However, such compounds were not identified in the GC-MS analysis. A possible explanation may be related to the high reactivity of the furan-derived compounds under the reaction conditions (i.e., high temperatures and hydrogen pressures). In the presence of the Pd-based catalysts, these compounds can be transformed into alkyl-phenols.

Although all these classes of compounds are produced regardless of the nature of the catalyst, the composition of the mixture differed from one catalyst to another. Thus, an increased amount of the Pd in the catalyst formulation results in an enrichment of the humins oil in alkyl-phenolic compounds.

The results showed that the nature of the support and the palladium amount have a major effect on the conversion of the humins but also on the yield in the humins oil. A possible explanation may be related to the textural properties of the zeolites. These can greatly influence the accessibility of large molecules to the active metal sites and the subsequent reactions. Another effect can be induced by the different acid strengths of the investigated zeolites, which may initiate a repolymerization of reactive intermediates, i.e., an effect also observed in previous lignin depolymerization studies [53].

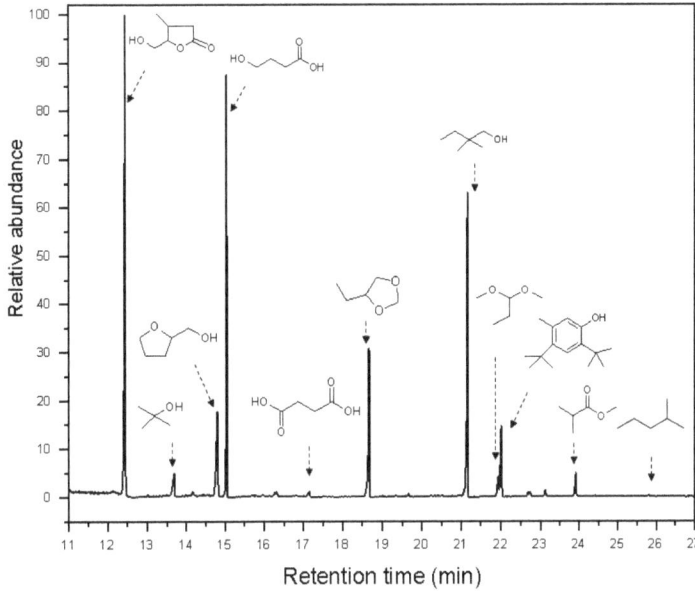

**Figure 15.** GC-MS spectrum of the humins oil (from the hydrodeoxygenation reaction in the presence of 1.5Pt@USY, at 250 °C, 6 h, and 30 bar of $H_2$).

**Table 5.** Major component classes and representative individual components in the liquid phase analysis by GC-MS.

| | |
|---|---|
| Alkanes | 2-methyl-penthane |
| Alcohols | 2.2-dimethylbutanol    tetrahydro-furfurylalcohol    4-hydroxy-5-(hydroxymethyl) oxolan-2-one    1.1-dimethyl ethanol |
| Acids | 4-hydroxybutanoic acid    butanedioic acid |
| Esters | 1.1-dimethoxypropane    4-ethyl-1.3-dioxolane |
| Phenolics | 4.6-di-tert-butyl-m-cresol |
| Esters | methyl isobutyrate |

## 3. Materials and Methods

### 3.1. Humins Synthesis and Characterization

Humins were prepared in accordance with a recently reported hydrothermal methodology [19]. An aqueous solution containing D-glucose (36.0 g D-glucose in 200 mL water, 1.0 M) and $H_2SO_4$ (1.078 g, 5.5 mM) was added to an autoclave and heated at 180 °C, for 7 h. The solid (i.e., humins) was isolated by filtration and washed with an excess of water (300 mL), dried for 12 h at 80 °C, grounded, and purified via a Soxhlet extraction.

### 3.2. Synthesis of Bi-Functional Pd@zeolite Catalysts

Zeolites (purchased from ZEOLYST International Company) in this study had a $SiO_2/Al_2O_3$ ratio of 25.0 ($NH_3$-Beta CP814E), 5.2 (H-Y CBV 600), and 30.0 (H-USY CBV 720), respectively. The surface areas of the zeolites, provided by the company are 680 $m^2/g$ (CP814E), 660 $m^2/g$ (CBV 600), and 780 $m^2/g$ (CBV 720) [54]. Before use, the $NH_3$-Beta zeolite was calcined in an oven at 450 °C, static air atmosphere, for 10 h, when $NH_4^+$ was decomposed to $NH_3(g)$ and $H^+$, generating the H-Beta zeolite. The calcination temperature was raised with a temperature ramp of 2 °C/min. The other zeolites were dried in a vacuum for 6 h at 110 °C. The preparation of the catalysts was performed as follows: over 1.5 g zeolite was added to a solution containing 16.0, 32.0, or 48.0 mg $Pd(CH_3COO)_2$ (for a final content of 0.5 wt% Pd, 1.0 wt% Pd and 1.5 wt% Pd, respectively) in benzene (15 mL). The mixture was stirred for 24 h at room temperature, after which the solvent was removed by evaporation. The obtained solids were dried at 80 °C for 8 h, calcined at 500 °C for 4 h (heating rate 1 °C/min), and reduced in a stream of hydrogen (flow rate of 30 mL/min) at 400 °C, for 4 h (heating rate 5 °C/min). The final catalysts were denoted as: 0.5Pd@Beta, 1Pd@Beta, 1.5Pd@Beta, 0.5Pd@Y, 1Pd@Y, 1.5Pd@Y, 0.5Pd@USY, 1Pd@USY, and 1.5Pd@USY.

### 3.3. Pd@zeolite Catalysts Characterization

The synthesized catalysts were characterized by various techniques such as adsorption-desorption isotherms of nitrogen at −196 °C, X-ray diffraction (XRD), thermogravimetry (TG-DTA), diffuse reflectance infrared Fourier transform spectroscopy (DRIFT), temperature programmed desorption ($CO_2$- and $NH_3$-TPD), temperature programmed reduction ($H_2$-TPR) and X-ray photoelectron spectroscopy (XPS).

The textural properties were determined from nitrogen adsorption-desorption isotherms of nitrogen at −196 °C using a Micrometrics Tristar 3020 apparatus. Prior to adsorption, all samples were systematically degassed at 200 °C under primary vacuum for 4 h. The surface area was calculated from the BET equation and the pore size distribution was determined based on the Barret-Joyner-Halenda (BJH) approach.

XRD diffractograms were recorded with a Schimadzu XRD-7000 diffractometer with K$\alpha$ radiation ($\lambda$ = 1.5418 Å, 40 kV, 40 mA) with steps of 0.02° in the 2θ range of 5–80 degrees.

TG-DTA analyses were performed with a Shimadzu apparatus. The heating rate of the sample was maintained at a value of 10 °C $min^{-1}$, starting from room temperature up to 800 °C, in a nitrogen stream of 10 mL $min^{-1}$.

DRIFT spectra were recorded with a Thermo Electron Nicolet 4700 FTIR spectrometer equipped with a Smart accessory for diffuse reflectance measurements. IR spectra were scanned in the range of 4000–400 $cm^{-1}$. The final spectra corresponded to an accumulation of 400 scans. The baseline was collected taking KBr as reference.

$NH_3$-TPD experiments were carried out in an AutoChem II 2920 station from Micromeritics. The samples, placed in a U-shaped quartz reactor with an inner diameter of 0.5 cm, were pretreated under He (Purity 5.0, from Linde) at 120 °C for 1 h and then exposed to a flow of $NH_3$ (10.01% in He, SIAD) for 1 h. After that, the samples were purged with a flow of He (50 mL·$min^{-1}$) for 30 min at 25 °C in order to remove the weakly adsorbed species. TPD was then started, with a heating rate of 3 °C·$min^{-1}$ till 500 °C. The desorbed products were analyzed with a TC detector/by GC-TCD chromatography. The desorbed $NH_3$ expressed as μmols per gram of catalyst was determined using a calibration curve.

H$_2$-TPR was performed in the same station. The samples were kept under a flow of H$_2$ (5% in He) with a heating rate of 3 °C·min$^{-1}$ till 500 °C. The consumed H$_2$ was detected by a thermo-conductivity detector (TCD) and expressed as mmols per gram of catalyst.

The X-ray photoelectron spectroscopy (XPS) analysis of the samples was performed in an AXIS Ultra DLD (Kratos Surface Analysis) setup using Mg K$\alpha$ (1253.6 eV) radiation produced by a non-monochromatized X-Ray source at operating power of 144 W (12 kV × 12 mA). The base pressure in the analysis chamber was at least $1 \times 10^{-8}$ mbar. All core level spectra were deconvoluted with the use of Voigt functions (Lorentzian and Gaussian widths) with a distinct inelastic background for each component [55,56]. A minimum number of components is used to obtain a convenient fit. The binding energy scale was calibrated to the C 1 s standard value of 284.6 eV (measured at the beginning of XPS spectra).

*3.4. Catalytic Tests*

Catalytic tests (i.e., humins depolymerization followed by hydrodeoxygenation reactions) were performed under the following conditions: 0.01–0.02 g of catalyst was added to a solution of 0.1 g humins in 8 mL of isopropanol (IPA). After closing the autoclave, it was pressurized with 30 bars of molecular hydrogen and the mixture was heated at 250 °C, under stirring, for 6–24 h. After the reaction, the reactor was cooled to room temperature, the pressure was released, the catalyst and the untransformed humins were separated by filtration, and the collected reaction products from the liquid phase were recovered by vacuum distillation of the solvent. The reaction liquid product was denoted as *"humins oil"*. The solid residue (unreacted humins and catalyst) was dried at 70 °C for 12 h, under vacuum, and weighed for mass balance calculations. Solid and liquid yields were calculated on the base of the formula:

$$Solid\ yield(\%) = \frac{mass\ of\ solid\ product - catalyst\ intake}{humins\ intake} \times 100$$

$$Liquid\ yield\ (\%) = \frac{mass\ of\ liquid\ phase}{humins\ intake + solvent\ intake} \times 100$$

The conversion was calculated based on humins intake and the solid products isolated after the reaction. It assumes that the solid residue consists of unconverted humins and, as such, the solid form due to repolymerization reactions of reactive intermediates was not taken into account.

$$Humins\ conversion\ (\%) = \frac{humins\ intake - (mass\ of\ solid\ after\ reaction - catalyst\ intake)}{humins\ intake} \times 100$$

*3.5. Products Analysis*

The obtained products (i.e., humins oil) were analyzed by GC-FID chromatography (GC-Shimadzu) and identified by GC-MS analysis (THERMO Electron Corporation equipped with TG-5SilMS column, 30 m × 0.25 mm × 0.25 µm). The non-transformed humins were analyzed by GPC-SEC chromatography to reveal its degree of decomposition during the catalytic process. An Agilent Technologies instrument (model 1260) equipped with Agilent PLgel MIXED-E column (7.5 × 300 mm, 3 µm) and a multi-detection unit (260 GPC/SEC MDS with RID, LS, and vs. detectors) was used in this scope. The analyses were performed under the following conditions: THF flow—1 mL/min, injection volume—100 µL, and temperature 35 °C. Calibration of the GPC system was performed in the range of 162–1,000,000 g mol$^{-1}$, with good accuracy of measurements for MW > 1000. GPC chromatograms allowed the calculation of the average molecular weight of the humins fragments recovered from the hydrodeoxygenation reaction.

## 4. Conclusions

In conclusion, a series of Pd@zeolite catalysts with loadings of 0.5 wt%, 1.0 wt% and 1.5 wt% Pd (zeolites: Beta (Si/Al = 12.5), Y (Si/Al = 2.6) and USY (Si/Al = 15)) were prepared using palladium acetate ($Pd(CH_3COO)_2$) as a precursor.

The resulting catalysts were characterized by XRD, adsorption-desorption isotherms of nitrogen, XPS, TG-DTA, $H_2$-TPR, DRIFT, and $NH_3$-TPD techniques. The obtained results suggest the formation of the hierarchically micro-mesoporous architectures in which both oxide and metallic palladium species exist preponderantly encapsulated within the channels or cavities of zeolites. During the activation (i.e., calcination and reduction) process the migration of palladium species and their aggregation as larger particles take place. The higher the amount of palladium, the larger the formed particles and the lower the reduction degree. On the other hand, the higher the Si/Al ratio, the stronger interaction of the palladium species with the zeolite framework. This leads to small particles highly dispersed on the zeolite surface.

The humins obtained from glucose were converted by hydrodeoxygenation to humins oil using the prepared Pd@zeolite catalysts, at only 250 °C. In the series of Beta-based catalysts, the highest level of conversion was 57.8% (1.5Pd@Beta, 24 h, 250 °C) while the highest percentage in the humins oil was 27.3%, on the 1Pd@Beta catalyst (12 h, 250 °C). For the series of USY-based-catalysts, the variation in the productivity of the humins oil showed a similar trend, i.e., higher amounts were obtained on the 1Pd@USY catalyst. However, in this case, these were significantly smaller (17.5%) than on 1Pd@Beta. Unfortunately, for high humins conversions, the amount of humins oil is lower. These results clearly show a high influence of the features of the catalyst (nature of the Pd species and physico-chemical features of the zeolite carrier) upon the reaction products.

These findings confirm that the solid useless humins waste may be (partly) depolymerized to liquid oil, which may further serve as a source for the production of valuable biobased chemicals such as alcohols, hydroxyacids, ketones, and alkyl-phenolics. The advantage of the process consists of a much lower reaction temperature by comparing with those reported in the literature (250 °C versus 400 °C). However, increased productivity of the process requires additional investigations focusing on the optimization of the catalytic system.

**Supplementary Materials:** The following supporting information can be downloaded at: https://www.mdpi.com/article/10.3390/catal12101202/s1, Figure S1: Deconvolutions of the HR XPS spectra for sample 1Pd@Y and the survey spectra; Figure S2: Deconvolutions of the HR XPS spectra for sample 1Pd@Beta and the survey spectra; Figure S3: Deconvolutions of the HR XPS spectra for sample 1.5Pd@Beta and the survey spectra; Figure S4: Deconvolutions of the HR XPS spectra for sample 1Pd@USY and the survey spectra; Figure S5: TG-DTA profiles of 0.5Pd@Beta (A) and 1Pd@Beta (B); Figure S6: TG-DTA profiles of 0.5Pd@Y (A) and 1.5Pd@Y (B); Figure S7: TG-DTA profiles of 0.5Pd@USY (A), 1Pd@USY (B) and 1.5Pd@USY (C); Table S1: Binding energies and the fractions of the $Pd_{3d5/2}$; Table S2: Si 2p peak components of Pd@zeolites samples; Table S3: Al 2p peak components of the Pd@Zeolite samples.

**Author Contributions:** Conceptualization, S.M.C.; methodology, N.C., M.E.F. and I.P.; validation, M.T., C.M.T., A.N. and B.C.; formal analysis, M.T., B.C., A.N. and N.C.; investigation, N.C., M.E.F., A.N. and I.P.; writing—original draft preparation, S.M.C.; writing review and editing, S.M.C. and V.I.P.; visualization, S.M.C. and V.I.P.; supervision, S.M.C., C.M.T. and V.I.P.; funding acquisition, N.C. All authors have read and agreed to the published version of the manuscript.

**Funding:** This study was funded by the Government of Romania, Ministry of Research and Innovation, project PN-III-P1-1.1-TE-2019-1933, Nr.69/2020.

**Conflicts of Interest:** The authors declare no conflict of interest.

## References

1. Jing, Y.; Guo, Y.; Xia, Q.; Liu, X.; Wang, Y. Catalytic production of value-Added chemicals and liquid fuels from lignocellulosic biomass. *Chem* **2019**, *5*, 2520–2546. [CrossRef]
2. Covinich, L.G.; Clauser, N.M.; Felissia, F.E.; Vallejos, M.E.; Area, M.C. The challenge of converting biomass polysaccharides into levulinic acid through heterogeneous catalytic processes. *Biofuels Bioprod. Biorefin.* **2020**, *14*, 417–444. [CrossRef]
3. Ding, D.; Wang, J.; Xi, J.; Liu, X.; Lu, G.; Wang, Y. High-yield production of levulinic acid from cellulose and its upgrading to γ-valerolactone. *Green Chem.* **2014**, *16*, 3846–3853. [CrossRef]
4. Ding, D.; Xi, J.; Wang, J.; Liu, X.; Lu, G.; Wang, Y. Production of methyl levulinate from cellulose: Selectivity and mechanism study. *Green Chem.* **2015**, *17*, 4037–4044. [CrossRef]
5. Mija, A.; van der Waal, J.C.; Pin, J.M.; Guigo, N.; de Jong, E. Humins as promising material for producing sustainable carbohydrate-derived building materials. *Constr. Build. Mater.* **2017**, *139*, 594–601. [CrossRef]
6. Suganuma, S.; Nakajima, K.; Kitano, M.; Yamaguchi, D.; Kato, H.; Hayashi, S.; Hara, M. Hydrolysis of cellulose by amorphous carbon bearing SO3H, COOH, and OH Groups. *J. Am. Chem. Soc.* **2008**, *130*, 12787–12793. [CrossRef]
7. Pye, E.K. Lignin Line and Lignin-Based Product Family Trees. In *Biorefineries-Industrial Processes and Products*; Kamm, B., Gruber, P.R., Kamm, M., Eds.; Wiley-VCH Verlag GmbH.: Weinheim, Germany, 2008; Volume 2, pp. 165–199.
8. Hoang, T.M.C.; Eck, E.R.H.V.; Bula, W.P.; Gardeniers, J.G.E.; Lefferts, L.; Seshan, K. Humin based by-products from biomass processing as a potential carbonaceous source for synthesis gas production. *Green Chem.* **2015**, *17*, 959–972. [CrossRef]
9. Wang, Y.; Agarwal, S.; Kloekhorst, A.; Heeres, H.J. Catalytic hydrotreatment of humins in mixtures of formic acid/2-Propanol with supported ruthenium catalysts. *ChemSusChem* **2016**, *9*, 951–961. [CrossRef]
10. Agarwal, S.; Es, D.V.; Heeres, H.J. Catalytic pyrolysis of recalcitrant, insoluble humin byproducts from C6 sugar biorefineries. *J. Anal. Appl. Pyrol.* **2017**, *123*, 134–143. [CrossRef]
11. Sangregorio, A.; Guigo, N.; Waal, J.C.V.N.; Sbirrazzuoli, N. Humins from biorefineries as thermoreactive macromolecular systems. *ChemSusChem* **2018**, *11*, 4246–4255. [CrossRef]
12. Hoang, T.M.C.; Lefferts, L.; Seshan, K. Valorization of humin-based byproducts from biomass processing—A route to sustainable hydrogen. *ChemSusChem* **2013**, *6*, 1651–1658. [CrossRef]
13. Maerten, S.G.; Voß, D.; Liauw, M.A.; Albert, J. Selective catalytic oxidation of humins to low-chain carboxylic acids with tailor-made polyoxometalate catalysts. *ChemistrySelect* **2017**, *2*, 7296–7302. [CrossRef]
14. Wang, Y.; Agarwal, S.; Heeres, H.J. Catalytic liquefaction of humin substances from sugar biorefineries with Pt/C in 2-Propanol. *ACS Sustain. Chem. Eng.* **2017**, *5*, 469–480. [CrossRef]
15. Cheng, Z.; Saha, B.; Vlachos, D.G. Catalytic hydrotreatment of humins to bio-oil in methanol over supported metal catalysts. *ChemSusChem* **2018**, *11*, 3609–3617. [CrossRef] [PubMed]
16. Wang, Y.; Agarwal, S.; Tang, Z.; Heeres, H.J. Exploratory catalyst screening studies on the liquefaction of model humins from C6 sugars. *RSC Adv.* **2017**, *7*, 5136–5147. [CrossRef]
17. Sun, J.; Cheng, H.; Zhang, Y.; Zhang, Y.; Lan, X.; Zhang, Y.; Xia, Q.; Ding, D. Catalytic hydrotreatment of humins into cyclic hydrocarbons over solid acid supported metal catalysts in cyclohexane. *J. Energy Chem.* **2021**, *53*, 329–339. [CrossRef]
18. Sumerskii, I.V.; Krutov, S.M.; Zarubin, M.Y. Humin-like substances formed under the conditions of industrial hydrolysis of wood. *Russ. J. Appl. Chem.* **2010**, *83*, 320–327. [CrossRef]
19. Fergani, M.E.; Candu, N.; Tudorache, M.; Bucur, C.; Djelal, N.; Granger, P.; Coman, S.M. From useless humins by-product to Nb@graphite-like carbon catalysts highly efficient in HMF synthesis. *Appl. Catal. A Gen.* **2021**, *618*, 118130. [CrossRef]
20. Zandvoort, I.V.; Koers, E.J.; Weingarth, M.; Bruijnincx, P.C.A.; Baldus, M.; Weckhuysen, B.M. Structural characterization of 13C-enriched humins and alkali-treated 13C humins by 2D solid-state NMR. *Green Chem.* **2015**, *17*, 4383–4392. [CrossRef]
21. Filiciotto, L.; Balu, A.M.; Romero, A.A.; Angelici, C.; Waal, J.C.V.D.; Luque, R. Reconstruction of humins formation mechanism from decomposition products: A GC-MS study based on catalytic continuous flow depolymerizations. *Mol. Catal.* **2019**, *479*, 110564. [CrossRef]
22. Fodor, D.; Ishikawa, T.; Krumeich, F.; Bokhoven, J.A.V. Synthesis of single crystal nanoreactor materials with multiple catalytic functions by incipient wetness impregnation and ion exchange. *Adv. Mater.* **2015**, *27*, 1919–1923. [CrossRef]
23. Chai, Y.; Shang, W.; Li, W.; Wu, G.; Dai, W.; Guan, N.; Li, L. Noble metal particles confined in zeolites: Synthesis, characterization, and applications. *Adv. Sci.* **2019**, *6*, 1900299. [CrossRef]
24. Okumura, K.; Niwa, M. Regulation of the Dispersion of PdO through the interaction with acid sites of zeolite studied by extended X-ray absorption fine structure. *J. Phys. Chem. B* **2000**, *104*, 9670–9675. [CrossRef]
25. Fergani, M.E.; Candu, N.; Coman, S.M.; Parvulescu, V.I. Nb-Based zeolites: Efficient bi-functional catalysts for the one-pot synthesis of succinic acid from glucose. *Molecules* **2017**, *22*, 2218. [CrossRef]
26. Beyerlein, R.A.; Choi Feng, C.; Hall, J.B.; Huggins, B.J.; Ray, G.J. Effect of steaming on the defect structure and acid catalysis of protonated zeolites. *Top. Catal.* **1997**, *4*, 27–42. [CrossRef]
27. Remy, M.J.; Stanica, D.; Poncelet, G.; Feijen, E.J.P.; Grobet, P.J.; Martens, J.A.; Jacobs, P.A. Dealuminated H-Y zeolites: Relation between physicochemical properties and catalytic activity in heptanes and decane isomerization. *J. Phys. Chem.* **1996**, *100*, 12440–12447. [CrossRef]
28. Jamalzadeh, Z.; Haghighi, M.; Asgari, N. Synthesis, physicochemical characterizations and catalytic performance of Pd/carbon-zeolite and Pd/carbon-CeO2 nanocatalysts used for total oxidation of xylene at low temperatures. *Front. Environ. Sci. Eng.* **2013**, *7*, 365–381. [CrossRef]

29. Jabłon'ska, M.; Król, A.; Kukulska-Zajac, E.; Tarach, K.; Chmielarz, L.; Góra-Marek, K. Zeolite Y modified with palladium as effective catalyst for selective catalytic oxidation of ammonia to nitrogen. *J. Catal.* **2014**, *316*, 36–46. [CrossRef]
30. Tian, F.; Wu, Y.; Shen, Q.; Li, X.; Chen, Y.; Meng, C. Effect of Si/Al ratio on mesopore formation for zeolite beta via NaOH treatment and the catalytic performance in α-pinene isomerization and benzoylation of naphthalene. *Micropor. Mesopor. Mat.* **2013**, *173*, 129–138. [CrossRef]
31. Penner, S.; Wang, D.; Jenewein, B.; Gabasch, H.; Klotzer, B.; Knop-Gericke, A.; Schlogl, B.; Hayek, K. Growth and decomposition of aligned and ordered PdO nanoparticles. *J. Chem. Phys.* **2006**, *125*, 94703. [CrossRef]
32. Fleisch, T.H.; Hicks, R.F.; Bell, A.T. An XPS study of metal-support interactions on $PdSiO_2$ and $PdLa_2O_3$. *J. Catal.* **1984**, *87*, 398–413. [CrossRef]
33. Chichova, D.; Mäki-Arvela, P.; Heikkilä, T.; Kumar, N.; Väyrynen, J.; Salmi, T.; Murzin, D.Y. X-ray Photoelectron Spectroscopy Investigation of Pd-Beta Zeolite Catalysts with Different Acidities. *Top. Catal.* **2009**, *52*, 359–379. [CrossRef]
34. Stakheev, A.Y.; Kustov, L.M. Effects of the support on the morphology and electronic properties of supported metal clusters: Modern concepts and progress in 1990s. *Appl. Catal. A Gen.* **1999**, *188*, 3–35. [CrossRef]
35. Klie, R.F.; Browning, N.D.; Chowdhuri, A.R.; Takoudis, C.G. Analysis of ultrathin $SiO_2$ interface layers in chemical vapor deposition of $Al_2O_3$ on Si by in situ scanning transmission electron microscopy. *Appl. Phys. Lett.* **2003**, *83*, 1187. [CrossRef]
36. Guillemot, D.; Polisset-Thfoin, M.; Fraissard, J.; Bonnin, D. New Method for Obtaining Palladium Particles on Y Zeolites. Characterization by TEM, $^{129}$Xe NMR, $H_2$ Chemisorption, and EXAFS. *J. Phys. Chem. B* **1997**, *101*, 8243–8249. [CrossRef]
37. Wagner, C.D.; Davis, L.E.; Zeller, M.V.; Taylor, J.A.; Raymond, R.H.; Gale, L.H. Empirical atomic sensitivity factors for quantitative analysis by electron spectroscopy for chemical analysis. *Surf. Interface Anal.* **1981**, *3*, 211–225. [CrossRef]
38. Gallagher, P.K.; Gross, M.E. The thermal decomposition of palladium acetate. *J. Therm. Anal.* **1986**, *31*, 1231–1241. [CrossRef]
39. Shen, W.J.; Okumura, M.; Matsumura, Y.; Haruta, M. The influence of the support on the activity and selectivity of Pd in CO hydrogenation. *Appl. Catal. A Gen.* **2001**, *213*, 225–232. [CrossRef]
40. Pieterse, J.A.Z.; Van den Brink, R.W.; Booneveld, S.; de Bruijn, F.A. Influence of zeolite structure on the activity and durability of Co-Pd-zeolite catalysts in the reduction of NOx with methane. *Appl. Catal. B-Environ.* **2003**, *46*, 239–250. [CrossRef]
41. Chang, T.-C.; Chen, J.-J.; Yeh, C.-T. Temperature-Programmed Reduction and Temperature-Resolved Sorption Studies of Strong Metal-Support Interaction in supported palladium catalysts. *J. Catal.* **1985**, *96*, 51–57. [CrossRef]
42. Byun, M.Y.; Park, D.W.; Lee, M.S. Effect of Oxide Supports on the Activity of Pd Based Catalysts for Furfural Hydrogenation. *Catalysts* **2020**, *10*, 837. [CrossRef]
43. Zheng, Q.; Farrauto, R.; Deeba, M. Part II: Oxidative Thermal Aging of Pd/Al2O3 and Pd/CexOy-ZrO2 in Automotive Three Way Catalysts: The Effects of Fuel Shutoff and Attempted Fuel Rich Regeneration. *Catalysts* **2015**, *5*, 1797–1814. [CrossRef]
44. Flanigen, E.M.; Khatami, H.; Szymanski, H.A. Infrared Structural Studies of Zeolite Frameworks. In *Molecular Sieve Zeolites, Advances in Chemistry 101*; Flanigen, E.M., Sand, L.B., Eds.; American Chemical Society: Washington, DC, USA, 1971; pp. 201–229. [CrossRef]
45. Banse, B.A.; Koel, B.E. Interaction of oxygen with Pd(111): High effective O2 pressure conditions by using nitrogen dioxide. *Surf. Sci.* **1990**, *232*, 275–285. [CrossRef]
46. Sacchetto, V.; Gatti, G.; Paul, G.; Braschi, I.; Berlier, G.; Cossi, M.; Marchese, L.; Bagatin, R.; Bisio, C. The interactions of methyl tert-butyl ether on high silicazeolites: A combined experimental and computational study. *Phys. Chem. Chem. Phys.* **2013**, *15*, 13275–13287. [CrossRef]
47. Othman, I.; Mohamed, R.M.; Ibrahim, I.A.; Mohamed, M.M. Synthesis and modification of ZSM-5 with manganese and lanthanum and their effects on decolorization of indigo carmine dye. *Appl. Catal. A-Gen.* **2006**, *299*, 95–102. [CrossRef]
48. Topsøe, N.Y.; Pedersen, K.; Derouane, E.G. Infrared and temperature-programmed desorption study of the acidic properties of ZSM-5-type zeolites. *J. Catal.* **1981**, *70*, 41–52. [CrossRef]
49. Bläkera, C.; Pasel, C.; Luckas, M.; Dreisbach, F.; Bathen, D. Investigation of load-dependent heat of adsorption of alkanes and alkenes on zeolites and activated carbon. *Micropor. Mesopor. Mat.* **2017**, *241*, 1–10. [CrossRef]
50. Zhang, J.; Wang, L.; Zhu, L.; Wu, Q.; Chen, C.; Wang, X.; Ji, Y.; Meng, X.; Xiao, F.S. Solvent-Free Synthesis of Zeolite Crystals Encapsulating Gold–Palladium Nanoparticles for the Selective Oxidation of Bioethanol. *ChemSusChem* **2015**, *8*, 2867–2871. [CrossRef]
51. Panagiotopoulou, P.; Vlachos, D.G. Liquid phase catalytic transfer hydrogenation of furfural over a Ru/C catalyst. *Appl. Catal. A Gen.* **2014**, *480*, 17–24. [CrossRef]
52. Jae, J.; Zheng, W.; Lobo, R.F.; Vlachos, D.G. Production of Dimethylfuran from Hydroxymethylfurfural through Catalytic Transfer Hydrogenation with Ruthenium Supported on Carbon. *ChemSusChem* **2013**, *6*, 1158–1162. [CrossRef]
53. Jongerius, A.L.; Copeland, J.R.; Foo, G.S.; Hofmann, J.P.; Bruijnincx, P.C.A.; Sievers, C.; Weckhuysen, B.M. Stability of Pt/γ-Al2O3 catalysts in lignin and lignin model compound solutions under liquid phase reforming reaction conditions. *ACS Catal.* **2013**, *3*, 464–473. [CrossRef]
54. Zeolyst International. Available online: https://www.zeolyst.com/our-products/standard-zeolite-powders.html (accessed on 1 June 2021).
55. Teodorescu, C.M.; Esteva, J.M.; Karnatak, R.C.; Afif, A.E. An approximation of the Voigt I profile for the fitting of experimental x-ray absorption data. *Nucl. Instrum. Methods Phys. Res. Sect. A Accel. Spectrom. Dect. Assoc. Equip.* **1994**, *345*, 141–147. [CrossRef]
56. Mardare, D.; Luca, D.; Teodorescu, C.M.; Macovei, D. On the hydrophilicity of nitrogen-doped $TiO_2$ thin films. *Surf. Sci.* **2007**, *601*, 4515–4520. [CrossRef]

*Article*

# Use of Photocatalytically Active Supramolecular Organic–Inorganic Magnetic Composites as Efficient Route to Remove β-Lactam Antibiotics from Water

Sabina G. Ion [1,2], Octavian D. Pavel [1,2], Nicolae Guzo [1,2], Madalina Tudorache [1,2], Simona M. Coman [1,2], Vasile I. Parvulescu [1,2,*], Bogdan Cojocaru [1,2,*] and Elisabeth E. Jacobsen [3,*]

[1] Faculty of Chemistry, University of Bucharest, 4-12 Regina Elisabeta Av., 030018 Bucharest, Romania
[2] Research Center for Catalysts & Catalytic Processes, Faculty of Chemistry, University of Bucharest, 4-12, Blv. Regina Elisabeta, 030018 Bucharest, Romania
[3] Department of Chemistry, Norwegian University of Science and Technology, Høgskoleringen 5, 7491 Trondheim, Norway
* Correspondence: vasile.parvulescu@chimie.unibuc.ro (V.I.P.); bogdan.cojocaru@chimie.unibuc.ro (B.C.); elisabeth.e.jacobsen@ntnu.no (E.E.J.); Tel.: +40-213051464 (V.I.P. & B.C.); +47-98843559 (E.E.J.)

**Abstract:** Considerable efforts have been made in recent years to identify an optimal treatment method for the removal of antibiotics from wastewaters. A series of supramolecular organic-inorganic magnetic composites containing Zn-modified MgAl LDHs and Cu-phthalocyanine as photosensitizers were prepared with the aim of removing β-lactam antibiotics from aqueous solutions. The characterization of these materials confirmed the anchorage of Cu-phthalocyanine onto the edges of the LDH lamellae, with a negligible part inserted in the interlayer space. The removal of the β-lactam antibiotics occurred via concerted adsorption and photocatalytic degradation. The efficiency of the composites depended on (i) the LDH: magnetic nanoparticle (MP) ratio, which was strongly correlated with the textural properties of the catalysts, and (ii) the phthalocyanine loading in the final composite. The maximum efficiency was achieved with a removal of ~93% of the antibiotics after 2 h of reaction.

**Keywords:** layered double hydroxides; magnetic nanoparticles; phthalocyanines; antibiotics removal

## 1. Introduction

Antibiotics are by far the most successful class of drugs used to fight against bacterial infections and, therefore, they are extensively used in both the human and veterinary medicine. They are produced by both microorganisms and synthetic or semi-synthetic routes, where, regardless of the pathway, the building of the complex chemical structures generates pollutant residues [1]. Furthermore, their excretion after their use and the disposal of unused medicinal compounds are additional sources of pollution. Some antibiotics, such as penicillin, can be easily degraded, while others, such as fluoroquinolones or tetracyclines, are considerably more persistent. Accordingly, these predominate for a longer time in the environment and accumulate in higher concentrations [1]. Thus, their potential role in promoting the development of a high level of resistance to human and animal pathogens enhances the concern about their presence in the environment.

Considerable efforts have been made in recent years to identify an optimal treatment method for the removal of antibiotics from wastewaters. With this aim, advanced oxidation processes (AOP) emerged as a more promising tool compared to traditional routes, such as adsorption on activated carbon, air stripping, or reverse osmosis [2]. Indeed, many of these techniques only transfer pollutants from one phase to another, without complete destruction up to mineralization. Biological treatment is not toxic to organic cultures, but it has some limitations [3]. The most important is related to its applicability to wastewaters containing biodegradable compounds.

Among the various advanced oxidation processes, photocatalysis has emerged as a promising technology for wastewater treatment in general, and antibiotic removal in particular [4–6]. The main advantages of the process are the possibility of the simultaneous removal of antibiotics and antibiotic-resistant bacteria, the lack of mass-transfer limitations under operational environmental conditions, and the price, the commercially availability, the non-toxicity and the photochemical stability of the catalyst [7].

Recent research focused on organic photocatalysts mimicking natural supramolecular systems, in which the triplet states of the excited organic molecule can participate in electron-transfer processes in a manner that is quite similar to semiconductors [8,9]. Metallophthalocyanines (Pcs) can act as electron donors upon a photochemical excitation in the visible region [10–12]. However, the use of bare (metallo)phthalocyanines is limited due to the recovery and stability problems [13]. Therefore, their immobilization represents a viable alternative through which to solve this. $TiO_2$ immobilized phthalocyanines [14] or mechanical mixtures of metallophthalocyanines and zeolites [13] have already been found to be effective for the photocatalytic degradation of organic compounds. The encapsulation of these complexes in porous supports led to an increase in activity compared with the free complex, as a result of two factors: (*i*) the concentration of substrates around the active sites, and (*ii*) protection against degradation [15].

The encapsulation of metallo-phthalocyanines inside layered double hydroxides (LDHs) is another suggested solution to create hybrid organic–inorganic supramolecular systems [16]. In complement to their adsorptive and oxidation catalytic properties, LDHs were utilized in photocatalysis under both UV and Vis light irradiation [17,18]. These studies demonstrated that the immobilization of macrocyclic complexes may enhance their catalytic behavior for environmental pollution control under Vis irradiation, as well as the use of anionic phthalocyanines (sulfonated phthalocyanines) as charge-neutralizing anions [16,19–24]. However, conventional catalyst-separation methods, such as filtration or centrifugation, are not always suitable for nanoparticles due to the incomplete separation they produce. To solve this practical problem, magnetic nanoparticles (MNPs) were considered as alternative catalytic supports due to their properties, such as high surface areas and chemical stability, non-swelling properties in organic solvents [25], easy recovery by an external magnet, reusability [26], regular shapes, uniform size, low cost of production, non-toxicity, and biocompatibility [27].

$\beta$-Lactam antibiotics, such as amoxicillin (AMO), ampicillin (AMP), penicillin G (PEG), penicillin V (PEV), and cloxacillin (CLX), have antimicrobial properties through the presence of a $\beta$-lactam ring. These drugs are widely used in medicinal treatments and a high content of their metabolites ends in waste waters, which leads to several negative effects. Therefore, their removal from the environment is imperative, and various procedures involving UV irradiation [28], radiolysis [29], ozonolysis [30], Fenton- and photo-Fenton processes [28,31], photocatalysis with semiconductors [7,32–35], or a combination of these [36] have already been reported. Among these, techniques involving solar light as a source of irradiation in the photodegradation of $\beta$-lactam antibiotics are the most appropriate.

In this respect, it was already reported that AMO, AMP, PEV, PEG, or CLX can be degraded under visible irradiation by ZnS nanoparticles as photocatalysts. In a large domain of concentrations (0.5–1000 mg L$^{-1}$), the rate of degradation increased as the stability of the antibiotic molecules decreased [37]. ZnO nanoparticles and ZnO/polyaniline core-shell nanocomposite also degraded ampicillin under sunlight irradiation [38], while photocatalytic membranes composed of hybrid polylactic acid (PLA)/$TiO_2$ nanofibers deposited on fiberglass supports exhibited a high rate of degradation of AMP, which was virtually eliminated in the first 30 min of the process. Furthermore, metallic nanoparticles (Ag-NPs) were used in the photocatalytic degradation of AMP under natural sunlight irradiation with an efficiency of 96.5% [39].

However, to the best of our knowledge, MPc-LDH-MNP organic–inorganic supramolecular hybrid systems have not yet been investigated as photocatalysts for the removal of $\beta$-lactam antibiotics. Therefore, this study reports the preparation of such materials, along with their characterization and catalytic behavior, for the photodegradation of β-lactam antibiotics under visible light. Each component in the system contributes to the photocatalysts' efficiency in the removal of antibiotics: LDH serves as a support for Pc, and is also responsible for the adsorption of the antibiotics from the environment; Pc serves as photosensitizer, and the MNPs make possible the rapid extraction of the composite from the reaction environment.

## 2. Results and Discussion

The coding of the catalysts based on their preparation conditions is presented in Table 1.

**Table 1.** The preparation details for the catalysts prepared using different LDH:MNP ratios and from different MNP precursors.

| Sample | Precursor | Conditions |
|---|---|---|
| P1 | $FeCl_2 + Fe(NO_3)_3$ | LDH:$Fe_3O_4$ = 1:3 (10 min ageing) |
| P2 | $FeCl_2 + Fe(NO_3)_3$ | LDH:$Fe_3O_4$ = 1:3 (30 min ageing) |
| P3 | $FeCl_2 + Fe(NO_3)_3$ | LDH:$Fe_3O_4$ = 3:1 (60 min ageing) |
| P4 | $FeCl_2 + Fe(NO_3)_3$ | LDH:$Fe_3O_4$ = 3:2 (60 min ageing) |
| P5 | $FeCl_2 + Fe(NO_3)_3$ | LDH:$Fe_3O_4$ = 1:1 (60 min ageing) |
| P6 | $FeCl_2 + Fe(NO_3)_3$ | LDH:$Fe_3O_4$ = 1:2 (60 min ageing) |
| P7 | $FeSO_4 + FeCl_3$ (2:1 mol) | LDH:$Fe_3O_4$ = 3:1 (60 min ageing) |
| P8 | $FeSO_4 + FeCl_3$ (1:2 mol) | LDH:$Fe_3O_4$ = 3:1 (60 min ageing) |

### 2.1. Characterization

#### 2.1.1. X-ray Diffraction (XRD)

As previously reported [40], the X-ray diffractogram of the bare LDH shows (Figure 1a), at low angles, narrow and intense diffraction lines and, at high diffraction angles, wide and less intense lines characteristic of the LDH materials [41]. In addition to these lines, additional lines corresponding to the zincite phase (ZnO) appeared, as minor impurities, in the 2 theta range 31–38° [40]. The reconstruction by memory effect occurred with the preservation of the lines of the stable ZnO phase (ICDD 005-0664).

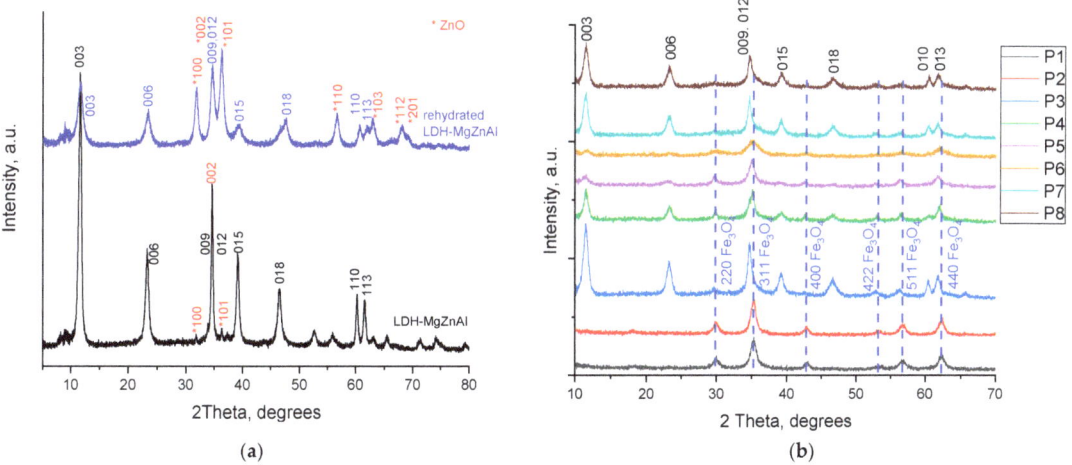

**Figure 1.** X-ray diffractograms corresponding to (**a**) as-prepared and rehydrated LDH after deposition of CuPc and (**b**) P1–P8 catalysts.

The X-ray diffractograms corresponding to P1–P8 catalysts are presented in Figure 1b. The magnetite phase was observed in almost all the samples and was most predominant for P1, P2, P5, and P6. Apart from LDH, Fe$_3$O$_4$ (ICDD 19-0629), and small impurities of ZnO, no other phases or impurities were detected.

The network parameter "**a**" (Table 2), calculated as **a** = 2·d$_{110}$, indicating the average cation—cation distance in the layered network, showed a decrease in inter-layer space, as an effect of the hydration associated with the elimination, by calcination, of the larger carbonate anions that were larger, and the replacement of these by hydroxyl anions, which are much smaller in size. The network parameter "**c**", calculated as **c** = 3/2·(d$_{003}$ + 2d$_{006}$), keeps the trend of IFS variation caused by an increase in the electrostatic forces between the layers and the interlayer anions. According to the literature, the presence of heavy interlayer molecular species led to a change in the electron density of the 00l harmonics, i.e., organic anions [42], metal nanoparticles [43], or oxometalates [44]. Considering the "**a**" and "**c**" parameters, we can conclude that the majority of phthalocyanine is anchored on the edge of the hydroxide lamellae, while a smaller part is placed in the interplanar space. The crystallite size, which in this case was an average of the entire composite, depended on the LDH: MP ratio.

**Table 2.** Network parameters determined from the X-ray diffractograms for the bare MgZnAl LDH and P1–P8 catalysts.

| Sample | a (Å) | c (Å) | IFS (Å) [1] | $I_{003}/I_{006}$ | $I_{003}/I_{110}$ | D(Å) [2] |
|---|---|---|---|---|---|---|
| LDH–MgZnAl | 3.0717 | 22.8689 | 2.82 | 2.96 | 5.54 | 132 |
| P1 | 3.0462 | 22.5822 | 2.73 | 1.00 | 0.67 | 111 |
| P2 | 3.0557 | 22.6262 | 2.74 | 1.08 | 0.68 | 115 |
| P3 | 3.0616 | 22.9058 | 2.84 | 2.21 | 4.89 | 96 |
| P4 | 3.0539 | 22.8707 | 2.82 | 2.21 | 6.73 | 105 |
| P5 | 3.0644 | 23.1147 | 2.90 | 1.95 | 4.56 | 79 |
| P6 | 3.0663 | 23.5299 | 3.04 | 3.00 | 1.29 | 200 |
| P7 | 3.0612 | 22.9415 | 2.85 | 2.17 | 3.98 | 97 |
| P8 | 3.0589 | 22.8956 | 2.83 | 2.20 | 4.04 | 95 |

[1] IFS represents the distance of the interplanar space (the values using Miyata's reported sheet thickness of 4.8 Å [45]).
[2] The crystallite size (obtained with the Debye–Scherrer equation) and determined from the FWHM value of the reflection 003.

2.1.2. Diffuse Reflectance UV-Vis Spectroscopy (DR-UV-Vis)

The bare LDH showed a wide absorption band in the range of 240–380 nm, with maxima at 348 and 359 nm corresponding to a zincite phase in the layered structure (Figure 2a) [40]. The additional band at 212 nm corresponds to Mg(OH)$_2$/MgO [40,46], while those in the region 550–750 nm to the strong Q bands, assigned to the CuPc [47]. The DR-UV-Vis spectra of the phthalocyanine-LDH composites are presented in Figure 2b. These disclose B-bands of CuPc (300–350 nm) [48] and magnetite 500 nm [49].

2.1.3. Diffuse-Reflectance Infrared Spectroscopy (DRIFT)

The intercalation of phthalocyanine was also confirmed by the DRIFT analysis (Figure 3). Regarding the LDH (Figure 3a), the wide absorption band located in the 3700–3400 cm$^{-1}$ range corresponds to the vibrations of the OH, ν(O–H) groups in the inter-planar space, while the band at 3000 cm$^{-1}$ was assigned to the hydrogen bonds between the water and the carbonate anion [50]. The band at 1638–1650 cm$^{-1}$ corresponded to the vibrations of the water molecules in the layered structure, that in the range 1100 and 650 cm$^{-1}$ to the vibrations of the CO$_3^{2-}$ groups, and those before 600 cm$^{-1}$ to the Mg-O, Zn-O and Al-O bonds. The infrared spectra also showed some bands associated with the presence of LDH-anchored phthalocyanine [51] (Figure 3b), although their intensity was very low. For example, the band at approximately 1650 cm$^{-1}$ was be assigned to the C=N bond, that at 1120 cm$^{-1}$ to the S=O bond, and those at 730, 1033 and 1090 cm$^{-1}$ to the C-H bond.

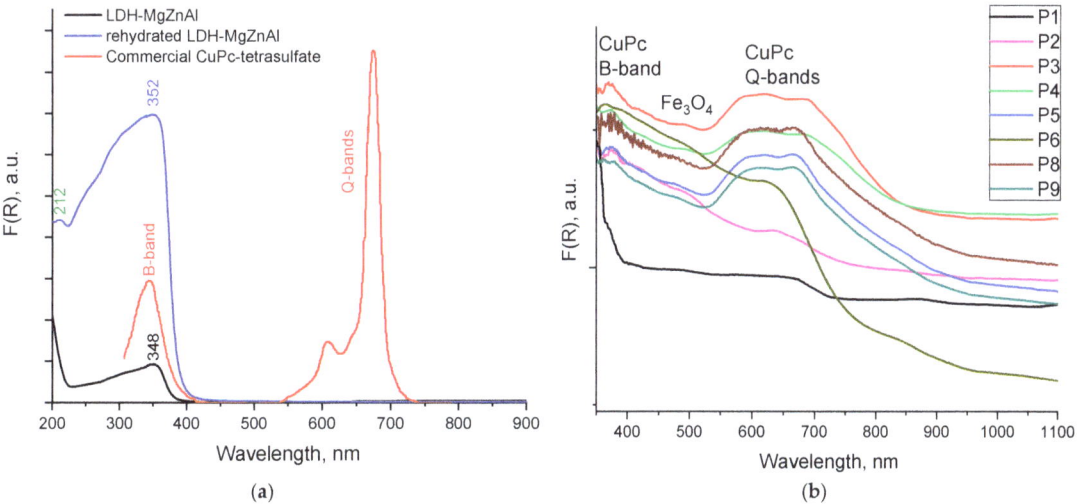

**Figure 2.** DR-UV-VIS spectra collected for (**a**) as-prepared and rehydrated LDH, commercial CuPc tetrasulfate, and (**b**) P1–P8 catalysts.

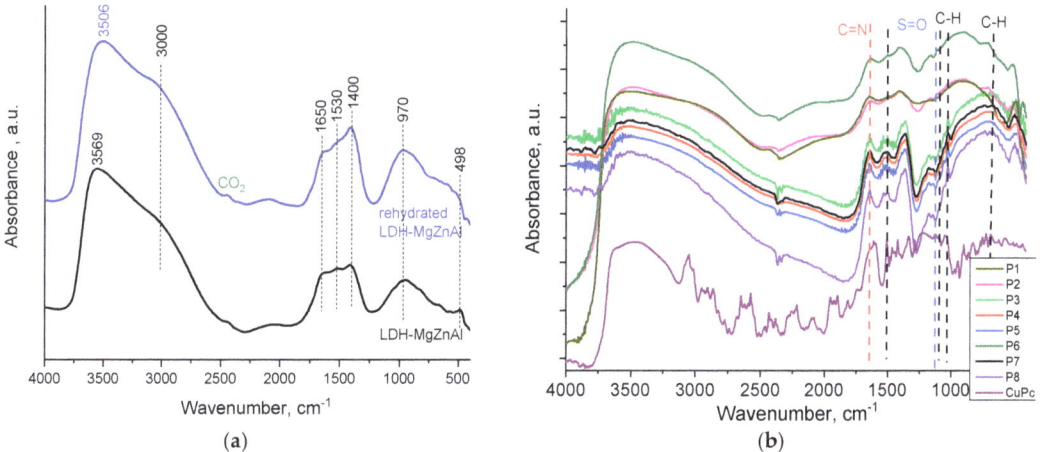

**Figure 3.** DRIFT spectra collected for (**a**) as-prepared and rehydrated LDH and (**b**) P1–P8 catalysts and commercial CuPc tetrasulfate.

### 2.1.4. Textural Properties

The surface areas of the P1–P8 catalysts (Table 2) varied in a rather narrow range, i.e., from 110 (P4) to 160 m$^2 \cdot$g$^{-1}$ (P5). The increase in the ageing time, which also meant an increase in the MNP size, led to a decrease in the surface area (samples P1 and P2). The decrease of the surface area compared to the parent LDH was due to the tendency to clusterize in large conglomerates systems confirmed by TEM (see Supplementary Materials) [52]. All the samples presented a Type IV adsorption–desorption isotherm, characteristic of mesoporous materials (Figure 4). However, depending on the LDH: MNP ratio and the nature of MNP precursors, the variation in the pore size was larger (Table 3, Figure 4).

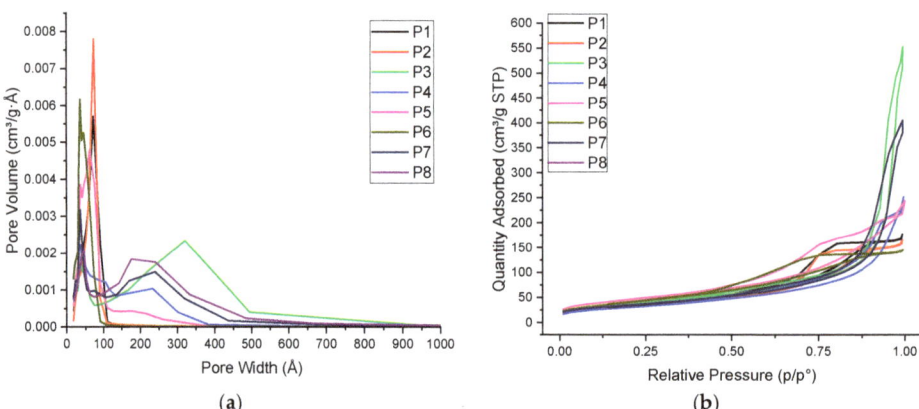

**Figure 4.** Pore-size distribution (a) and adsorption–desorption isotherms (b) for the P1–P8 catalysts.

**Table 3.** Surface area, pore volume, and main pore size for LDH, rehydrated LDH, and P1–P8 catalysts.

| Sample | $S_s$ (BET) (m$^2$·g$^{-1}$) | Pore Volume (cm$^3$·g$^{-1}$) | Average Pore Size (Å) |
| --- | --- | --- | --- |
| LDH-MgZnAl | 69 | 0.387 | 222 |
| Rehydrated LDH-MgZnAl | 25 | 0.186 | 238 |
| P1 | 129 | 0.277 | 72 |
| P2 | 116 | 0.258 | 73 |
| P3 | 137 | 0.858 | 34, 317 |
| P4 | 111 | 0.392 | 34, 90, 231 |
| P5 | 163 | 0.336 | 37, 63, 182 |
| P6 | 148 | 0.230 | 36, 45 |
| P7 | 124 | 0.628 | ~200 |
| P8 | 124 | 0.498 | ~200 |

## 2.2. Degradation of Antibiotics

The experiments carried out in this study for the removal of antibiotics from water using the organic–inorganic magnetic composite photocatalysts indicated two parallel processes, namely, adsorption and photocatalytic degradation (Figures 5 and 6, respectively).

Simple adsorption measurements (i.e., in the absence of photo-irradiation) indicated a removal of ampicillin of 9–87%. Depending on the catalyst, the removal of the ampicillin varied in the following order: P7 > P3 > P8 > P5 > P6 > P4 > P2~P1, i.e., in good concordance with their textural properties (pore volume and interlayer diameter). Although P3 had the highest pore volume in the series, it presented some interlayer limitation disfavoring the adsorption of ampicillin. For mesoporous supports with larger pores (200–700 nm), the loading of the adsorbed ampicillin does not depend on the pore size, and it was related only to the surface charge density (σ) of the sorbent surface [53]. For the LDHs, the ampicillin-related amoxicillin was found to adsorb predominantly by a chemisorption mechanism, following pseudo-second-order kinetics [54]. Indeed, as shown in Table 3, the interlayer distance did not influence the mass transfer or the access to the chemisorption sites.

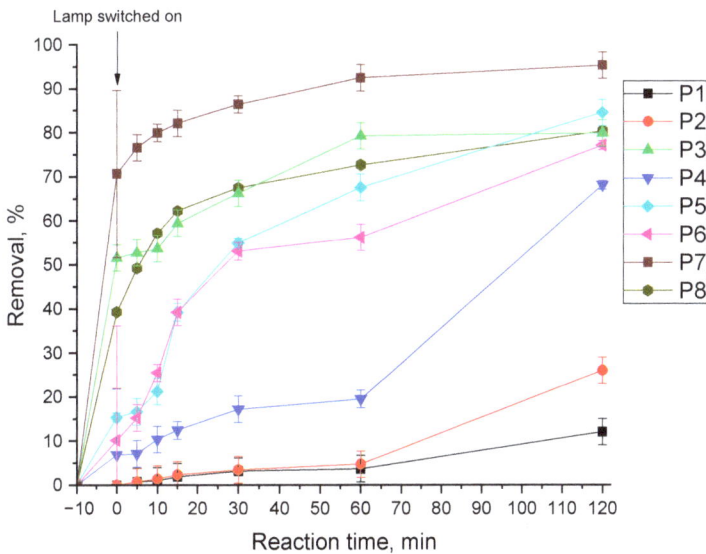

**Figure 5.** Removal of ampicillin from water solution using a blue LED (445–465 nm) as irradiation source.

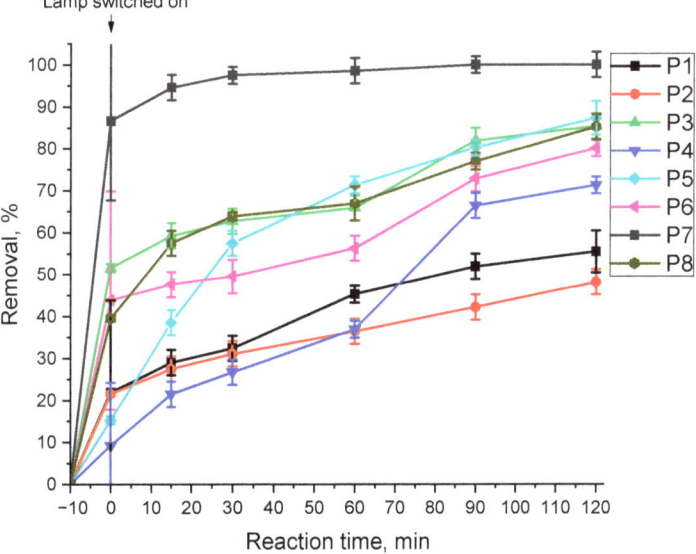

**Figure 6.** Removal of ampicillin from water solution using a solar simulator as irradiation source.

The photo-irradiation of the system changed the order established by the adsorption. However, the photo-catalytic degradation could not be separated from the adsorption of the ampicillin and degraded fragments, and it was also dependent on the photo-irradiation energy. Under the LED irradiation (the initial concentration of ampicillin in solution at t = 0), the maximum degradation was achieved for P7, with a removal of ~93% after 2 h of reaction. Overall, summarizing the effects of the adsorption and photocatalytic degradation, the removal decreased in the following order: P7 > P5 > P8~P3 > P6 > P4 > P2 > P1 (Figure 5).

The performances of these catalysts depended both on the LDH:MNP ratio (which was also associated with the amount of CuPc) and the textural properties of the composite.

Moving to a solar simulator as an irradiation source (Figure 6) afforded a higher removal efficiency, which was correlated with the higher photocatalytic efficiency attributed to the absorption of the light via the Q bands of the phthalocyanine.

The stability and reusability of catalysts are very important factors in the characterization of their efficiency and the limits of their applicability in real applications. Based on these considerations, the composite P7 demonstrated an enhanced efficiency. After three runs (the catalyst was magnetically removed from the reaction environment, washed thoroughly with deionized water, and then dried at 120 °C for 4 h) the conversion of the ampicillin was conserved at 98%. After each run, the DRIFT analysis confirmed no leaching of the phthalocyanine (Figure 7).

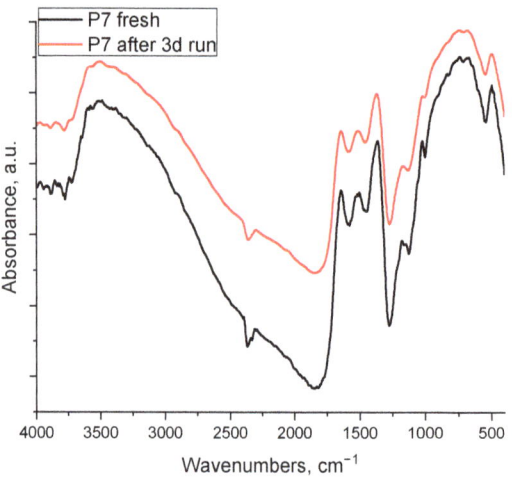

**Figure 7.** DRIFT spectra of fresh and spent P7 sample.

Similar tests on the removal of amoxicillin under simulated solar-light irradiation (Figure 8) using selected composites confirmed the order of activity determined for the ampicillin. However, both the absorption part and the photocatalytic behavior showed some differences. These were directly related to the structural features of these antibiotics (Figure 8), which can affect both the adsorption of molecules and photocatalytic degradation processes. Thus, the degradation of β-lactam antibiotics can follow two routes: hydroxylation and opening of the β-lactam ring [55]. The ratio between these routes is influenced by the structure of the antibiotic and the nature of the catalyst. For example, hydroxylation can be enhanced for amoxicillin compared to ampicillin by the population and the strength of the hydroxyl groups able to activate the aromatic ring, making it more susceptible to their attachment. However, the mechanism of the photo-degradation of β-lactam antibiotics is still under discussion.

FTIR analysis of the amoxicillin solution in the presence of P7 sample at each reaction time shows an increase in the amount of bonds involving oxygen (Figure 9a–c), which back up both reaction pathways.

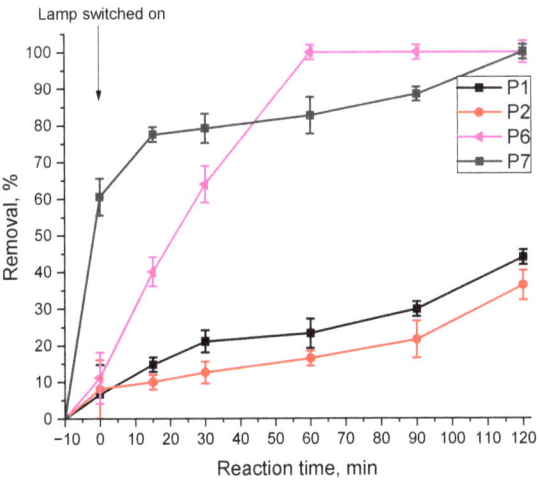

**Figure 8.** Example of removal of amoxicillin from water solution using a solar simulator as an irradiation source.

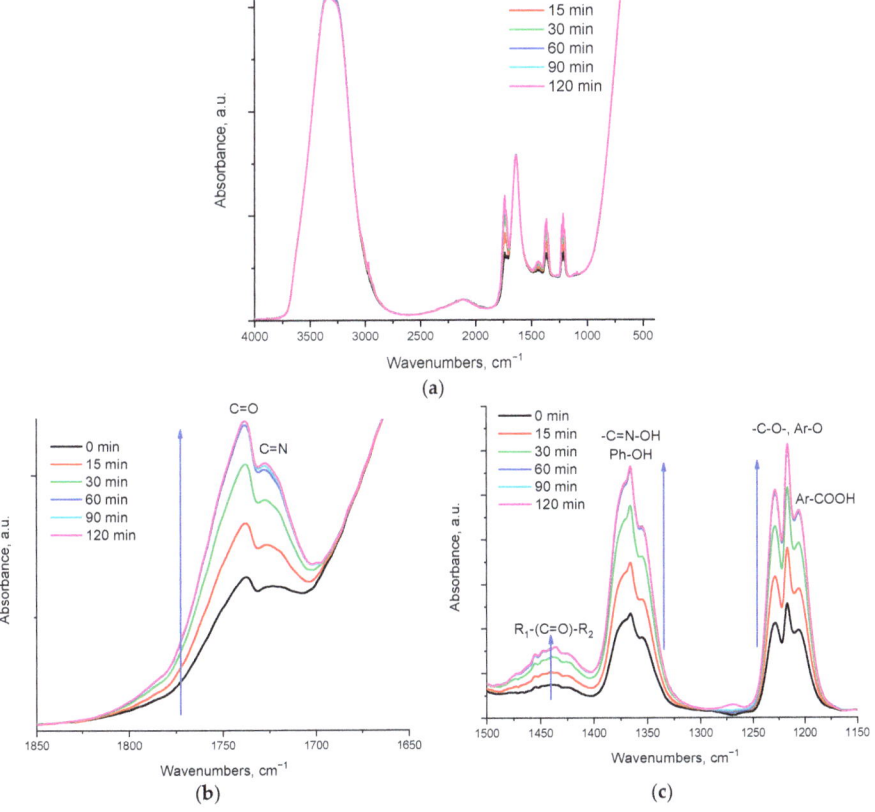

**Figure 9.** FTIR spectra of amoxicillin solution in the presence of P7 sample at each reaction time: (a) full spectra, (b) 1850–1650 cm$^{-1}$ region, (c) 1500–1150 cm$^{-1}$ region.

## 3. Materials and Methods

### 3.1. Synthesis

The synthesis of LDH $Mg_{0.375}Zn_{0.375}Al_{0.25}$ was carried out following a previously reported procedure [40]. Therefore, two solutions (A and B), were mixed at an inlet flow of 1 cm$^3$·min$^{-1}$, at RT under stirring (600 rpm). Solution A of a concentration of 1.5M contained the nitrate-like precursors of Mg, Zn, and Al in the molar ratio Mg/Zn/Al = 0.375/0.375/0.25. The base solution contained 0.23 moles NaOH and 0.092 moles $Na_2CO_3$ at a concentration of 1M in $Na_2CO_3$. Co-precipitation of these took place for 1.5 h at the pH of 10 ± 0.2 (continuously monitored by a pH-meter, pH315i), after which the resulted precipitate was aged for 18 h, at 80 °C, in air atmosphere at the same stirring rate. The suspension was then cooled at room temperature, filtered under vacuum, and washed with distilled water until reaching a pH of 7. The precipitate was dried at 120 °C for 24 h in air atmosphere, resulting in LDH-MgZnAl. The insertion of phthalocyanine into the structure of LDH started from the calcination of LDH for 18 h at 460 °C (10 °C/min), followed by stirring of the resulting mixed oxide with a solution of Co-phthalocyanine 3,4′,4″,4‴-tetrasulfonic acid (0.1 M) for 12 h, filtration of the composite phthalocyanine-LDH, washing of the precipitate, and drying at 120 °C for 24 h in air. The amount of CuPc determined by elemental analysis was ~10 × 10$^{-5}$ moles CuPc/g of LDH.

The attachment of the magnetic particles (MPs) was achieved by mixing 100 mg of modified LDH with Cu-phthalocyanine in 100 mL of water. After vigorous stirring for 10 min, the corresponding amounts of $FeCl_2$ and $Fe(NO_3)_3$ were added, resulting in a MP@CuPc@LDH-MgZnAl composite. The mixture was then heated at 70 °C in an oil bath and treated, by a slow addition, with 7.5 mL of 5M NaOH. The resulting suspension was then left to age for 60 min. Finally, the solid was magnetically removed, washed with water by decantation, and dried at 70 °C for 12 h.

In order to verify the structural influence of the LDH: MNPs ratio and the nature of the MNP precursors, a series of materials containing CuPc@LDH-MgZnAl were prepared using a similar protocol (Table 1).

### 3.2. Characterization

#### 3.2.1. Specific Surface Area (BET) and Pore-Size Measurements

The investigation of the textural characteristics was performed using the 77K nitrogen adsorption–desorption isotherms collected using a Micromeritics ASAP2020 Surface Area and Porosity Analyzer. The samples were outgassed under vacuum for 24 h at 120 °C.

#### 3.2.2. X-ray Diffraction (XRD)

XRD data were collected at room temperature using a Shimadzu XRD-7000 apparatus taking the monochromatic Cu K$_\alpha$ radiation (λ = 1.5406 Å, 40 kV, 40 mA), with a scanning rate of 0.1 degrees per minute, in the range 2θ = 5–80 degrees.

The size of the crystallites was estimated using Sherrer's equation:

$$D = k \cdot \lambda / FWHM \cdot \sin\theta \tag{1}$$

where:

k = 0.9 (shape factor)
λ = wavelength of the X-ray radiation source (for Cu = 1.54 Å)
FWHM = full width at half height (in radians)
θ = maximum position (in radians)

#### 3.2.3. Diffuse-Reflectance UV-Vis Spectroscopy (DR-UV-Vis)

The DR-UV-Vis spectra were collected under ambient conditions using Specord 250 equipment (Analytic Jena) with an integrating sphere as a measuring device in the reflectance mode. MgO was used as reference material.

### 3.2.4. Diffuse-Reflectance Infrared Spectroscopy (DRIFT)

The infrared spectra were collected with a Brucker Tensor II device equipped with a Harrick Praying Mantis Diffuse Reflection accessory. The final spectra averaged 128 scans with a resolution of 4 cm$^{-1}$. KBr was used as a reference.

### 3.3. Photocatalytic Tests

The irradiation was generated by a sunlight simulator using a Luzchem LZC-4b source (LED 445–465 nm), or a Sciencetech SF150-A Small Collimated Beam Solar Simulator (equipped with Air Mass AM1.5G Filter and Light-Tight Reaction Chamber). The evolution of the photocatalyzed reaction was followed by means of a liquid chromatograph (HPLC) equipped with Zorbax SB-C18 column (4.6 × 150 mm, 5 microns), mobile phase 25 mM $KH_2PO_4$:ACN = 60:40, mobile phase flow 0.5 mL·min$^{-1}$; analysis time 30 min; column temperature 60 °C, DAD detection 204 nm). In a typical experiment, 15 mL of a prepared 0.015 M solution of antibiotic and 60 mg of catalyst were added into a quartz test tube. To establish the absorption–desorption equilibrium; before turning on the lamp, the samples were kept in the dark, under stirring, for 10 min.

## 4. Conclusions

A series of supramolecular organic–inorganic magnetic composites containing Zn-modified MgAl-LDHs and Cu-phthalocyanine as photosensitizers were prepared with the aim of removing β-lactam antibiotics from aqueous solutions. The resulting composites were characterized using textural, spectral, and diffractometric techniques. These confirmed the anchorage of Cu-phthalocyanine onto the edges of the LDH lamellae, without any insertion in the interlayer space. The removal of these antibiotics occurred via concerted adsorption and photocatalytic degradation. The efficiency of the composites depended on (*i*) the LDH:MP ratio, which was strongly correlated with the textural properties of the catalysts, and (*ii*) the phthalocyanine loading in the final composite. The maximum efficiency was achieved for the P8 photocatalyst, with a removal of ~93% ampicillin after 2 h of reaction.

**Supplementary Materials:** The following supporting information can be downloaded at: https://www.mdpi.com/article/10.3390/catal12091044/s1, Figure S1: TEM image of fresh LDH; Figure S2: TEM image of rehydrated LDH; Figure S3: TEM image of MNP@ rehydrated LDH. Reference [52] is cited in the Supplementary Materials.

**Author Contributions:** Conceptualization, B.C.; methodology, B.C. and O.D.P.; investigation, S.G.I., N.G., O.D.P., M.T. and S.M.C.; resources, B.C. and E.E.J.; writing—original draft preparation, B.C., O.D.P. and E.E.J.; writing—review and editing, V.I.P., B.C., O.D.P. and E.E.J.; supervision, B.C. and M.T.; funding acquisition, M.T. and E.E.J. All authors have read and agreed to the published version of the manuscript.

**Funding:** This work was supported by a grant of The Romanian Ministry of Education and Research, CNCS–UEFISCDI, project number PN-III-P4-ID-PCE-2020-2207, within PNCDI III, and by The Education, Scholarship, Apprenticeships and Youth Entrepreneurship Programme–EEA, grants 2014–2021, project no. 18-Cop-0041.

**Data Availability Statement:** The data are available on request from the corresponding author.

**Conflicts of Interest:** The authors declare no conflict of interest.

## References

1. Martinez, J.L. Environmental pollution by antibiotics and by antibiotic resistance determinants. *Environ. Pollut.* **2009**, *157*, 2893–2902. [CrossRef] [PubMed]
2. Zhang, G.; Ji, S.; Xi, B. Feasibility study of treatment of amoxillin wastewater with a combination of extraction, Fenton oxidation and reverse osmosis. *Desalination* **2006**, *196*, 32–42. [CrossRef]
3. Surenjan, A.; Pradeep, T.; Philip, L. Application and performance evaluation of a cost-effective vis- LED based fluidized bed reactor for the treatment of emerging contaminants. *Chemosphere* **2019**, *228*, 629–639. [CrossRef] [PubMed]

4. Liu, C.; Mao, S.; Wang, H.; Wu, Y.; Wang, F.; Xia, M.; Chen, Q. Peroxymonosulfate-assisted for facilitating photocatalytic degradation performance of 2D/2D WO$_3$/BiOBr S-scheme heterojunction. *Chem. Eng. J.* **2022**, *430*, 132806. [CrossRef]
5. Liu, C.; Mao, S.; Shi, M.; Wang, F.; Xia, M.; Chen, Q.; Ju, X. Peroxymonosulfate activation through 2D/2D Z-scheme CoAl-LDH/BiOBr photocatalyst under visible light for ciprofloxacin degradation. *J. Hazard. Mater.* **2021**, *420*, 126613. [CrossRef] [PubMed]
6. Liu, C.; Mao, S.; Shi, M.; Hong, X.; Wang, D.; Wang, F.; Xia, M.; Chen, Q. Enhanced photocatalytic degradation performance of BiVO$_4$/BiOBr through combining Fermi level alteration and oxygen defect engineering. *Chem. Eng. J.* **2022**, *449*, 137757. [CrossRef]
7. Elmolla, E.S.; Chaudhuri, M. Photocatalytic degradation of amoxicillin, ampicillin and cloxacillin antibiotics in aqueous solution using UV/TiO$_2$ and UV/H$_2$O$_2$/TiO$_2$ photocatalysis. *Desalination* **2010**, *252*, 46–52. [CrossRef]
8. Bobo, M.V.; Kuchta, J.J.I.; Vannucci, A.K. Recent advancements in the development of molecular organic photocatalysts. *Org. Biomol. Chem.* **2021**, *19*, 4816–4834. [CrossRef]
9. Romero, N.A.; Nicewicz, D.A. Organic Photoredox Catalysis. *Chem. Rev.* **2016**, *116*, 10075–10166. [CrossRef]
10. Schmidt, A.M.; Calvete, M.J.F. Phthalocyanines: An Old Dog Can Still Have New (Photo)Tricks! *Molecules* **2021**, *26*, 2823. [CrossRef]
11. Vallejo Lozada, W.A.; Diaz-Uribe, C.; Quiñones, C.; Lerma, M.; Fajardo, C.; Navarro, K. Phthalocyanines: Alternative Sensitizers of TiO2 to be Used in Photocatalysis. In *Phthalocyanines and Some Current Applications*; Yilmaz, Y., Ed.; IntechOpen: London, UK, 2017. [CrossRef]
12. Ion, R. Porphyrins and Phthalocyanines: Photosensitizers and Photocatalysts. In *Phthalocyanines and Some Current Applications*; Yilmaz, Y., Ed.; IntechOpen: London, UK, 2017. [CrossRef]
13. Corma, A.; Garcia, H. Zeolite-based photocatalysts. *Chem. Commun.* **2004**, 1443–1459. [CrossRef] [PubMed]
14. Ranjit, K.T.; Willner, I.; Bossmann, S.; Braun, A. Iron(III) phthalocyanine-modified titanium dioxide: A novel photocatalyst for the enhanced photodegradation of organic pollutants. *J. Phys. Chem. B* **1998**, *102*, 9397–9403. [CrossRef]
15. Zsigmond, A.; Notheisz, F.; Bäckvall, J.-E. Rate enhancement of oxidation reactions by the encapsulation of metal phthalocyanine complexes. *Catal. Lett.* **2000**, *65*, 135–139. [CrossRef]
16. Perez-Bemal, E.; Ruano-Casero, R.; Pinnavaia, T.J. Catalytic autoxidation of 1-decanethiol by cobalt(II) phthalocyaninetetrasulfonate intercalated in a layered double hydroxide. *Catal. Lett.* **1991**, *11*, 55. [CrossRef]
17. Mohapatra, L.; Parida, K. A review on the recent progress, challenges and perspective of layered double hydroxides as promising photocatalysts. *Mater. Chem. A* **2016**, *4*, 10744. [CrossRef]
18. Gao, L.-G.; Gao, Y.-Y.; Song, X.-L.; Ma, X.-R. A novel La$^{3+}$-Zn$^{2+}$-Al$^{3+}$-MoO$_4{}^{2-}$ layered double hydroxides photocatalyst for the decomposition of dibenzothiophene in diesel oil. *Pet. Sci. Technol.* **2018**, *36*, 850–855. [CrossRef]
19. Barbosa, C.A.S.; Ferrira, A.M.D.C.; Constantino, V.R.L.; Coelho, A.C.V. Preparation and Characterization of Cu(II) Phthalocyanine Tetrasulfonate Intercalated and Supported on Layered Double Hydroxides. *J. Incl. Phenom. Macro. Chem.* **2002**, *42*, 15. [CrossRef]
20. Barbosa, C.A.S.; Dias, P.M.; Ferrira, A.M.C.; Constantino, V.R.L. Mg–Al hydrotalcite-like compounds containing iron-phthalocyanine complex: Effect of aluminum substitution on the complex adsorption features and catalytic activity. *Appl. Clay. Sci.* **2005**, *28*, 147. [CrossRef]
21. Barbosa, C.A.S.; Ferrira, A.M.D.C.; Constantino, V.R.L. Synthesis and Characterization of Magnesium-Aluminum Layered Double Hydroxides Containing (Tetrasulfonated porphyrin)cobalt. *Eur. J. Inorg. Chem.* **2005**, 1577. [CrossRef]
22. Kanan, S.; Awate, S.V.; Agashe, M.S. Incorporation of anionic copper phthalocyanine complexes into the intergallery of Mg-Al layered double hydroxides. *Stud. Surf. Sci. Catal.* **1998**, *113*, 927. [CrossRef]
23. Parida, K.M.; Baliarsingh, N.; Sairam Patra, B.; Das, J. Copperphthalocyanine immobilized Zn/Al LDH as photocatalyst under solar radiation for decolorization of methylene blue. *J. Mol. Catal. A Chem.* **2007**, *267*, 202–208. [CrossRef]
24. Maretti, L.; Carbonell, E.; Alvaro, M.; Scaiano, J.C.; Garcia, H. Laser flash photolysis of dioxo iron phthalocyanine intercalated in hydrotalcite and its use as a photocatalyst. *J. Photochem. Photobio. A Chem.* **2009**, *205*, 19–22. [CrossRef]
25. Abu-Reziq, R.; Alper, H.; Wang, D.; Post, M.L. Metal Supported on Dendronized Magnetic Nanoparticles: Highly Selective Hydroformylation Catalysts. *J. Am. Chem. Soc.* **2006**, *128*, 5279–5282. [CrossRef]
26. Karaoğlu, E.; Baykal, A.; Erdemi, H.; Alpsoy, L.; Sozeri, H. Synthesis and characterization of dl-thioctic acid (DLTA)–Fe$_3$O$_4$ nanocomposite. *J. Alloys Compd.* **2011**, *509*, 9218–9225. [CrossRef]
27. Naeimi, H.; Nazifi, Z.S. A highly efficient nano-Fe$_3$O$_4$ encapsulated-silica particles bearing sulfonic acid groups as a solid acid catalyst for synthesis of 1,8-dioxo-octahydroxanthene derivatives. *J. Nanopart. Res.* **2013**, *15*, 2026–2032. [CrossRef] [PubMed]
28. Brillas, E. A review on the photoelectro-Fenton process as efficient electrochemical advanced oxidation for wastewater remediation. Treatment with UV light, sunlight, and coupling with conventional and other photo-assisted advanced technologies. *Chemosphere* **2020**, *250*, 126198. [CrossRef]
29. Szabó, L.; Tóth, T.; Engelhardt, T.; Rácz, G.; Mohácsi-Farkas, S.; Takács, E.; Wojnárovits, L. Change in hydrophilicity of penicillinsduring advanced oxidation by radiolytically generated OH compromises the elimination of selective pressure on bacterial strains. *Sci. Total Environ.* **2016**, *551–552*, 393–403. [CrossRef]
30. Andreozzi, R.; Canterino, M.; Marotta, R.; Paxeus, N. Antibiotic removal from wastewaters: The ozonation of amoxicillin. *J. Hazard. Mater.* **2005**, *122*, 243–250. [CrossRef]

31. Rozas, O.; Contreras, D.; Mondaca, M.A.; Pérez-Moya, M.; Mansilla, H.D. Experimental design of Fenton and photo-Fenton reactions for the treatment of ampicillin solutions. *J. Haz. Mat.* **2010**, *177*, 1025–1030. [CrossRef]
32. Belhacova, L.; Bibova, H.; Marikova, T.; Kuchar, M.; Zouzelka, R.; Rathousky, J. Removal of Ampicillin by Heterogeneous Photocatalysis: Combined Experimental and DFT Study. *Nanomaterials* **2021**, *11*, 1992. [CrossRef]
33. Dimitrakopoulou, D.; Rethemiotaki, I.; Frontistis, Z.; Xekoukoulotakis, N.P.; Venieri, D.; Mantzavinos, D. Degradation, mineralization and antibiotic inactivation of AMX by UV-A/TiO$_2$ photocatalysis. *J. Environ. Manag.* **2012**, *98*, 168–174. [CrossRef] [PubMed]
34. Klauson, D.; Babkina, J.; Stepanova, K.; Krichevskaya, M.; Preis, S. Aqueous photocatalytic oxidation of amoxicillin. *Catal. Today* **2010**, *151*, 39–45. [CrossRef]
35. Shaykhi, Z.M.; Zinatizadeh, A.A.L. Statistical modeling of photocatalytic degradation of synthetic AMX wastewater in an immobilized TiO$_2$ photocatalytic reactor using response surface methodology. *J. Taiwan Inst. Chem. Eng.* **2014**, *45*, 1717–1726. [CrossRef]
36. Hou, J.; Chen, Z.; Gao, J.; Xie, Y.; Li, L.; Qin, S.; Wang, Q.; Mao, D.; Luo, Y. Simultaneous removal of antibiotics and antibioticresistance genes from pharmaceutical wastewater using the combinations of up-flow anaerobic sludge bed, anoxic-oxic tank, and advanced oxidation technologies. *Water Res.* **2019**, *159*, 511–520. [CrossRef] [PubMed]
37. Pouretedala, H.R.; Hasanali, M.A. Photocatalytic degradation of some b-lactam antibiotics in aqueous suspension of ZnS nanoparticles. *Desalination Water Treat.* **2013**, *51*, 2617–2623. [CrossRef]
38. Nosrati, R.; Olad, A.; Maramifar, R. Degradation of ampicillin antibiotic in aqueous solution by ZnO/polyaniline nanocomposite as photocatalyst under sunlight irradiation. *Environ. Sci. Pollut. Res.* **2012**, *19*, 2291–2299. [CrossRef]
39. Jassal, P.S.; Khajuria, R.; Sharma, R.; Debnath, P.; Verma, S.; Johnson, A.; Kumar, S. Photocatalytic degradation of ampicillin using silver nanoparticles biosynthesised by Pleurotus ostreatus. *BioTechnologia* **2020**, *101*, 5–14. [CrossRef]
40. Zăvoianu, R.; Mihăilă, S.-D.; Cojocaru, B.; Tudorache, M.; Parvulescu, V.I.; Pavel, O.D.; Oikonomopoulos, S.; Jacobsen, E.E. An advanced approach for MgZnAl-LDH catalysts synthesis used in Claisen-Schmidt condensation. *Catalysts* **2022**, *12*, 759. [CrossRef]
41. Pavel, O.D.; Stamate, A.-E.; Zăvoianu, R.; Bucur, I.C.; Bîrjega, R.; Angelescu, E.; Pârvulescu, V.I. Mechano-chemical versus co-precipitation for the preparation of Y-modified LDHs for cyclohexene oxidation and Claisen-Schmidt condensations. *Appl. Catal. A Gen.* **2020**, *605*, 117797. [CrossRef]
42. Prévot, V.; Casala, B.; Ruiz-Hitzky, E. Intracrystalline alkylation of benzoate ions into layered double hydroxides. *J. Mater. Chem.* **2001**, *11*, 554–560. [CrossRef]
43. Mastalir, Á.; Király, Z. Pd nanoparticles in hydrotalcite: Mild and highly selective catalysts for alkyne semihydrogenation. *J. Catal.* **2003**, *220*, 372–381. [CrossRef]
44. Carja, G.; Delahay, G. Mesoporous mixed oxides derived from pillared oxovanadates layered double hydroxides as new catalysts for the selective catalytic reduction of NO by NH$_3$. *Appl. Catal. B Environ.* **2004**, 47. [CrossRef]
45. Miyata, S. The Syntheses of Hydrotalcite-Like Compounds and Their Structures and Physico-Chemical Properties—I: The Systems Mg$^{2+}$-Al$^{3+}$-NO$_3^-$, Mg$^{2+}$-Al$^{3+}$-Cl$^-$, Mg$^{2+}$-Al$^{3+}$-ClO$_4^-$, Ni$^{2+}$-Al$^{3+}$-Cl$^-$ and Zn$^{2+}$-Al$^{3+}$-Cl$^-$. *Clays Clay Miner.* **1975**, *23*, 369–375. [CrossRef]
46. Yousefi, S.; Ghasemi, B.; Tajally, M.; Asghari, A. Optical properties of MgO and Mg(OH)$_2$ nanostructures synthesized by a chemical precipitation method using impure brine. *J. Alloys Compd.* **2017**, *711*, 521–529. [CrossRef]
47. Sakamoto, K.; Ohno-Okumura, E. Syntheses and Functional Properties of Phthalocyanines. *Materials* **2009**, *2*, 1127–1179. [CrossRef]
48. Zhang, C.; Tong, S.W.; Jiang, C.; Kang, E.T.; Chan, D.S.H.; Zhu, C. Simple tandem organic photovoltaic cells for improved energy conversion efficiency. *Appl. Phys. Lett.* **2008**, *92*, 68. [CrossRef]
49. Rajendran, K.; Balakrishnan, G.S.; Kalirajan, J. Synthesis of Magnetite Nanoparticles for Arsenic Removal from Ground Water Pond. *Int. J. PharmTech Res.* **2015**, *8*, 670–677.
50. Pavel, O.D.; Zăvoianu, R.; Bîrjega, R.; Angelescu, E.; Pârvulescu, V.I. Mechanochemical versus co-precipitated synthesized lanthanum-doped layered materials for olefin oxidation. *Appl. Catal. A Gen.* **2017**, *542*, 10–20. [CrossRef]
51. Huang, F.; Tian, S.; Qi, Y.; Li, E.; Zhou, L.; Qiu, Y. Synthesis of FePcS-PMA-LDH Cointercalation Composite with Enhanced Visible Light Photo-Fenton Catalytic Activity for BPA Degradation at Circumneutral pH. *Materials* **2020**, *13*, 1951. [CrossRef]
52. Hur, T.-B.; Phuoc, T.X.; Chyu, M.K. New approach to the synthesis of layered double hydroxides and associated ultrathin nanosheets in de-ionized water by laser ablation. *J. Appl. Phys.* **2010**, *108*, 114312. [CrossRef]
53. Nairi, V.; Medda, L.; Monduzzi, M.; Salis, A. Adsorption and release of ampicillin antibiotic from ordered mesoporous silica. *J. Colloid Interface Sci.* **2017**, *497*, 217–225. [CrossRef] [PubMed]
54. Elhaci, A.; Labed, F.; Khenifi, A.; Bouberka, Z.; Kameche, M.; Benabbou, K. MgAl-Layered double hydroxide for amoxicillin removal from aqueous media. *J. Environ. Anal. Chem.* **2020**, *101*, 2876–2898. [CrossRef]
55. Dogan, S.; Kidak, R. A Plug flow reactor model for UV-based oxidation of amoxicillin. *Desalin. Water Treat.* **2016**, *57*, 13586–13599. [CrossRef]

Article

# Green Chemo-Enzymatic Protocols for the Synthesis of Enantiopure β-Blockers (S)-Esmolol and (S)-Penbutolol

Susanne Hansen Trøøyen, Lucas Bocquin, Anna Lifen Tennfjord, Kristoffer Klungseth and Elisabeth Egholm Jacobsen *

Department of Chemistry, Norwegian University of Science and Technology, Høgskoleringen 5, 7491 Trondheim, Norway
* Correspondence: elisabeth.e.jacobsen@ntnu.no

**Abstract:** The β-blocker (S)-esmolol, has been synthesized in 97% enantiomeric excess and 26% total yield in a four-step synthesis, with a transesterification step of the racemic chlorohydrin methyl 3-(4-(3-chloro-2-hydroxypropoxy)phenyl)propanoate, catalysed by lipase B from *Candida antarctica* from Syncozymes, Shanghai, China. The β-blocker (S)-penbutolol, has been synthesized in 99% enantiomeric excess and in 22% total yield. The transesterification step of the racemic chlorohydrin 1-chloro-3-(2-cyclopentylphenoxy)propan-2-ol was catalyzed by the same lipase as used for the esmolol building block. We have used different bases for the deprotonation step of the starting phenols, and vinyl butanoate as the acyl donor in the transesterification reactions. The reaction times for the kinetic resolution steps catalysed by the lipase varied from 23 to 48 h, and were run at 30–38 °C. Specific rotation values confirmed the absolute configuration of the enantiopure drugs, however, an earlier report of the specific rotation value of (S)-esmolol is not consistent with our measured specific rotation values, and we here claim that our data are correct. Compared to the previously reported syntheses of these two enantiopure drugs, we have replaced toluene or dichloromethane with acetonitrile, and replaced the flammable acetyl chloride with lithium chloride. We have also reduced the amount of epichlorohydrin and bases, and identified dimeric byproducts in order to obtain higher yields.

**Keywords:** (S)-esmolol; (S)-penbutolol; enantiopure building blocks; characterisation of a dimeric by-product; *Candida antarctica* lipase B; chiral chromatography

Citation: Trøøyen, S.H.; Bocquin, L.; Tennfjord, A.L.; Klungseth, K.; Jacobsen, E.E. Green Chemo-Enzymatic Protocols for the Synthesis of Enantiopure β-Blockers (S)-Esmolol and (S)-Penbutolol. *Catalysts* 2022, 12, 980. https://doi.org/10.3390/catal12090980

Academic Editor: Takeshi Sugai

Received: 1 August 2022
Accepted: 28 August 2022
Published: 31 August 2022

**Publisher's Note:** MDPI stays neutral with regard to jurisdictional claims in published maps and institutional affiliations.

**Copyright:** © 2022 by the authors. Licensee MDPI, Basel, Switzerland. This article is an open access article distributed under the terms and conditions of the Creative Commons Attribution (CC BY) license (https://creativecommons.org/licenses/by/4.0/).

## 1. Introduction

We have previously developed efficient synthesis protocols for the syntheses of single enantiomers of several β-adrenergic receptor blockers and their chiral building blocks [1–3]. We present here the syntheses of enantiopure (S)-esmolol and (S)-penbutolol (Figure 1), with focus on green chemistry principles in every reaction step. By the use of kinetic resolution, high enantiomeric excess of the corresponding secondary alcohols as building blocks for such compounds can be obtained. The only drawback with this method is that the enantiopure product can only be obtained in a 50% yield. Dynamic kinetic resolution can be used to improve the yields of the kinetic resolutions, which will lower the amount of waste produced, and thus give an even greener synthesis of the enantiopure drugs [4].

Figure 1. (S)-Esmolol (**left**) and (S)-penbutolol HCl salt (**right**).

We also give more in-depth information about these syntheses, which we have found to be missing in previous reports. This includes an accurate measurement of the specific rotation of (S)-esmolol and precursors of both (S)-esmolol and (S)-penbutolol, and also the characterisation of a dimeric by-product formed in the synthesis path to (S)-esmolol. To limit formation of by-products, knowledge of their identity is essential, which is why we also provide plausible mechanisms for the formation of these dimers.

Esmolol is a hydrophilic $\beta_1$-adrenergic receptor blocker, which has a rapid onset and a short duration of action [5,6]. The drug is administered intravenously, and is widely used in the treatment of hypertension, cardiac arrhythmia, and angina pectoris. The most potent enantiomer (eutomer) of esmolol is (S)-esmolol [7]. The drug is manufactured as Brevibloc®, with a racemic active pharmaceutical ingredient (API). Two methods for the synthesis of (S)-esmolol have been reported. Narsaiah and Kumar obtained the epoxide (S)-methyl 3-(4-(oxiran-2-ylmethoxy)phenyl)propanoate in 94% enantiomeric excess (*ee*) by kinetic resolution of the racemic epoxide, catalysed by Jacobsens catalyst. Subsequent amination gave (S)-esmolol, however, the authors do not report any *ee* value of the product [8]. Banoth and Banerjee have reported a non-enzymatic and a chemo-enzymatic route to (S)-esmolol. Commercially available (R)-epichlorohydrin was used as a starting material, however, by this method, (S)-esmolol was obtained only in 93% *ee*. In a kinetic resolution of the chlorohydrin methyl 3-(4-(3-chloro-2-hydroxypropoxy)phenyl)-propanoate with vinyl acetate in toluene, lipase from *Pseudomonas cepacia* was used as the catalyst. The chlorohydrin was obtained in 98% *ee* and is reported to have *R*-configuration. Amination of the chlorohydrin gave (S)-esmolol in 98% *ee* [7]. Substituting toluene for a safer and more sustainable solvent in the enzymatic step is desired for a greener synthesis. Acetonitrile has lower toxicity and environmental impact than toluene, and would be a preferred alternative [9], however, Banoth and Banerjee did not find acetonitrile to give high enantioselectivity when using lipase from *Pseudomonas cepacia* [7].

Penbutolol is a non-selective $\beta$-blocker used in the treatment of hypertension. The drug inhibits both $\beta_1$- and $\beta_2$-adrenergic receptors in the heart and in the kidneys. Betapressin® is manufactured with the racemic API. Penbutolol sulphate is manufactured as Levatol® with enantiopure (S)-penbutolol sulphate as a prodrug giving (S)-penbutolol when it enters the body. There are several methods reported for the synthesis of (S)-penbutolol, however, many of these methods use expensive enzymes and resolving agents, and suffer from low yields and low enantiomeric purity. As early as 1984, Hamaguchi et al. obtained (S)-penbutolol in 100 % *ee* with lipase Amano 3 as a catalyst in the hydrolysis of the racemic building block 3-(*tert*-butyl)-5-(hydroxymethyl)oxazolidin-2-one. However, this method has many steps, and although the lipase is efficient, it is no longer listed on the market [10]. In a kinetic resolution of the corresponding chlorohydrin using lipase from *Pseudomonas* sp., Ader et al. obtained (S)-penbutolol, however, only in 91% *ee* [11]. (S)-Penbutolol has also been obtained in 95% *ee* by the use of Sharpless asymmetric dihydroxylation. The drawback of this protocol is the use of toxic and expensive catalysts, and the *ee* of the (S)-penbutolol obtained is not optimal. The authors also report an efficient synthesis of the starting material 2-cyclopentylphenol, which today is available from Merck, but is quite expensive. [12]. Klunder et al. reported in 1989, the synthesis of (S)-penbutolol in 86% *ee* by the addition of enantiopure (2S)-glycidyl tosylate to a penbutolol precursor [13]. Kan et al. have reported a synthesis of (S)-penbutolol hydrochloride from racemic 5-acyloxymethyl-3-alkyl-2-oxazolidinones resolved with lipases or microorganisms [14].

## 2. Results and Discussion

### 2.1. Synthesis of Racemic Chlorohydrin 3 (for Esmolol)

The chlorohydrin (R)-methyl 3-(4-(3-chloro-2-hydroxypropoxy)phenyl)-propanoate, (R)-**3**, which is a chiral building block in the synthesis of the $\beta$-blocker (S)-esmolol ((S)-**5**), has been synthesised in 97% *ee* (Scheme 1). A deprotonation of the commercial phenol methyl 3-(4-hydroxyphenyl)propanoate (**1a**) gave the corresponding alkoxide, which in reacting with epichlorohydrin, gave the chlorohydrin **3**. A transesterification reaction of **3**, catalysed by lipase B from *Candida antarctica*, gave the corresponding *S*-butanoate ester (*S*)-**4** and the *R*-chlorohydrin (*R*)-**3**. From the reaction of (*R*)-**3** with isopropylamine, (*S*)-esmolol was synthesised in 97% *ee* and 26 % overall yield. We have previously shown that the type and concentration of base used in the reaction of phenolic starting materials with epichlorohydrin strongly influences the ratio of epoxide vs. chlorohydrin, and also the ratio of by-products formed in these reactions [3]. In the synthesis of epoxide **2** and chlorohydrin **3**, sodium hydroxide favors the formation of chlorohydrin **3** over the epoxide. However, full conversion of the starting material **1a** was not obtained. We have previously used catalytic amounts of the base in the syntheses of similar compounds, but in the deprotonation of the phenolic starting material **1a**, this did not give sufficient conversion. By the use of potassium carbonate, epoxide **2** was obtained in 68% yield and with the full conversion of **1a**. Ring opening of epoxide **2** was performed by the protonation with acetic acid and opening with lithium chloride in acetonitrile to give chlorohydrin **3** in 96% yield for this step. We have managed to increase the yield from the previously reported 92% [7] in the epoxide ring opening, with the use of LiCl and acetic acid in acetonitrile instead of the highly flammable acetyl chloride and less preferable solvent dichloromethane, at the expense of slightly longer reaction times.

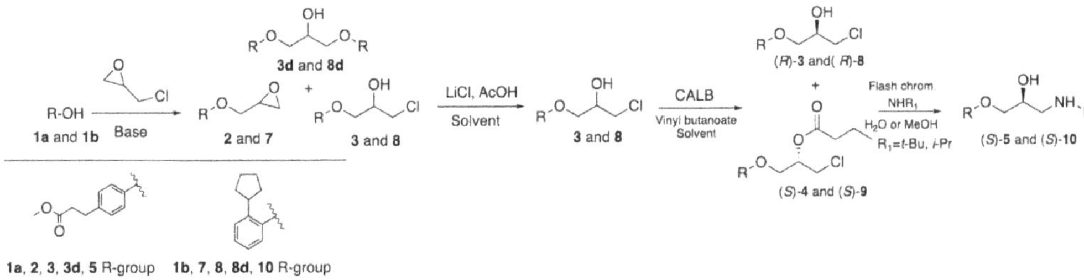

**Scheme 1.** Building blocks (*R*)-**3** and (*R*)-**8** synthesised in 97-99% *ee* for use in synthesis of the (*S*)-enantiomers of the β-blockers ((*S*)-esmolol ((*S*)-**5**) and (*S*)-penbutolol ((*S*)-**10**) with the same enantiopurity as the respective building blocks.

*2.2. Synthesis of Racemic Chlorohydrin **8** (for Penbutolol)*

Chlorohydrin (*R*)-1-chloro-3-(2-cyclopentylphenoxy)propan-2-ol, (*R*)-**8**, which is used as a chiral building block for the β-blocker (*S*)-penbutolol ((*S*)-**10**), was synthesised in 99% *ee* by a similar chemo-enzymatic method as for (*R*)-**3** (Scheme 1). From the reaction of (*R*)-**8** with isopropylamine, (*S*)-penbutolol was synthesised in 99% *ee* and 27% overall yield. The reaction between 2-cyclopentylphenol (**1b**) and epichlorohydrin gave the highest conversion of the starting material into the products epoxide 2-((2-cyclopentylphenoxy)methyl) oxirane (**7**) and the chlorohydrin 1-chloro-3-(2-cyclopentylphenoxy)propan-2-ol (**8**), when 1.5 equivalents of sodium hydroxide were used, and the reaction time was 48 h. $^1$H NMR analysis of the obtained mixture before opening of the epoxide **7** revealed the presence of around 13% of the starting material 2-cyclopentylphenol (**1b**), 36% of the epoxide **7** and 39% of chlorohydrin **8**. Other impurities, most likely a dimer, similar to the by-product **3d** described for esmolol synthesis, makes up approximately 12% of the reaction mixture. The by-products have not been further characterised. From this mixture, chlorohydrin **8** was obtained in 70% yield by the same method as described for the esmolol building block **3**.

*2.3. Characterisation of by-Products in Synthesis of Esmolol Precursors*

A dimeric by-product was observed in the reaction between deprotonated phenol **1a**, epichlorohydrin, and potassium carbonate to form epoxide **2** (Schemes 1 and 2). LC-MS analysis gave a peak with $m/z$ = 439.2, molecular formula $C_{23}H_{28}O_7Na$, which corresponds to 3,3'-(((2-hydroxypropane-1,3-diyl)bis(oxy))bis(4,1-phenylene))dipropanoate (**3d**) with a molecular mass of 416.45 g/mol. By purification of the reaction mixture using flash chromatography in order to obtain pure epoxide **2**, dimer **3d** was isolated in 6% yield with a purity of 99%. Characterisation of **3d** was performed by $^1$H-, $^{13}$C-, H,H COSY-, HSQC- and HMBC NMR spectroscopy, with deuterated chloroform as the solvent. $^1$H- and $^{13}$C-NMR spectra for dimer **3d** are given in the Supplementary Materials, in addition to the achiral HPLC chromatogram. We have observed the same type of dimers in the synthesis of similar β-blocker precursors [3], and here we propose plausible mechanisms for the **3d** dimer formation, see Scheme 2. We suggest that using a larger amount of epichlorohydrin could limit the amount of dimer formed, as the alkoxide **1a**$_{\text{Anion}}$ would have better access to epichlorohydrin as opposed to the products **2** and **3**, which could improve the yield of the first reaction step. However, we have managed to reduce the epichlorohydrin from 5 equivalents in similar syntheses reported by Bevinakatti and Banerji in 1992 [15] to 2 equivalents, also used in our previous report on the syntheses of several β-blockers [3].

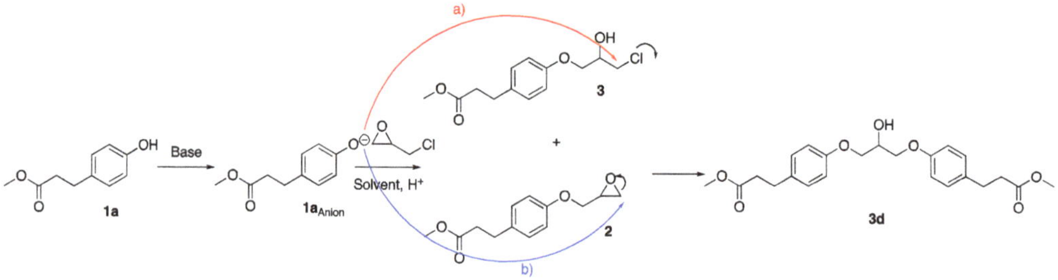

**Scheme 2.** Two suggested mechanistic pathways for the formation of dimer **3d**. Reaction mechanism (a) shows a nucleophilic attack on chlorohydrin **3** by alkoxide **1a**$_{\text{Anion}}$, mechanism (b) shows a nucleophilic attack on epoxide **2** by alkoxide **1a**$_{\text{Anion}}$.

### 2.4. Lipase-Catalysed Kinetic Resolution of Chlorohydrins **3** and **8**

Kinetic resolution of the chlorohydrins **3** and **8** was catalysed by lipase B from *Candida antarctica* (CALB) in dry acetonitrile with vinyl butanoate as the acyl donor. This gave E-values of 157 and 183, respectively (calculated by E&K Calculator, 2.1b0 PPC) [14] (Figures 2 and 3, respectively), and ee-values of 87% for S-ester (S)-**4** and 97-99% of the R-chlorohydrins (R)-**3** and (R)-**8**, as described above. The ee-values were retained upon the conversion of the enantiopure chlorohydrins to the respective drugs, see Table 1. The reaction times for the transesterification reactions for obtaining (R)-**3** and (R)-**8** were 23 and 48 h, respectively. The use of acetonitrile as the solvent in these kinetic resolutions makes the syntheses greener than the previous reports using toluene [7], and here we have shown that acetonitrile gives high selectivity for CALB in the synthesis of R-chlorohydrins (R)-**3** and (R)-**8**.

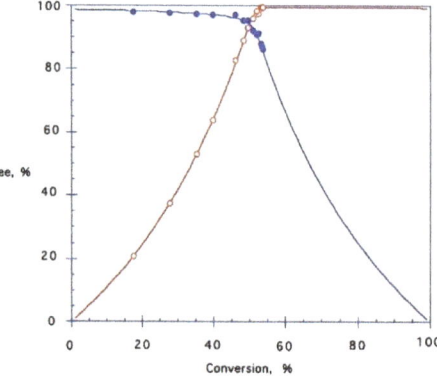

**Figure 2.** Graphical representation of the kinetic resolution of chlorohydrin **3** with CALB in dry acetonitrile

with vinyl butanoate as the acyl donor. Enantiomeric excess of the remaining substrate ($ee_S$, red circles) and enantiomeric excess of the product ester ($ee_P$, blue circles) is shown in percent plotted against conversion in percent. The red and the blue curves are generated from the experimental values of $ee_S$ and $ee_P$, respectively. The E-value was calculated to be 157. E-values were calculated from E&K Calculator 2.1b0 PPC [16].

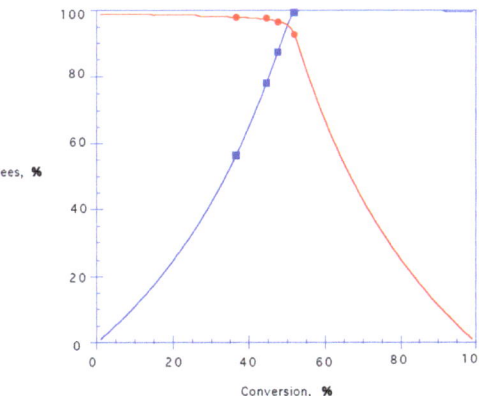

**Figure 3.** Graphical representation of the kinetic resolution of chlorohydrin **8** with CALB in dry acetonitrile with vinyl butanoate as the acyl donor. $ee_S$ (blue squares) and $ee_P$ (red circles) in percent plotted against conversion in percent. The blue and the red curves are generated from the experimental values of $ee_S$ and $ee_P$, respectively. The E-value was calculated to be 183. E-values were calculated from E&K Calculator 2.1b0 PPC [16].

**Table 1.** E-values, ee values and yields of the enantiopure chlorohydrins (R)-**3** and (R)-**8**, ester (S)-**4**, and the drugs (S)-**5**, (S)-**10** and (S)-**10**·HCl. The kinetic resolutions were catalysed by CALB from Syncozymes in dry acetonitrile. Specific rotations $[\alpha]_D^T$ were determined at 20–23 °C in different solvents with c = 1. For additional parameters, see Materials and Methods. The yields for each compound in the table are for each reaction step. The overall yield for (S)-esmolol ((S)-**5**) is 26%, and for (S)-penbutolol hydrochloride ((S)-**10**·HCl) the overall yield is 20%.

| Chloro-Hydrin | E-Value | Chlorohydrin ee, Yield, % | Specific Rotation | Ester, ee, Yield, % | Specific Rotation | Drug, ee, Yield, % | Specific Rotation |
|---|---|---|---|---|---|---|---|
| (R)-**3** | 157 | (R)-**3a**, 97, 43 | $[\alpha]_D^{20} = -5.33$ (c 1.6, i-PrOH) | (S)-**4**, 87, 41 | $[\alpha]_D^{20} = +30.71$ (c 1.4, i-PrOH) | (S)-**5**, 97, 92 | $[\alpha]_D^{20} = -6.80$ (c 1.03, CHCl$_3$) |
| (R)-**8** | 183 | (R)-**8**, 99, 39 | $[\alpha]_D^{25} = -14.00$ (c 1.6, MeOH), | | | (S)-**10**·HCl, 99, 89 | $[\alpha]_D^{20} = -23.00$ (c 1.0, MeOH), |
| | | | | | | (S)-**10**, 99, 82 | $[\alpha]_D^{20} = -14.00$ (c 1.0, MeOH), |

### 2.5. Synthesis of (S)-Esmolol ((S)-**5**)

The R-Chlorohydrin (R)-**3** was converted to (S)-esmolol ((S)-**5**) in 92% yield by amination with isopropylamine in methanol. The ee was retained in the conversion. For the ee to be retained, it is important that the S-ester ((S)-**4**) and R-chlorohydrin (R)-**3** are completely separated during the flash chromatography separation of the crude mixture from the enzymatic kinetic resolution step. If the separation was not complete, we observed a lowering of the ee. We suggest that this lowering of the ee could be a result of aminolysis of the S-ester, followed by amination of the resulting S-chlorohydrin to give the unwanted (R)-esmolol ((R)-**5**).

## 2.6. Synthesis of (S)-Penbutolol ((S)-10)

The R-Chlorohydrin (R)-8 was converted to (S)-penbutolol ((S)-10) in 82% yield by amination with *tert*-butylamine in methanol and with an *ee* of 99%.

## 2.7. Specific Rotation of Pure Enantiomers

Specific rotation values for the enantiopure chlorohydrin (R)-3 or the corresponding ester (S)-4 have not been reported previously. The absolute configuration of compounds (R)-3, (S)-4, and (S)-5 shown in Table 1 was determined by the known enantioselectivity of CALB towards similar compounds, which has been previously reported [3,17]. In 2011, Narsaiah and Kumar reported a specific rotation for the S-enantiomer of esmolol of $[\alpha]_D^{20} = +4.50$ (c 1, CHCl$_3$) [8], while we report $[\alpha]_D^{20} = -6.80$ (c 1.04, CHCl$_3$) for the S-enantiomer of esmolol ((S)-5) in 97% *ee*. Narsaiah and Kumar did not report the enantiomeric excess of their (S)-esmolol, nor how the absolute configuration was determined. We claim that our measurements are correct, and that (S)-esmolol is levorotatory. The specific rotation value of (R)-8 has not been reported previously. The absolute configuration of compound (R)-8 was determined by the enantioselectivity of CALB, which we have reported previously [3,17]. The specific rotation of (S)-penbutolol as a free base has been reported by Phukan and Sudalai to be $[\alpha]_D^{20} = -10.90$ (c 0.8, MeOH) in 95% *ee*. We here report $[\alpha]_D^{20} = -14.00$ (c 1.0, MeOH) for (S)-10 in 99% *ee*. The specific rotation of the hydrochloric salt of (S)-penbutolol ((S)-10·HCl) has been reported by Kan et al. to be $[\alpha]_D^{20} - 26.40$ (c 1.0, MeOH). We have synthesized (S)-10·HCl from (S)-10 and hydrochloric acid in isopropanol, in 89% yield, with an *ee* of 99% and a specific rotation of $[\alpha]_D^{20} = -23.00$ (c 1.0, MeOH), which is consistent with Kan et al. [14].

## 3. Materials and Methods

### 3.1. Chemicals and Solvents

All chemicals used in this project are commercially available, of analytical grade and were purchased from either Sigma-Aldrich Norway AS (Oslo, Norway), or VWR International AS, Norway (Oslo, Norway). HPLC-grade solvents were used for the HPLC analyses. Dry MeCN was acquired from a solvent purifier, MBraun MD-SPS800 (München, Germany), and stored in a flask containing molecular sieves (4Å).

### 3.2. TLC Analyses and Column Chromatography

TLC analyses were performed on Merck silica 60 F$_{254}$ and detection with UV at $\lambda$ = 254 nm. Flash chromatography was performed using silica gel from Sigma-Aldrich Norway AS (Oslo, Norway) (pore size 60 Å, 230–400 mesh particle size, 40–63 µm particle size).

### 3.3. Enzymes

*Candida antarctica* Lipase B (CALB) (activity $\geq$ 10,000 PLU/g, lot#20170315) immobilised on highly hydrophobic macro porous resin, and produced in fermentation with genetically modified *Pichia pastoris*, was a gift from SyncoZymes Co. Ltd. (Shanghai, China). The enzyme reactions were performed in a New Brunswick G24 Environmental Incubator Shaker from New Brunswick Co. (Edison, NJ, USA).

### 3.4. Chiral HPLC Analyses

All chiral analyses were performed on Agilent 1100 and 1200 HPLC systems using a manual injector (Rheodyne 77245i/Agilent 10 mL loop (Agilent 1100), an autosampler (Agilent 1200), and a variable wavelength detector (VWD) set to 254 nm). Separations of enantiomers were performed on a Chiralcel OD-H column (250 mm × 4.6 mm ID, 5 µm particle size, Daicel, Chiral Technologies Europe, Gonthier d'Andernach, Illkirch, France). Chlorohydrin 3 enantiomers: (*n*-hexane:*i*-PrOH, 80:20), flow 1 mL/min, 10 µL injection, $t_R((S)$-3) = 9.4 min, $t_R((R)$-3) = 10.3 min, $R_s((S)/(R)$-3) = 1.86. Ester 4 enantiomers: (*n*-hexane:*i*-PrOH, 97:3), flow 1 mL/min, 10 µL injection, $t_R((S)$-4) = 15.6 min, $t_R((R)$-4) = 17.3 min, $R_s((S)/(R)$-4) = 2.08. Esmolol (5) enantiomers: (*n*-hexane:*i*-PrOH:Et$_2$NH

(80:19.6:0.4), flow 1 mL/min, 10 µL injection, $t_R((R)\text{-}5)$ = 5.8 min and $t_R((R)\text{-}5)$ = 9.3 min, $R_s((S)/(R)\text{-}5)$ = 11.5. Chlorohydrin **8** enantiomers: (hexane:*i*-PrOH, 90:10), flow 1 mL/min, 10 µL injection. $t_R((S)\text{-}8)$ = 6.63 min and $t_R((R)\text{-}8)$ = 7.31 min. $R_s((S)/(R)\text{-}8)$ = 2.41. Ester **9** enantiomers: (hexane:*i*-PrOH, 99.4:0.6), flow 1 mL/min, 10 µL injection. $t_R((S)\text{-}9)$ = 8.05 min and $t_R((R)\text{-}9)$ = 9.37 min. $R_s((S)/(R)\text{-}9)$ = 1.56. Penbutolol **10** enantiomers: (hexane:*i*-PrOH:Et$_2$NH, 90:9.8:0.2), flow 1 mL/min, 10 µL injection. $t_R((R)\text{-}10\cdot\text{HCl})$ = 4.67 min and $t_R((S)\text{-}10\cdot\text{HCl})$ = 7.39 min. $R_S((S)/(R)\text{-}10\cdot\text{HCl})$ = 10.5.

*3.5. Achiral HPLC Analysis of Dimer* **3d**

Achiral HPLC analysis of dimer **3d** was performed on an Agilent 1290 system from Matriks AS (Oslo, Norway), equipped with an auto injector (4 µL). Detection was performed by a diode array detector (DAD, $\lambda$ = 254 nm). Also used was an ACE Excel 5 C18 column from Matriks AS (Oslo, Norway) (150 mm × 4.6 mm ID; 5 µm particle size), with an isocratic eluent (H$_2$O:MeCN, 50:50) over 12 min, flow = 1 mL/min. Dimer **3d** $t_R$ = 9.1 min.

*3.6. Liquid Chromatography-Mass Spectroscopy (LC-MS) of Esmolol Dimer* **3d**

LC-MS analysis of by-product **3d** was performed on an AQUITY UPLC I-Class system (Waters, Milford, CT, USA) coupled to a quadrupole time-of-flight mass analyzer (QTOF; SYNAPT-G2S) with a ZSpray EIS ion source (Waters, Milford, CT, USA). A AQUITY UPLC BEH C18 column (100 mm × 2.1 mm ID, 130Å, 1.7 µm particle size) with a mobile phase composition of H$_2$O and MeCN, both with 0.1% formic acid. Method: Isocratic (H$_2$O:MeCN, 80:20) over 12 min, then gradient (100% MeCN) from 12-13.5 min, and then back to (H$_2$O:MeCN, 80:20) for 15 min, flow 0.25 mL/min. Molecular mass of **3d** 3,3'-(((2-hydroxypropane-1,3-diyl)bis(oxy))bis(4,1-phenylene))dipropanoate is 416.45 g/mol. LC-MS analysis gave a peak with $m/z$ = 439.2, molecular formula C$_{23}$H$_{28}$O$_7$Na.

*3.7. Mass Spectrometry Analysis of by-Product* **3d**

Exact mass of **3d** was determined with a Synapt G2-S Q-TOF mass spectrometer from Waters[TM] (Waters Norway, Oslo, Norway). Ionization of the sample was performed with an ASAP probe (APCI), and the calculation of exact masses and spectra processing were performed with Waters[TM] Software (Masslynxs V4.1 SCN871). See Supplementary Materials for spectra.

*3.8. Optical Rotation*

Optical rotation values were performed with an Anton Paar MCP 5100 polarimeter from Dipl.Ing. Houm AS (Oslo, Norway), and a wavelength of 589 nm (D), for values, see single enantiomers for specific rotation values.

*3.9. Absolute Configurations*

The absolute configuration of (S)-esmolol ((S)-**5**) was determined by the enantioselectivity of CALB which we have reported previously [3,17]. We report here a specific rotation of (S)-esmolol ((S)-**5**), disputing the previously reported value [8]. Specific rotation values of (R)-**3**, (S)-**4** and (R)-**8** have not been reported previously and were determined by the enantioselectivity of CALB, which we have reported previously [3,17]. The absolute configuration of (S)-penbutolol (S)-**10** and (S)-**10**·HCl were determined by comparing the specific rotation with previously reported data [12,14].

*3.10. NMR Analyses*

NMR analyses were recorded on a Bruker 400 MHz Avance III HD instrument equipped with a 5 mm SmartProbe Z-gradient probe operating at 400 MHz for $^1$H and 100 MHz for $^{13}$C, respectively, or on a Bruker 600 MHz Avance III HD instrument equipped with a 5 mm cryogenic CP-TCI Z-gradient probe operating at 600 MHz for $^1$H and 150 MHz for $^{13}$C (Bruker, Rheinstetten, Germany). Chemical shifts are in ppm relative to TMS (or

CHCl$_3$ shift), and coupling constants are in hertz (Hz). $^1$H- and $^{13}$C NMR spectra can be found in the Supplementary Materials.

*3.11. Synthesis Protocols*

3.11.1. Methyl 3-(4-(Oxiran-2-ylmethoxy)phenyl)propanoate (**2**)

To a stirred solution of methyl 3-(4-hydroxyphenyl)propanoate (**1a**) (0.25 g, 2.84 mmol) in dry MeCN (30 mL), K$_2$CO$_3$ (1.01 g, 7.40 mmol) and epichlorohydrin (0.44 mL, 5.68 mmol) were added. The mixture was heated under reflux for 45 h. Full conversion was detected by TLC (CH$_2$Cl$_2$:MeCN, 11:1, $v/v$), $R_f$ (**1a**) = 0.39, $R_f$ (**2**) = 0.66. The reaction mixture was filtered, and the filtrate was concentrated under reduced pressure. EtOAc (25 mL) was added, and the resulting solution was washed with distilled H$_2$O (10 mL). The water phase was extracted with EtOAc (3 × 15 mL). The organic phases were combined, washed with saturated NaCl solution (20 mL) and dried over MgSO$_4$. The crude mixture (0.62 g) was purified by flash chromatography (CH$_2$Cl$_2$:MeCN, 11:1, $v/v$) to afford **2** as a clear liquid in 68% yield (0.45 g, 1.92 mmol) and 99% purity ($^1$H NMR). $^1$H NMR (600 MHz, CDCl$_3$) δ: 7.10–7.12 (m, 2H, Ar-**H**), 6.83–6.86 (m, 2H, Ar-**H**), 4.19 (dd, 1H, $^2J$ = 11.01 Hz, $^3J$ = 3.20 Hz, **CH**$_2$-O), 3.94 (dd, 1H, $^2J$ = 11.01 Hz, $^3J$ = 5.65 Hz, **CH**$_2$-O), 3.66 (s, 3H, **CH**$_3$), 3.34 (ddt, 1H, $^2J$ = 2.70 Hz, $^3J$ = 3.20 Hz, $^3J$ = 5.65 Hz, **CH**-O), 2.88–2.91 (m, 3H, Ar-**CH**$_2$/**CH**$_2$-O), 2.75 (dd, 1H, $^2J$ = 4.94, $^3J$ = 2.70, **CH**$_2$-O), 2.59 (t, 2H, **CH**$_2$COOR). $^{13}$C NMR (150 MHz, CDCl$_3$) δ: 173.4, 157.0, 133.2, 129.3, 114.7, 68.8, 51.6, 50.2, 44.7, 35.9, 30.1.

3.11.2. Methyl 3-(4-(3-Chloro-2-hydroxypropoxy)phenyl)propanoate (**3**)

To a stirred solution of methyl 3-(4-(oxiran-2-ylmethoxy)phenyl)propanoate (**2**) (0.44 g, 1.86 mmol) and LiCl (0.16 g, 3.72 mmol) in MeCN (10 mL), glacial AcOH (350 µL, 9.30 mmol) was added. The solution was stirred at rt for 47 h. The reaction was monitored by TLC (CH$_2$Cl$_2$:MeCN, 11:1, $v/v$), $R_f$ (**2**) = 0.66, $R_f$ (**3**) = 0.44. The reaction was quenched with Na$_2$CO$_3$ (aq) and extracted with CH$_2$Cl$_2$ (3 × 20 mL). The organic phase was washed with saturated NaCl solution (10 mL) and dried over MgSO$_4$ before the solvent was removed under reduced pressure. Chlorohydrin **3** was obtained as a light-yellow oil in 96% yield (0.49 g, 1.79 mmol) and 96% purity ($^1$H NMR). $^1$H NMR (600 MHz, CDCl$_3$) δ: 7.11–7.14 (m, 2H, Ar-**H**), 6.83–6.85 (m, 2H, Ar-**H**), 4.18–4.23 (m, 1H, **CH**-OH), 4.04–4.09 (m, 2H, **CH**$_2$-O), 3.71–3.79 (m, 2H, **CH**$_2$-Cl), 3.66 (s, 3H, **CH**$_3$), 2.90 (t, 2H, $^3J$ = 7.92 Hz, **CH**$_2$-Ar), 2.60 (t, 2H, $^3J$ = 7.92 Hz, **CH**$_2$COOR), 2.49–2.52 (m, 1H, -O**H**). $^{13}$C NMR (150 MHz, CDCl$_3$) δ: 173.4, 156.8, 133.5, 129.4, 114.7, 68.9, 68.6, 51.6, 46.0, 36.0, 30.1.

3.11.3. 3,3′-(((2-Hydroxypropane-1,3-diyl)bis(oxy))bis(4,1-phenylene))dipropanoate (**3d**)

By the purification of epoxide **2** by flash chromatography (CH$_2$Cl$_2$:MeCN, 11:1, $v/v$), dimer **3d** was isolated in 6% yield (33.2 mg, 0.08 mmol), purity 99% (HPLC). $^1$H NMR (600 MHz, CDCl$_3$) δ: 7.10–7.13 (m, 4H, Ar-**H**), 6.84–6.89 (m, 4H, Ar-**H**), 4.34–4.37 (quint., 1H, $^3J$ = 5.40 Hz), 4.09–4.15 (m, 4H, -**CH**$_2$-O) 3.66 (s, 6H, -**CH**$_3$), 2.88–2.91 (t, 4H, $^3J$ = 7.85 Hz, -**CH**$_2$-), 2.58–2.61 (t, 4H, $^3J$ = 7.85 Hz, **CH**$_2$), 2.55-2.56 (d, 1H, $^3J$ = 5.10 Hz, -O**H**). $^{13}$C NMR (150 MHz, CDCl$_3$) δ: 173.4 (2C), 156.0 (2C), 133.2 (2C), 129.4 (4C), 114.6 (4C), 68.9 (4C), 68.8 (1C), 51.6 (2C), 36.0 (4C), 30.1 (4C).

3.11.4. Synthesis of Chlorohydrin (*R*)-**3** and Ester (*S*)-**4** by CALB Catalysed Kinetic Resolution of Methyl 3-(4-(3-Chloro-2-hydroxypropoxy)phenyl)propanoate (**3**)

To a solution of methyl 3-(4-(3-chloro-2-hydroxypropoxy)phenyl)propanoate (**3**) (0.13 g, 0.49 mmol) in dry MeCN (10 mL) containing activated 4Å molecular sieves, vinyl butanoate (249 µL, 1.96 mmol) and CALB (150 mg) was added. The reaction vial was capped and placed in an incubator at 37 °C and stirred at 200 rpm for 48 h. The reaction mixture was filtered and the solvent was removed under reduced pressure before the separation of (*R*)-methyl 3-(4-(3-chloro-2-hydroxypropoxy)phenyl)propanoate ((*R*)-**3**) and (*S*)-1-chloro-3-(4-(3-methoxy-3-oxopropyl)phenoxy)propan-2-yl butanoate ((*S*)-**4**) by flash chromatography (n-pentane:EtOAc, 4:1, $v/v$). TLC (n-pentane:EtOAc, 4:1, $v/v$): $R_f$ (**3**) = 0.15, $R_f$ (**4**) = 0.51.

Chlorohydrin (*R*)-**3** was isolated as a clear oil in 43% yield (57 mg, 0.21 mmol), 99% purity ($^1$H NMR) and 97% *ee* (HPLC), $[\alpha]_D^{20} = -5.33$. (*c* 1.6, *i*-PrOH), ($^1$H NMR of (*R*)-**3** as for **3**). Ester (*S*)-**4** was obtained as a clear oil in 41% yield (68 mg, 0.20 mmol), 99% purity ($^1$H NMR) and 87% *ee*, $[\alpha]_D^{20} = +30.71$ (*c* 1.4, *i*-PrOH). (*S*)-**4**: $^1$H-NMR (600 MHz, CDCl$_3$) $\delta$: 7.13–7.11 (m, 2H, Ar-**H**), 6.85–6.82 (m, 2H, Ar-**H**), 5.35–5.31 (quint., 1H, $^3J$ = 5.16 Hz, C**H**), 4.16-4.11 (m, 2H, C**H$_2$**-O-), 3.86–3.76 (m, 2H, C**H$_2$**-Cl), 3.66 (s, 3H, C**H$_3$**), 2.90–2.88 (t, 2H, $^3J$ = 7.77 Hz, C**H$_2$**-Ar), 2.61–2.58 (t, 2H, $^3J$ = 7.77 Hz, C**H$_2$**-CO$_2$-), 2.34–2.32 (m, 2H, C**H$_2$**-CO$_2$), 1.71–1.64 (m, 2H, C**H$_2$**-CH$_3$), 0.99–0.95 (m, 3H, C**H$_3$**); $^{13}$C NMR (150 MHz CDCl$_3$) $\delta$: 173.4, 172.9, 156.8, 133.5, 129.4 (2C), 114.7 (2C), 70.9, 66.2, 51.6, 42.6, 36.1, 36.0, 30.1, 13.6, 18.4.

3.11.5. Esmolol (**5**)

Methyl 3-(4-(3-chloro-2-hydro-xypropoxy)phenyl)propanoate (**3**) (24.7 mg, 0.09 mmol) was dissolved in MeOH (3 mL) and *i*-PrNH$_2$ (157 µL, 1.84 mmol) was added. The mixture was stirred under reflux for 24 h. The reaction was monitored by TLC (CH$_2$Cl$_2$:MeCN, 11:1, *v/v*), R$_f$ (**5**) = 0.16. The solvent was removed under reduced pressure, and the residue was diluted with EtOAc (50 mL) and washed with distilled H$_2$O (2 × 20 mL) and NaHCO$_3$ (aq). The organic phase was dried over MgSO$_4$ and the solvent was removed under reduced pressure to afford racemic esmolol (**5**) as a pale-yellow solid in 86% yield (23.4 mg, 0.08 mmol), 88% purity ($^1$H NMR). $^1$H NMR (600 MHz, CDCl$_3$) $\delta$: 7.08–7.10 (m, 2H, Ar-**H**), 6.82–6.84 (m, 2H, Ar-**H**), 3.98–4.00 (m, 1H, C**H**-OH), 3.92–3.96 (m, 1H, C**H$_2$**O), 3.65 (s, 3H, C**H$_3$**), 2.85–2.89 (m, 3H, C**H$_2$**-Ar, C**H**-NH), 2.81 (p, 1H, $^3J$ = 6.30 Hz, C**H**Me$_2$), 2.70 (dd, 1H, $^3J$ = 8.03 Hz, $^2J$ = 12.14, C**H**-NH), 2.58 (t, 2H, $^3J$ = 7.60, C**H$_2$**COOR), 1.07 (d, 6H, $^3J$ = 6.30 Hz, C**H$_3$**). $^{13}$C NMR (150 MHz, CDCl$_3$) $\delta$: 173.5, 157.3, 133.0, 129.3, 114.7, 70.7, 68.6, 51.7, 49.5, 49.0, 36.0, 30.2, 23.2, 23.1.

3.11.6. (*S*)-Esmolol ((*S*)-**5**)

Following the procedure described above, and purification by preparative TLC (CH$_2$Cl$_2$: MeCN, 11:1, *v/v*), (*R*)-methyl 3-(4-(3-chloro-2-hydroxypropoxy)phenyl)propanoate ((*R*)-**3**) (38.0 mg, 0.14 mmol, 97% *ee*) was converted to (*S*)-esmolol (*S*)-**5**, as a clear oil in 92% yield (37.9 mg, 0.13 mmol), 99% purity ($^1$H NMR) and 97% *ee* (HPLC), $[\alpha]_D^{20} = -6.80$ (*c* 1.03, CHCl$_3$). ($^1$H NMR, as for **5**).

3.11.7. 1-Chloro-3-(2-cyclopentylphenoxy)propan-2-ol (**8**)

To a solution of NaOH (160 mg, 4.00 mmol) in distilled H$_2$O (4 mL), was added 2-cyclopentylphenol (**1b**) (432 mg, 2.66 mmol). The reaction mixture was stirred for 1 min, and 2-(chloromethyl)oxirane (epichlorohydrin) (431 µL, 509 mg, 5.50 mmol) was added. The mixture was stirred at rt for 48 h. Distilled H$_2$O (10 mL) was then added and the product was extracted with EtOAc (3 × 10 mL). The combined organic phases were washed with a saturated NaCl solution (10 mL), dried over anhydrous MgSO4, and the solvent was removed under reduced pressure, yielding 619 mg of a mixture of 2-cyclopentylphenol (**1**), 2-((2-cyclopentylphenoxy)methyl)oxirane (**7**) and 1-chloro-3-(2-cyclopentylphenoxy)propan-2-ol (**8**), as a slightly-yellow oil. A mixture of **7**/**8** (570 mg) was dissolved in THF (3 mL). AcOH (409 µL, 429 mg, 7.14 mmol) and LiCl (303 mg, 7.15 mmol) were added. The reaction mixture was stirred at rt for 24 h. The solution was then concentrated under reduced pressure. The obtained product was dissolved in EtOAc (10 mL) and washed with distilled H$_2$O (10 mL). The aqueous phase was extracted with EtOAc (10 mL). The combined organic phases were washed with a saturated NaCl solution (10 mL), dried over anhydrous MgSO$_4$, and the solvent was removed under reduced pressure. The product was purified by flash chromatography (n-pentane:EtOAc, 9:1, *v/v*), yielding chlorohydrin **8** as a colourless oil (435 mg, 1.71 mmol, 70% yield, 93% ($^1$H NMR)). TLC (*n*-pentane:EtOAC, 9:1, *v/v*) R$_f$ = 0.34 for product **8**. $^1$H NMR (600 MHz, CDCl$_3$) $\delta$ 7.24 (m, 1H, Ar-**H**), 7.16 (m, 1H, Ar-**H**), 6.96 (m, 1H, Ar-**H**), 6.86 (m, 1H, Ar-**H**), 4.25 (h, $^3J$ = 5.3 Hz, 1H, C**H**-OH), 4.13 (dd, $^2J$ = 9.4, $^3J$ = 5.1 Hz, 1H, C**H$_2$**O), 4.09 (dd, $^2J$ = 9.4, $^3J$ = 5.4 Hz, 1H, C**H$_2$**O), 3.82 (dd, $^2J$ = 11.2, $^3J$ = 5.2 Hz, 1H, C**H$_2$**Cl), 3.76 (dd, $^2J$ = 11.2, $^3J$ = 5.6 Hz, 1H, C**H$_2$**Cl), 3.35–3.26 (m, 1H, C**H**),

2.48 (d, $^3J$ = 6.2 Hz, 1H, **OH**), 2.04–1.56 (m, 8H). $^{13}$C NMR (151 MHz, CDCl$_3$) δ 156.01, 134.89, 127.05, 126.84, 121.52, 111.67, 70.20, 68.79, 46.35, 39.19, 33.12, 25.62.

3.11.8. Synthesis of Chlorohydrin (R)-**8** by CALB-Catalysed Kinetic Resolution of 1-Chloro-3-(2-cyclopentylphenoxy)propan-2-ol (**8**)

To a solution of 1-chloro-3-(2-cyclopentylphenoxy)propan-2-ol (**8**) (163 mg, 0.63 mmol) dissolved in dry MeCN (20 mL) activated molecular sieves (4Å), vinyl butanoate (408 µL, 364 mg, 3.18 mmol) and CALB (280 mg) were added. The mixture was placed in an incubator shaker (38 °C, 200 rpm) for 23 h. The enzymes and molecular sieves were filtered off and solvents were removed under reduced pressure. The obtained product was dissolved in EtOAc (10 mL) and was washed with distilled H$_2$O (3 × 15 mL) and a saturated NaCl solution (10 mL). The solution was dried over anhydrous MgSO$_4$, and the solvent was removed under reduced pressure. (R)-**8** and (S)-**9** were separated by flash chromatography (n-pentane:EtOAc, 9:1, v/v). (R)-**8** was obtained as a colourless oil (63 mg, 0.243 mmol, 39% yield, 95% purity ($^1$H NMR), ee = 99% (chiral HPLC)). $[\alpha]_D^{25}$= −14.00 (c 1.0, MeOH).

3.11.9. (S)-Penbutolol ((S)-**10**)

To a mixture of (R)-1-chloro-3-(2-cyclopentylphenoxy)propan-2-ol ((R)-**8**) (31 mg, 0.12 mmol) in MeOH (2 mL), was added *tert*-butylamine (0.18 mL, 0.13 g, 1.71 mmol). The mixture was stirred under reflux for 24 h, then concentrated under reduced pressure. The obtained product was dissolved in EtOAc (10 mL) and washed with distilled H$_2$O (5 mL). The organic phase was dried over anhydrous MgSO$_4$, and the solvent was removed under reduced pressure to give (S)-penbutolol ((S)-**10**) as a white solid (29 mg, 82% yield, 93% purity ($^1$H NMR), ee = 99% (chiral HPLC)). $[\alpha]_D^{20}$ = −14.00 (c 1.0, MeOH). $^1$H NMR (600 MHz, CDCl$_3$) δ 7.22 (m, 1H, Ar-**H**), 7.17–7.11 (m, 1H, Ar-**H**), 6.92 (m, 1H, Ar-**H**), 6.85 (m, 1H, Ar-**H**), 4.05–4.00 (m, 1H, **CH**-OH), 3.97 (m, 2H, **CH$_2$**O), 3.35–3.28 (m, 1H, **CH**), 2.88 (dd, $^2J$ = 11.9, $^3J$ = 3.7 Hz, 1H, **CH$_2$**NH), 2.78–2.72 (m, 1H, **CH$_2$**NH), 2.07–1.53 (m, 8H), 1.12 (s, 9H). $^{13}$C NMR (151 MHz, CDCl$_3$) δ 156.48, 134.62, 126.78, 126.62, 120.84, 111.39, 70.61, 68.78, 50.32, 44.78, 39.30, 32.94, 32.87, 29.14, 25.46, 25.45.

3.11.10. (S)-Penbutolol Hydrochloride ((S)-**10**·HCl)

(S)-Penbutolol ((S)-**10**) (10.0 mg) was dissolved in *i*-PrOH (40 µL), and a solution of HCl in *i*-PrOH (5 %, 80 µL) was added. The reaction was run for 1 h, and the solvent was removed under reduced pressure to give (S)-penbutolol·HCl ((S)-**10**·HCl) as a colourless solid (10.0 mg, 30.4 µmol, 89% yield, 93% purity ($^1$H-NMR), ee = 99% (chiral HPLC)). $[\alpha]_D^{20}$= −23.00 (c 1.0, MeOH). $^1$H NMR (600 MHz, CDCl$_3$) δ 9.81 (s, 1H), 8.32 (s, 1H), 7.22 (m, 1H, Ar-**H**), 7.13 (m, 1H, Ar-**H**), 6.93 (m, 1H, Ar-**H**), 6.80 (m, 1H, Ar-**H**), 4.69–4.60 (m, 1H, **CH**-OH), 4.12 (dd, $^2J$ = 9.5, $^3J$ = 4.3 Hz, 1H, **CH$_2$**O), 3.98 (dd, $^2J$ = 9.6, $^3J$ = 6.4 Hz, 1H, **CH$_2$**O), 3.34 (m, 2H, **CH$_2$**NH), 3.11 (m, 1H, **CH**), 2.04–1.54 (m, 8H), 1.50 (s, 9H). $^{13}$C NMR (151 MHz, CDCl$_3$) δ 155.98, 134.55, 126.74, 126.70, 121.19, 111.33, 69.66, 65.83, 57.57, 45.92, 39.15, 32.89, 32.88, 25.90, 25.32.

4. Conclusions

A four-step synthesis of (S)-esmolol ((S)-**5**) in 26% overall yield and 97% ee has been performed from the starting materials methyl 3-(4-hydroxyphenyl)propanoate and epichlorohydrin. We have reported a specific rotation for (S)-esmolol ((S)-**5**) of $[\alpha]_D^{20}$ = −6.80 (c 1.03, CHCl$_3$), disputing the previously reported positive specific rotation value. A five-step synthesis of (S)-penbutolol ((S)-**10**) in 99% ee and (S)-penbutolol hydrochloride ((S)-**10**·HCl) in 20% overall yield and 99% ee have been performed. Both specific rotation values for (S)-penbutolol ((S)-**10**) and (S)-penbutolol hydrochloride ((S)-**10**·HCl) have been determined and are consistent with previously reported data. CALB-catalysed kinetic resolution of chlorohydrin precursors **3** and **8** in acetonitrile is an efficient method to obtain the enantiopure building blocks (R)-**3** and (R)-**8**. The yield of these compounds is limited to 50% due to the kinetic resolution step, but it could be further increased up to 100% by using special techniques, such as dynamic kinetic resolution [4]. We have shown that the thorough monitoring of these processes, with a focus on reducing reaction chemicals and replacing hazardous chemicals, leads to greener processes. We have previously shown that CALB can be reused up to six times with no loss of activity or selectivity [18].

**Supplementary Materials:** The following supporting information can be downloaded at: https://www.mdpi.com/article/10.3390/catal12090980/s1; $^1$H and $^{13}$C NMR spectra, relevant MS spectra and chiral HPLC chromatograms.

**Author Contributions:** Investigation, writing, original draft preparation, E.E.J.; supervision and writing, review and editing, E.E.J.; investigation, data curation and partly writing of manuscript S.H.T., L.B., A.L.T., K.K. All authors have read and agreed to the published version of the manuscript.

**Funding:** This work was financially supported by The Education, Scholarship, Apprenticeships and Youth Entrepreneurship Programmer—EEA Grants 2014-2021, Project No. 18-Cop-0041.

**Data Availability Statement:** The data presented in this study are available online.

**Acknowledgments:** This publication was realised with the EEA Financial Mechanism 2014–2021 financial support, project no 18-COP-0041. Its content (text, photos, videos) does not reflect the official opinion of the Programme Operator, the National Contact Point and the Financial Mechanism Office. Responsibility for the information and views expressed therein lies entirely with the authors. Syncozymes Co LTD, Shanghai, China is thanked for gift of CALB.

**Conflicts of Interest:** The authors declare no conflict of interest.

# References

1. Lund, I.T.; Bøckmann, P.L.; Jacobsen, E.E. Highly enantioselective CALB-catalyzed kinetic resolution of building blocks for β-blocker atenolol. *Tetrahedron* **2016**, *72*, 7288–7292. [CrossRef]
2. Blindheim, F.H.; Hansen, M.B.; Evjen, S.; Zhu, W.; Jacobsen, E.E. Chemoenzymatic Synthesis of Synthons as Precursors for Enantiopure Clenbuterol and Other β$_2$-Agonists. *Catalysts* **2018**, *8*, 516. [CrossRef]
3. Gundersen, M.A.; Austli, G.B.; Løvland, S.S.; Hansen, M.B.; Rødseth, M.; Jacobsen, E.E. Lipase Catalyzed Synthesis of Enantiopure Precursors and Derivatives for β-Blockers Practolol, Pindolol and Carteolol. *Catalysts* **2021**, *11*, 503. [CrossRef]
4. Verho, O.; Bäckvall, J.-E. Chemoenzymatic Dynamic Kinetic Resolution: A Powerful Tool for the Preparation of Enantiomerically Pure Alcohols and Amines. *J. Am. Chem. Soc.* **2015**, *137*, 3996–4009. [CrossRef] [PubMed]
5. Maisel, W.H.; Friedman, P.L. Esmolol and Other Intravenous Beta-Blockers. *Card. Electrophysiol. Rev.* **2000**, *4*, 240–242. [CrossRef]
6. Esmolol. Available online: https://www.legemiddelhandboka.no (accessed on 1 August 2022).
7. Banoth, L.; Banerjee, U.C. New chemical and chemo-enzymatic synthesis of (*RS*)-, (*R*)-, and (*S*)-esmolol. *Arab. J. Chem.* **2017**, *10*, S3603–S3613. [CrossRef]
8. Narsaiah, A.V.; Kumar, J.K. Novel Asymmetric Synthesis of (*S*)-Esmolol Using Hydrolytic Kinetic Resolution. *Synth. Commun.* **2011**, *41*, 1603–1608. [CrossRef]
9. Byrne, F.P.; Jin, S.; Paggiola, G.; Petchey, T.H.M.; Clark, J.H.; Farmer, T.J.; Hunt, A.J.; McElroy, C.R.; Sherwood, J. Tools and techniques for solvent selection: Green solvent selection guides. *Sustain. Chem. Process.* **2016**, *4*, 7. [CrossRef]
10. Hamaguchi, S.; Asada, M.; Hasegawa, J.; Watanabe, K. Asymmetric hydrolysis of racemic 2-oxazolidinone esters with lipases. *Agric. Biol. Chem.* **1984**, *48*, 2331–2337. [CrossRef]
11. Ader, U.; Schneider, M.P. Enzyme assisted preparation of enantiomerically pure β-adrenergic blockers III. Optically active chlorohydrin derivatives and their conversion. *Tetrahedron Asymmetry* **1992**, *3*, 521–524. [CrossRef]
12. Phukan, P.; Sudalai, A. Regioselective alkylation of phenol with cyclopentanol over montmorillonite k10: An efficient synthesis of 1-(2-cyclopentylphenoxy)-3-[(1,1-dimethylethyl) amino] propan-2-ol {(*S*)-penbutolol}. *J. Chem. Soc. Perkin Trans.* **1999**, *20*, 3015–3018. [CrossRef]
13. Klunder, J.M.; Onami, T.; Sharpless, K.B. Arenesulfonate derivatives of homochiral glycidol: Versatile chiral building blocks for organic synthesis. *J. Org. Chem.* **1989**, *54*, 1295–1304. [CrossRef]
14. Kan, K.; Miyama, A.; Hamaguchi, S.; Ohashi, T.; Watanabe, K. Synthesis of (*S*)-β-Blockers from (*S*)-5-Hydroxymethyl-3-tert butyl-2-oxazolidinone or (*S*)-5-Hydroxymethyl-3-isopropyl-2-oxazolidinone. *Agric. Biol. Chem.* **1985**, *49*, 207–210. [CrossRef]
15. Bevinakatti, H.S.; Banerji, A.A. Lipase Catalysis in Organic Solvents. Application to the Synthesis of (*R*)- and (*S*)-Atenolol. *J. Org. Chem.* **1992**, *57*, 6003–6005. [CrossRef]

16. Anthonsen, H.W.; Hoff, B.H.; Anthonsen, T. Calculation of enantiomer ratio and equilibrium constants in biocatalytic ping-pong bi-bi resolutions. *Tetrahedron Asymmetry* **1996**, *7*, 2633–2638. [CrossRef]
17. Jacobsen, E.E.; Hoff, B.H.; Anthonsen, T. Enantiopure derivatives of 1,2-alkanediols: Substrate requirements of lipase B from *Candida antarctica*. *Chirality* **2000**, *12*, 654–659. [CrossRef]
18. Moen, A.R.; Hoff, B.H.; Hansen, L.K.; Anthonsen, T.; Jacobsen, E.E. Absolute configurations of monoesters produced by enz-yme catalyzed hydrolysis of diethyl 3-hydroxyglutarate. *Tetrahedron Asymmetry* **2004**, *15*, 1551–1554. [CrossRef]

Review

# Properties and Recyclability of Abandoned Fishing Net-Based Plastic Debris

Anna Kozioł [1], Kristofer Gunnar Paso [2,*] and Stanisław Kuciel [3]

1 Faculty of Natural Sciences, Norwegian University of Science and Technology, 7491 Trondheim, Norway
2 Department of Chemical Engineering, Norwegian University of Science and Technology, 7491 Trondheim, Norway
3 Faculty of Materials Engineering and Physics, Cracow University of Technology, Jana Pawla II 37, 37-864 Cracow, Poland
* Correspondence: kristofer.g.paso@ntnu.no

**Abstract:** Plastics in marine environments undergo molecular degradation via biocatalytic and photocatalytic mechanisms. Abandoned, lost, or discarded fishing gear (ALDFG) damages marine and coastal environments as well as plant and animal species. This article reviews ghost fishing, ecological damage from marine plastics, recommended recycling practices and alternative usages of derelict fishing gear. Material mixing techniques are proposed to counteract the effect of biocatalytic and photocatalytic biodegradation within the context of plastic fish net recycling. There is a need for a new and rapid "multidimensional molecular characterization" technology to quantify, at a batch level, the extent of photocatalytic or biocatalytic degradation experienced on each recovered fishing net, comprising molecular weight alteration, chemical functional group polydispersity and contaminant presence. Rapid multidimensional molecular characterization enables optimized conventional material mixing of recovered fishing nets. In this way, economically attractive social return schemes can be introduced for used fishing nets, providing an economic incentive for fishers to return conventional fishing nets for recycling.

**Keywords:** ghost fishing; fish net recycling; material blending; multidimensional; molecular; characterization

## 1. Introduction

Marine litter has been defined as 'any persistent, manufactured or processed solid material discarded, disposed or abandoned in the marine environment [1]. It was estimated that the total amount of plastic that has been produced between 1950 and 2017 equals approx. 9.2 billion tons. Specifically, more than half of this plastic has been produced since 2004 and less than 10% of it has been recycled [2]. The general amount of plastic debris in the ocean varies. However, it has been estimated that on average, around 300,000 items of plastic debris are present per km$^2$ of ocean surface [3].

It is important to notice that plastic domination of marine litter occurs because of its longevity and density—some plastic types, like, for example, polypropylene, are less dense than seawater or, like polyester, are slightly denser. Among plastic waste in the ocean, it was indicated by the US National Marine Debris Monitoring Program that 17.7% of plastic litter found on beaches came from ocean fishing activity [4]. Taking into consideration abandoned, lost or discarded fishing gear's (ALDFG) behavior and impact on ocean habitats, it is considered one of the greatest threats to ocean biodiversity.

It has been estimated that in 2018, global fish production will exceed 179 million tons. Unfortunately, this great number is directly connected to the increasing amount of ALDFG compared to the other components of marine debris. Every year, the global fishing gear losses include 5.7% of all fishing nets, 8.6% of all traps and 29% of all lines used [5]. This is why the majority of plastic waste was ALDFG, with 37% of fishing lines and 34% of fishing nets [6].

Habitat degradation due to plastic debris has a far-reaching impact on ocean biodiversity. The movement of plastic litter by tides and storms can result in severe physical damage to marine habitats and animals. Floating plastic is able to carry life-threatening bacteria and transport pharmaceuticals and toxins into coastal areas. The plastic may also contain several other chemicals and toxins added during the production process [2]. Plastic accumulated in water is often interpreted as food by marine animals. This will intoxicate their bodies. Many of these fish contaminated by plastic will eventually be consumed by people. Additionally, plastic debris that is passively moved by tides and wind transports non-indigenous species to new locations where they can become invasive and endanger local fauna and flora [2].

Plastic enters the ocean through several pathways. Mainly from rivers, directly from the land, and to a lesser extent, the input is atmospheric or biological. For example, birds that consume plastic particles on the land and excrete them into the ocean. However, the amount of plastic transported this way is smaller compared to other paths [7]. It is predicted that annual plastic flow to the oceans will nearly triple between 2016 and 2040. Additionally, more than 1000 rivers are responsible for 80% of the annual release of plastic, with small urban rivers being the most polluting. It has been estimated that the input from rivers ranges between 0.8 and 2.7 million tons per year [8]. Moreover, the annual direct input from the industry since 1950 has been estimated to be between 108 and 480 million tons [9].

It has been estimated that from 1.15 to 2.41 million tons of plastic enter the ocean every year. More than half of this plastic is less dense than water. This means that over 50% of this plastic is not going to sink to the seabed but will float closer to the surface once it enters the sea [10]. Buoyant plastic is transported by converging over extended distances and finally accumulating in the path. Once plastic particles enter the gyre, they are unlikely to leave until full degradation. More and more plastic is getting accumulated as one shows the mechanism behind the Great Pacific Garbage Patch (GPGP), the largest of five reported plastic accumulation zones in the world's oceans [11].

It has been estimated that the GPGP covers 1.6 million square kilometers. The mass was revealed to be significantly greater than the first assumptions. The weight is approx. 80,000 tons, which is from 4 to 16 times more plastic than previously reported [12]. Based on coastal clean-up data, fishing, aquaculture and shipping are responsible for 28.1% of the plastic in the ocean. However, from the observation of components of GPGP and other plastic accumulation places, it is estimated that the impact of these industries is significantly higher [13].

The type of floating plastic can be differentiated depending on the material or size. The most commonly found plastics include polyethylene (PE) or polypropylene (PP). Other systems of classification divide GPGP components by size or general categories. The following four size categories were established: microplastic (0.05–0.5 cm), mesoplastic (0.5–5 cm), macroplastic (5–50 cm) and megaplastic (>50 cm). 92% of the mass is made of debris larger than 5 mm. Out of this, at least 46% is comprised of fishing nets [12]. Figure 1 shows photographs of abandoned fishing nets on a beach in Norway.

Plastic migration is one of the causes of the huge GPGP. Available barcode system tracing for the plastic shows that some elements of the debris can be found 10 years later, and 10,000 km away from its origin [14]. A good example of this phenomenon is plastic debris found on beaches in Brazil, where at least 21.4% of the plastic came from Europe [15]. Thanks to satellite buoys, it was possible to track the movement of plastic waste in the marine environment. It was observed that plastic nets, which enter the atolls of the Hawaiian Islands from the northeast, tend to move southwest at a slow pace of 0.35 km per day. Eventually, those fishing nets remained stationary on reefs. The longer they stay, the greater the hazard they pose to the closest habitat. Nets moving with the winds will be moving towards the center of the atolls. Thus, remaining stationary for years and causing environmental issues by passively catching and entangling fauna and flora [16].

**Figure 1.** Parts of abandoned fishing nets found on a beach in southern Norway (Hessness, August 2022).

## 2. Ghost Fishing
*2.1. What Is Ghost Fishing?*

Ghost nets are fishing nets that have been discarded or lost in the ocean by fishing vessels [11]. Issues directly connected with that are bycatch and ghost fishing. Bycatch is observed while fishing. Other species of animals are entangled and caught in the net. This phenomenon can occur during a regular fishing procedure when any other marine species gets involved, except for the fishing species we intended to catch. Bycatch can be minimized by using suitable material, which does not absorb water and, as a consequence, does not change the mesh size. Ghost fishing is a completely passive action when ALDFG is floating and independently continues to catch fish or entangle around body parts of the water animals. This phenomenon seems to be directly connected to the quality of the fishing gear. The use of low-quality fishing gear often results in frequent losses when entangled with obstructions or in rough weather [17].

The problem of losing fishing gear consists of several aspects. Gear is considered abandoned when it is not possible for a fisher to retrieve it. That can happen when the gear is snagged on marine obstructions. Snagging fishing gear, including all types of fishing nets, is identified as a major cause of loss in many fisheries [18]. Additionally, during the fishing operation, it is possible to lose control over the gear and not be able to locate it. This situation can occur when tides or wave action carry the gear from the deployment location. Considerable gear loss can also be caused by interactions with other active fishing gear [19]. Other identified causes are long soak times, fishing in deep habitats, and deploying more gear than can be hauled in regularly [18]. An additional issue resulting in ALDFG is illegal, unreported and unregulated fishing; however, the exact connection is difficult to quantify [20].

It has been estimated that from 500,000 to 1 million tons of fishing gear are likely to enter the ocean and become fishing gear every year [19]. Additional attempts to qualify the problem revealed results with both local and global influence. In South Korea, 11,436 tons of traps and 38,535 tons of gillnets are abandoned every year [21]. What is more? Over 70 km of gillnets were lost in Canada's Greenland Halibut fishery in five years [22]. All those studies were conducted to indicate the significant problem of ghost gear. ALDFG that remains in the water for a sufficient period of time will eventually accumulate sessile organisms in a process called 'bio fouling'. It is suspected that this is the reason why the net becomes more visible to animals and ghost fishing efficiency decreases with time [23]. Monofilament nets have higher catch rates than multifilament nets. It is because the multifilament net is more visible in the water. Multifilament nets are made of materials buoyant in seawater. However, with time, it loses its properties and sinks down slowly. It has been suggested that stormy weather can accelerate the degradation and biofouling of fishing nets [24]. When the net loses the ghost catch, it can become buoyant again and rise back to the surface, starting the ghost fishing cycle again.

The topic of ghost fishing is relatively new and needs to be researched more, especially on its effects on population levels and available solutions or preventing actions. It has been suggested that ghost fishing nets should be treated separately from other components of marine plastic debris, as it requires a different managerial approach compared to debris originating from, for example, tourism [25].

The performance of a fishing net is highly dependent on the material. Among the most commonly used materials, polyamide (PA) tends to have the highest tensile properties compared to polypropylene or polyethylene. Based on the properties of PA, it was concluded that this material is more suitable for fishing to prevent ghost fishing. Nevertheless, it is possible to change the performance of fishing gear slightly by increasing the yarn diameter or the mesh size. Choosing proper parameters results in higher breaking load, tensile strength, and increases the drag coefficient as well as bending stiffness and breaking strength [26].

Studies on the degradation of nylon 6 showed surprising abilities for water absorption. What is more, the observed behavior after reaching the glass transition temperature suggests the use of this material for other specific applications. Once PA is at or above the glass transition temperature, it will tend to creep upon application of load [27]. That means, after only a few catches and not being used for a long time, the nylon fishing net is going to start losing its properties. As a result, being unable to catch the required number of fish.

*2.2. Degradation of Plastic in Marine Environment*

In 2016, NOfir reported on the effectiveness of the EUfir system related to ALDFG collection and recycling. The used methodology was called Life Cycle Assessment (LCA) and provided a systematic evaluation of the environmental aspect of the product through all stages of its life cycle. Results obtained in this method are reliable and helpful for achieving a life cycle economy. In this case, LCA was used to calculate the real environmental impact of a great system, from the availability of abandoned fishing equipment to the production of secondary materials after recycling operations. The most noticeable environmental effects of recycling were a decrease in the consumption of non-renewable resources and a decrease in carbon footprint (a decrease in carbon dioxide emissions). What is important to notice is that the most commonly found fishing equipment is made of nylon 6 (approx. 76%) [28].

Ropes and nets are exposed to the influence of oxygen, salt water, ultraviolet radiation and mechanical stress. Fishermen should ensure and care for the strength, elasticity, foldability and density of the material as well as the degradation rate. It has been determined that in the marine environment, the greatest impact on fiber degradation is due to the exposure to sunlight, the most abrasive condition that will degrade the fishing net. The marine environment reveals PA twines' weak resistance to sunlight for long periods at a high temperature. Additionally, nylon 6 fishing nets appear to absorb water and swell after some time [26].

The result of the effect of solar radiation on PA netting twines under ambient conditions revealed a decrease in breaking strength over time. Samples were exposed to direct solar radiation for 780 h, which caused significant changes in their properties. The degradation rate of untreated twines was three times higher than the treated ones. This leads to the conclusion that the exposure of the fishing nets to sunlight should be avoided. Synthetic nets do not need to be dried out as they do not rot and can be stored even when wet. Preventing fishing nets from exposure to sunlight extends the half-life of the net, reduces the capital cost of fishing and reduces the time spent on net maintenance and repair [29].

One of the biggest threats from abandoned fishing gear is the degradation of the material it is made of. During that process, plastic may decompose into dangerous chemical components. The degradation of large plastics is the direct source of microplastics. By observing the fragmentation of PP, PE and PA exposed to benthic conditions at 10 m depth over 12 months and by monitoring their weight, it was possible to estimate the behavior of abandoned plastic in a deep marine environment. Results revealed the presence of microplastic fibers and particles even though the photodegradation was reduced with time. This indicated that an alarming volume of microplastic is produced from the rope debris alone [30].

Microplastics generated from the degradation of macroplastics are called 'secondary microplastics'. Most of the microplastic present in the ocean has its source from land, but there is still a significant influence of marine activities on the amount. Marine-based sources include ALDFG, which releases microplastics during degradation in the water but also on beaches. Additionally, abrasion of aquaculture gear made of plastic and ships covered with synthetic paint releases plastic particles. As mentioned before, the primary causes of degradation are physical aberration and exposure to UV light. It is important to notice that big plastic particles release microplastic long before they themselves become small particles [31].

### 2.3. Bio-Based and Biodegradable Fishing Gear

Creating bio-fishing gear is one of the ideas for decreasing the amount of plastic waste in the ocean. Even though it has been marked as the solution, in some cases it is not beneficial for the environment and can cause similar harm to the environment as regular fishing gear.

Bioplastic production increased from 0.7 million tons in 2010 to 2.11 million tons in 9 years. More than 45% of production took place in Asia (European Bioplastics 2019). Nevertheless, in 2018, only around 0.6% of total plastic production was bioplastic. Increased demand for bioplastics is expected for the continued growth of this field. The most popular biodegradable bio-based plastics available on the market are the following: polylactic acid (PLA), polyhydroxyalkanoate (PHA) and polybutylene succinate (PBS)—the starch-based polymer. They are used as a substitution for polypropylene (PP), polyester (PET) and polystyrene (PS) [32].

To manufacture biodegradable nets, the following special material blend was synthesized: 82% of PBS mixed with 18% of polybutylene adipate-co-terephthalate (PBAT). The mechanical properties of the blend were compared to the properties of the nylon net. The nylon fishing net exhibited greater breaking strength and elongation when dry and better flexibility when wet—the biodegradable net appeared to be approx. 1,5-fold stiffer. Based on these results, it could be concluded that the bio-fishing net is going to have a lower catch efficiency than the nylon one. Nevertheless, a comparison revealed similar catch rates for yellow croakers. Degradation of the line started after two years, which made the net easy to be destroyed by potentially entangled organisms. The results of that experiment are promising and serve as a solution to the problem of ghost fishing [21].

However, the term bioplastic does not always mean that the material is bio-based or biodegradable. The meaning of the term is that we can find plant-based plastics that are be either biodegradable or non-biodegradable, or biodegradable fossil-based plastics [33]. The chemical and mechanical properties of materials highly differ. At the same time, different environments (soil or ocean) greatly influence the biodegradability of bioplastics [34]. Data collected during the last ten years shows that some problems connected with the influence on the environment are the same for bio as well as conventional plastic. Plant-based polymers are not necessarily biodegradable, can contain toxic additives and can degrade and persist as microplastics [2].

One of the most crucial disadvantages of bioplastic is the necessity of designing new recycling lines because it contaminates the recycling process of conventional plastic. Mostly, the sorting of plastic is based on visual examination, which does not distinguish bioplastics from non-bioplastics [35]. For example, PET and PLA (bio-based) plastic bottles look nearly identical, which makes it impossible to sort them based on their appearance. Mixing these two materials during recycling would cause problems for reprocessing because these materials have different melting points [36].

Discarded biodegradable plastic, including biodegradable plastic bags, poses the risk to aquatic life and the environment as those of non-biodegradable plastic. It has been found that they have a similar adverse impact on the infaunal abundance and biogeochemical processes. Based on a comparison of the specific examples of bio-HDPE and conventional HDPE, biodegradable plastic poses the same risk to biodiversity and the ecosystem. Both of them obstruct oxygen and light, decreasing the abundance of invertebrates and decreasing the flux of inorganic nutrients from the sediment [37].

There are still debates concerning the full environmental footprint of bioplastics. Most of the currently available analysis has been limited to carbon dioxide emissions [38]. However, there has been a standard ASTM D 6691 test method, which allows for the determination of the degradation of virgin and biodegradable plastics by aerobic mineralization. To check the behavior of the materials as future fishing materials, the marine environment was simulated in a laboratory. Out of the examined materials, the highest mineralization rate, which indicates degradation and biodegradation in the case of biodegradable plastics, was the highest for thermoplastic strath and plastic waste polymers. Thermoplastic strath showed a mineralization rate of 49.7% after 82 days and achieved 85% degradation after three months. That presents a much higher degradation rate than virgin polymers [39]. Moreover, the biodegradable fishing nets exposed to seawater show degradation after two years, resulting in abrasive changes in the surface [40].

## 3. The Wide-Ranging Impact of Marine Plastic

### 3.1. Impact of Marine Plastic on Mammals, Birds and Reptiles

Plastic has been proven to impose detrimental effects on at least 267 species around the world. This includes the following: 86% of sea turtle species, 44% of seabird species and 43% of all mammalian species. Animals are mostly harmed through ingestion (reducing stomach capacity, hindering growth, internal injuries, intestinal blockage), entanglement and subsequent strangulation [41]. Moreover, 340 original publications reported encounters between marine debris and marine animals [42]. At least 17% of those affected by entanglement and ingestion were listed as threatened or near threatened [43].

The number of species proven to be negatively affected by derelict plastic debris has doubled since 1997. Ghost gear is one of the most deadly forms of marine plastic debris [44]. It tends to continue to catch animals as long as it retains proper integrity [45]. This usually occurs during the first year after the loss of ghost gear, but there are observed types of fishing nets continuing to capture animals even decades after being lost [46]. Even though most fishing gear is designed to capture animals in a selective way, it is known by now that when lost, fishing gear can capture animals indiscriminately. It has been documented that in the Salish Sea, more than 260 species have been observed to get entangled and killed by lost salmon gillnets. It has also been estimated that the 4500 nets removed from 2002 to 2009 might have killed more than 2.5 million marine vertebrates, 800,000 fish and 20,000 seabirds [47]. Over 5400 animals from 40 different species of marine mammals, reptiles and elasmobranchs were entangled in ghost fishing nets [25].

Out of all marine mammals, seals and sea lions appeared to be the most endangered species by entanglement. In Australia, it has been estimated that 1500 sea lions die from entanglement every year [48]. In the Sea of Okhotsk, the most common victims of entanglement were young males, as a consequence of their natural curiosity and playful behavior. Additionally, the rotation of the body is a natural panic reaction that causes more entanglement for long periods. Most of the plastic debris found on sea lions was associated with nearby fishing [49]. There is evidence that even the relatively small entanglement rate of 0.4% of the northern fur seals is serious enough to affect the whole population. This is due to the disproportionate effect on individuals of fertile age [50].

Marine plastic in the form of net, rope, monofilament line and packaging bands can cause entanglement in a wide range of pinniped species. There is a noticeable potential for an acute impact on individuals by starvation and highly restrictive entanglement. Some animals live with chronic deep wounds for months or even years. Chronic wounds may cause a deep infection, leading to the premature death of an individual. The result of marine debris entanglement is the first and foremost suffering of animals through wounding, amputation or ingestion. This often goes hidden and unreported. Fur seals, monk seals, California sea lions, grey seals and common seals are the most likely species to be affected by entanglement [51]. It has been found that over the last two decades, entanglement records of seabirds have increased from 16% to 25%.

Lost fishing gear also damages important nearshore habitats, including seagrass beds, coral reefs and mangroves [52]. Lost fishing gear break corals, damage vegetation, build up sediments and impedes access to specific habitats [53]. It is considered likely that plastic on the seabed alters the dynamics of the entire ecosystem. Upon covering the seafloor, plastic sheets inhibit gas exchange, leading to low oxygen levels and the formation of artificial hand grounds, creating problems of burying creatures [54]. However, some organisms are able to adjust themselves to these conditions. Floating plastic debris was used by a variety of microorganisms as a newly created habitat [55]. Plastic debris also attracts fish or sea turtles to aggregate below its surface and follow the drifting material [56]. Damage to marine and coastal ecosystems [57] is challenging to calculate, but it has been proposed that a 1% decline in annual ecosystem services could equal a loss of USD 500 billion in global ecosystem benefits annually [58].

Plastic microparticles in the marine environment are being absorbed by small organisms at the base of the food chain. They are subsequently transported further up the food chain as the prey is eaten by the predator. Higher and higher concentrations are reached all the way to the top predator species. This process is called bioaccumulation and has an effect on human lives upon the consumption of fish and other seafood. Chemicals from oceanic plastics have been detected in human bodies as well [59].

On average, a human body absorbs approx. 52,000 particles of microplastic by ingestion per year. It is under investigation exactly where in the body it tends to accumulate the most and what kind of negative effect it would have on human health. Depending on the known impact of plastic on human beings, it is supposed that it may contribute to neurodevelopmental disorders, metabolic, respiratory and cardiovascular diseases as well as decreased antibody response to vaccines [60].

*3.2. Impact of Lost Fishing Gear on Fisheries*

Macroplastics have the potential to reduce the efficiency and productivity of commercial fisheries. The most important impact occurs through ghost fishing by ALDFG [5]. Ghost nets may get caught up and damage the machinery of the fishing boats [61]. Fishing operations near the coastline may have livestock as the nets are being picked up by the animals upon reaching shore.

According to experiments on abandoned and lost crab traps, an estimated 12,193 traps are lost annually in the Washington waters of the Salish Sea. Lost traps still show some catch rate, which results in animals being caught but never picked up. The annual Dungeness crab loss was estimated to be 4.5% of the value of harvest, translating into a value of USD 744,296. Unfortunately, the value of saved crabs is lower than the cost of removal. Nevertheless, the best solution could be to modify the trap design, which might reduce the mortality rate and negative impact on the abundance of crabs [62].

However, studies on the removal of derelict blue crab pots in the Chesapeake Bay showed more promising results. This may encourage fishers to organize an additional removal. Removing 34,408 derelict pots led to significant gains in gear efficiency and an additional 27% increase in income (USD 21.3 million). Global analysis shows that removing less than 10% of derelict pots and traps could result in a recovery of USD 831 million. Removing ALDFG will not only save marine biota but also appear to be profitable and sustainable for governments and communities whose livelihoods depend on income from the ocean [63].

In 2015 costs induced by derelict fishing gear on fisheries and aquaculture have been estimated at USD 1.47 billion. On transport and shipbuilding at USD 2.95 billion, which gave 13.4% and 27% of annual costs respectively [64]. In the Adriatic and Ionian Seas, the annual loss due to derelict fishing gear for the fishing sector was estimated at USD 21.86 million [65].

### 3.3. Tourism and Marine Port Operations

Marine plastic debris on beaches and in touristic marine environments (for example, coral reefs) presents a serious visual and aesthetic problem. The presence of litter has a significant negative impact on recreational experiences and overall beach enjoyment [66]. Visitors actively avoid spending time on polluted parts of the coast [67]. This generates lots of opportunities for industries because tourists favor alternative, less polluted locations, reducing income for businesses operated at less visited beaches [2].

The direct cost impact of marine debris on tourism has been estimated in 2015 at USD 6.41 billion, which is 59.2% of the total damage caused by derelict plastic [64]. In the region of the Adriatic Sea, the tourism sector lost an average of USD 6833 per year and harbors needed to spend USD 10,238 on managing marine litter [65]. In Orange Country, California, marine litter was reduced by 25%. This saved additional costs for visitors, who no longer needed to travel further in their search for non-polluted beaches [68].

Marine debris can present navigational hazards to ships at sea by entangled propellers, blocked water intakes and collisions with floating objects. Especially when the weather conditions are bad, the entanglement of propellers can significantly reduce stability and maneuverability [69]. Derelict fishing gear causes economic costs here as well, as sometimes changes in routes may be needed to avoid a collision. This may have a significant influence, especially in areas with heavy marine traffic [70].

### 3.4. Economic Costs

Different economic costs of pollution can be divided into prevention, remediation and damage costs. Prevention costs are the lowest and involve a range of actions organized by civil society organizations, governments and industries to reduce the amount of plastic litter entering the oceans to avoid damage and remediation costs in the future [2]. The annual global economic cost of marine plastic pollution is estimated to be at least USD 6–19 billion globally [71]. The cost of cleaning coasts could be reduced by a proper prevention policy [65]. The total cost of damage in 2015 in the region of the Asia-Pacific Economic Cooperation (APEC) has been estimated at USD 10.8 billion annually [64]. Moreover, the estimated cost of removing marine plastic from a remote atoll in the Seychelles was USD 4.68 million with 18,000 h of labor [72]. In the Republic of Korea, USD 282 million was sent over five years to remove plastic litter [73]. During a period of eight years, Japan spent USD 450 million on ocean plastic removal [74].

These damage costs, including lost opportunity costs and indirect costs, could be significantly reduced by preventing plastic from leaking into the environment. The worsening aesthetic of beaches polluted by waste reduces the number of tourists and income. Not only are fisheries affected (covering costs of damage caused by derelict fishing gear), but also land-based agricultural centers are affected by plastic litter blown onto beaches. Proper municipal clean-up practices are promising opportunities for the prevention of expenditure [2].

## 4. Recycling and Recommended Practices

*4.1. Recycling of Fishing Nets and Effective Actions*

The presence of ALDFG in marine environments is due to the following: irresponsible fishing practices, inadequate access to recycling facilities, low return prices for consumable plastic and a high cost of recycling [75]. Mechanical recycling is the simplest process. It involves following the following steps: sorting, cleaning, granulation, drying, melting, extrusion and pelletizing [35]. What is worth mentioning is that developing countries, such as Brazil, China and India, have high plastic recycling rates, between 20% and 60% [76]. In Australia and the United States, plastic recycling is low as follows: 10%–15%, whereas in Western Europe and Japan, recycling rates for plastic are around 25%–30% [76].

Technically, it is possible to separate most plastics into recognizable streams, but not all plastic streams are mechanically processable. It depends on the chemical and mechanical behavior as well as on the thermal properties. Only thermoplastic polymers (for example, polyethylene, polypropylene and polyester) are mechanically recyclable [77]. An alternative to mechanical recycling is chemical recycling, which produces plastic feedstock that can replace virgin plastic [78].

The main challenges for the circular fishing gear design are associated with the following: low utility of current materials, high level of mixing of different materials, lack of legal obligations for recycling from local authorities, lack of support and high cost of alternatives, low use of collection points in harbors and high organic contamination, which reduces the recyclability [79].

The most important practice for addressing the problem effectively is the prevention of gear loss. This is the ultimate goal of any progressive ghost gear program [80]. That is why it is aimed at the temporal separation of different gears, including the prohibition of high-risk types. For example, the Western Central Pacific Fisheries Commission prohibited large-scale driftnets. Additional separation of individual rope and net types is highly beneficial for all processing stages and the requirement to obtain uniform samples for material recycling [80].

Moreover, innovative solutions to end-of-life fishing gear promise to reduce the extent of lost fishing gear. Current actions taken by the European Commission have established a progressive goal of abandoned fishing net collection rate of 50% and a 15% recycling target, both to be met by 2025 [80]. There are many removal programs around the world that are focusing on different strategies of collection or cleaning the oceans. Some of them are highly specific. For example, the Northwest Straits Foundation's program is an initiative focused on the rapid removal of newly lost gillnets [80]. Other recommendations for the prevention of ALDFG are mostly focused on industry and governments. The great interest should be focused on solutions aiming at hot spot plastic areas. Mapping historic, ongoing and possible ALDFG data collection can significantly improve ocean cleaning practices and prevent the accumulation of plastic litter [81].

*4.2. Alternative Recycling Options*

Among the most important premises for establishing a recycling economy is creating international recycling standards, especially for mechanical recycling, as it is the most well-developed approach in terms of industrial feasibility [82]. One example is the creation of the European Strategy for Plastics in a Circular Economy where the design and production industry meet the needs of reuse, repair and recycling [79].

While eroding, polymeric chains decompose and release various chemical species. One of the most used materials is nylon 6, which was subjected to thermal analysis. The material was decomposed into volatile monomers at a temperature of approx. 400 °C at different heating rates (5, 15, 20 and 30 °C/min) [83]. Results showed that the decomposition of nylon 6 corresponds to a spectrum of caprolactam-based compounds during the most intense stage of decomposition. Pyrolysis of nylon 6 results in the reduction of the material into monomers, indicating the potential for the production of caprolactam. This also implies that waste nets can be converted to monomers via pyrolysis.

Available polymers have limited recyclability potential. Because of carbon-carbon backbone strength, depolymerization to monomers is prevented [84]. Polymers redesigned with ester backbones may be better suited for controlled chemical depolymerization. However, it may also be suitable for biological processing in managed systems, such as individual composting [84]. Even if the polymer satisfies the criteria for use and end-of-life, it is important to prepare a recoverable, sortable and separable product design. An example is the availability of the *APR Design Guide for Plastic Recyclability*, which is currently used in plastic-based packaging.

*4.3. Material Mixing Needs*

Promising R&D routes for establishing biodegradable fishing net materials often comprise blending mutually compatible biodegradable polymers. A unique R&D route for establishing a marine-degradable fishing net is the incorporation of photocatalyzable ether linkages along the polymer backbone architecture. Other R&D routes for establishing degradable fishing nets may promote biocatalytic degradation by various mechanisms.

However, designing photodegradable or biodegradable materials cannot be the sole solution as the environmental hazard remains for extended durations. Instead, a strong societal need exists for economically attractive "fishing net return schemes" (analogous to plastic bottle deposit schemes) for occupations fishers, providing an economic incentive to minimize abandoned, lost, or discarded fishing gear. The success of such economic return schemes would in the future enable the possibility of more conventional material mixing technologies for upgrading partially biodegraded fish nets. Such material mixing technologies would benefit from new rapid "multidimensional molecular characterization" technology to quantify, on a batch level, the amount of biodegradation experienced on each "homogenous" batch of recovered fishing nets. Such "multidimensional molecular characterization" would incorporate a quantified measure of chemical functional group polydispersity, enabling more accurate predictions of the mechanical properties of recycled polymer mixtures.

## 5. Alternative Usage of Derelict Fishing Gear

*5.1. Research Solutions*

Several research institutions have taken the challenge of finding new opportunities for recycled fishing nets and therefore getting them closer to the circular economy. Unfortunately, recycled polyolefin resins from fishing nets seem to have poor properties due to the presence of contamination. The blend of derelict PE nets with different types of virgin resins showed a potential for usage in packaging. Even though the created composites have certain limitations, it was possible to meet the required elongation at break as well as impact strength and environmental stress cracking resistance. With a properly chosen virgin resin, it is possible to use the plastic from fishing nets in packaging [85].

Interesting results were presented with the usage of recycled nylon fibers as tensile reinforcement of cementitious mortars. A significant increase in tensile strength and toughness was observed. Unreinforced material achieved approx. 35% lower tensile strength and up to 13 times lower toughness [86]. Moreover, it was discovered that the fibers of nylon fishing nets helped transfer stress through cracks and distribute stress by transforming a single wide crack into several smaller ones [87].

Obtaining the oil from the waste fishing net as a substitute for diesel fuel has been another, albeit uncommon investigation. This too, achieved some promising results. Oil from waste fishing nets possesses excellent fuel properties, with a calorific value of 44,450 kJ/kg (higher than diesel by 1.48%). Additionally, it works on a diesel engine without requiring any engine modifications. Nevertheless, the brake thermal efficiency decreased. Brake-specific fuel consumption increased, and so did engine emissions [88]. This is still, however, an idea worth further investigation as it may prove useful for retrieving fossil fuels.

Table 1 provides an overview of various market applications of ALDFG.

Table 1. Examples of different market applications for ALDFG.

| Type of Recycled Fiber | Market Application | Company |
|---|---|---|
| Polyethylene fishing nets mixed with different virgin resins | Packaging | Polymer Technology Laboratory, Spain |
| Nylon | Tensile reinforcement of cementitious mortars | University of Salerno, Italy |
| Nylon | Stress distribution in construction cracks | Hokkaido University, Japan |
| | Substitute for diesel fuel | College of Engineering, India |
| Nylon | Carpets and clothing | Gabriella, Poland |
| Mix of derelict fishing nets | Jewelry (New Stone design line) | Orska, Poland |
| Polypropylene waste | Material called Boomplastic used to make "Circula" bench | Studio Rygalik, Poland |
| Nylon fishing nets | Material called Econyl used to make spectacle frames | Karun, Chile |
| Different fishing fibers mixed with wooden fiber | Kelp Chair | Design Milk, Sweden |

*5.2. Solutions in Product Design*

In general, governments and international and local companies are aware of the negative impact of ALDFG on both the environment and the economy. To solve this problem, research on creating a new product by using the waste from fishing nets has already started. Currently, there are already being developed interesting solutions for transforming fishing ropes into nylon yarn for the production of clothes and carpets [89]. One of the many examples of using nylon fibers for clothing production is a Polish company named Gabriella. In 2021 designed tights consisting of 70% of oceanic wastes [90]. It has been proven that it is both possible and profitable to create sustainable and aesthetically pleasing products.

With the common initiative of the foundation MARE and the jewelry design company ORSKA, derelict fishing nets from the Baltic Sea were used for creating a new collection line. Ground fishing fibers were mixed with granules of recycled plastic that had undergone a thermal treatment, which resulted in the material used for creating the New Stone design line [91]. The Stone was created with the help of Tomasz Rygalik, the owner and designer of the Studio Rygalik company. His previous work shows the possibility of designing products out of plastic blends. He has created a material called Boomplastic, which has found its use in the creation of outdoor furniture—creating a garden bench called Circula. Boomplastic is a blend of polypropylene and colorful flakes obtained from polypropylene packages and bottle caps. The transparent matrix came from the grinding of damaged polypropylene bottles and cups [92].

Another brand called Karun started to create its products from plastic waste over 10 years ago. Their material is called Econyl, which is a nylon coming from ghost-fighting nets found in the ocean. Out of this plastic, there has been created spectacle frames, now available around the world [93]. Additionally, in 2022, the design company Design Milk from Sweden presented the innovative project of the 3D printed chair, using recycled fishing nets and wood fiber [94]. The creation of the Kelp Chair [94] prevented fishing nets from ending up in the depths of the Baltic Sea and instead turned them into new material. Promising results from research and usage of these materials create optimism for the further development of this field. The examples shown above indicate that recycling ALDFG and turning it into new material might become an economically profitable field. These new plastic types might be able to reduce the amount of virgin plastic entering the environment as well as limit costs connected to marine plastic debris.

## 6. Conclusions

Marine litter, and in particular plastic waste, including plastic from abandoned and derelict fishing gear, is a growing environmental concern. The influence of abandoned and derelict fishing gear is enormously threatening to natural marine and coastal habitats and endangered species. As well as being a burden on both the local and global economy. Designing materials, which are increasingly biodegradable, cannot be the sole solution to marine litter because the environmental hazard still remains for long stretches of time. It needs to be combined with a feasible plan for recovery and transformation into products that can compete in the open market. New research and businesses are already presenting alternative recycling paths and utilization of used fishing gear, which may be of benefit both environmentally and economically.

In the context of fishing net recycling, mixing pristine and partially degraded fishing net polymers marginally decreases new plastic production volumes. For re-usage as fishing nets, the mechanical properties of the recycled polymer must meet or exceed the pristine polymer's mechanical properties. In this manner, economically attractive return schemes can be implemented, reducing ecological harm caused by abandoned, lost, or discarded fishing nets.

**Author Contributions:** Conceptualization, A.K., K.G.P. and S.K.; writing—original draft preparation, A.K.; writing—review and editing, A.K., K.G.P. and S.K.; supervision, K.G.P. and S.K. All authors have read and agreed to the published version of the manuscript.

**Funding:** This research received no external funding.

**Data Availability Statement:** Not applicable.

**Acknowledgments:** The authors gratefully acknowledge Christian Karl and Anna-Maria Persson, both at SINTEF Industry in Oslo, Norway, for insightful conversations.

**Conflicts of Interest:** The authors declare no conflict of interest.

## References

1. United Nations Environment Programme, UNEP Annual Report 2005. Available online: https://www.unep.org/resources/synthesis-reports/unep-annual-evaluation-report-2005 (accessed on 6 June 2022).
2. United Nations Environment Programme. Drowning in Plastics—Marine Litter and Plastic Waste Vital Graphics. 2021. Available online: https://www.unep.org/resources/report/drowning-plastics-marine-litter-and-plastic-waste-vital-graphics (accessed on 6 June 2022).
3. NRC National Research Council. *Tackling Marine Debris in the 21st Century*; National Academies Press: Washington, DC, USA, 2008.
4. Sheavly, S.B. *National Marine Debris Monitoring Program: Final Program Report, Data Analysis and Summary*; Prepared for U.S. Environmental Protection Agency; Sheavly Consultants, Inc.: Virginia Beach, VA, USA, 2010.
5. Richardson, K.; Asmutis-Silvia, R.; Drinkwin, J.; Gilardi, K.V.; Giskes, I.; Jones, G.; O'Brien, K.; Pragnell-Raasch, H.; Ludwig, L.; Antonelis, K.; et al. Building evidence around ghost gear: Global trends and analysis for sustainable solutions at scale. *Mar. Pollut. Bull.* **2019**, *138*, 222–229. [CrossRef]
6. Gajanur, A.R.; Jaafar, Z. Abandoned, lost, or discarded fishing gear at urban coastlines. *Mar. Pollut. Bull.* **2022**, *175*, 113341. [CrossRef] [PubMed]
7. Stewart, L.G.; Lavers, J.L.; Grant, M.L.; Puskic, P.S.; Bond, A.L. Seasonal ingestion of anthropogenic debris in an urban population of gulls. *Mar. Pollut. Bull.* **2020**, *160*, 111549. [CrossRef] [PubMed]
8. Meijer, L.J.J.; van Emmerik, T.; van der Ent, R.; Schmidt, C.; Lebreton, L. More than 1000 rivers account for 80% of global riverine plastic emissions into the ocean. *Sci. Adv.* **2021**, *7*, eaaz5803. [CrossRef] [PubMed]
9. Li, W.C.; Tse, H.; Fok, L. Plastic waste in the marine environment: A review of sources, occurrence and effects. *Sci. Total Environ.* **2016**, *566–567*, 333–349. [CrossRef]
10. Lebreton, L.C.M.; Van Der Zwet, J.; Damsteeg, J.-W.; Slat, B.; Andrady, A.; Reisser, J. River plastic emissions to the world's oceans. *Nat. Commun.* **2017**, *8*, 15611. [CrossRef]
11. The Ocean Cleanup. The Great Pacific Garbage Patch. 12 May 2022. Available online: https://theoceancleanup.com/great-pacific-garbage-patch/ (accessed on 6 June 2022).
12. Lebreton, L.; Slat, B.; Ferrari, F.; Sainte-Rose, B.; Aitken, J.; Marthouse, R.; Hajbane, S.; Cunsolo, S.; Schwarz, A.; Levivier, A.; et al. Evidence that the Great Pacific Garbage Patch is rapidly accumulating plastic. *Sci. Rep.* **2018**, *8*, 4666. [CrossRef]

13. International Coastal Cleanup. Tracking Trash—25 Years of Action for the Ocean. 2011. Available online: https://oceanconservancy.org/wp-content/uploads/2017/04/2011-Ocean-Conservancy-ICC-Report.pdf (accessed on 6 June 2022).
14. Barnes, D.K.A.; Galgani, F.; Thompson, R.C.; Barlaz, M. Accumulation and fragmentation of plastic debris in global environments. *Philos. Trans. R. Soc. B* **2009**, *364*, 1985–1998. [CrossRef]
15. Santos, I.R.; Friedrich, A.C.; Barretto, F.P. Overseas garbage pollution on beaches of northeast Brazil. *Mar. Pollut. Bull.* **2005**, *50*, 783–786. [CrossRef]
16. McCoy, K.S.; Huntington, B.; Kindinger, T.L.; Morioka, J.; O'Brien, K. Movement and retention of derelict fishing nets in Northwestern Hawaiian Island reefs. *Mar. Pollut. Bull.* **2021**, *174*, 113261. [CrossRef]
17. Thomas, S.N.; Sandhya, K.M. Netting Materials for Fishing Gear with Special Reference to Resource Conservation and Energy Saving. In Proceedings of the ICAR Winter School: Responsible Fishing: Recent Advances in Resource and Energy Conservation, ICAR-CIFT, Kochi, India, 21 November–11 December 2019.
18. Brown, J.; Macfadyen, G. Ghost fishing in European waters: Impacts and management responses. *Mar. Policy* **2007**, *31*, 488–504. [CrossRef]
19. Macfadyen, G.; Huntington, T.; Cappell, R. *Abandoned, Lost or Otherwise Discarded Fishing Gear*; FAO Fisheries and Aquaculture Technical Paper 523; UNEP Regional Seas Reports and Studies 185; FAO: Rome, Italy, 2009.
20. Edyvane, K.S.; Penny, S.S. Trends in derelict fishing nets and fishing activity in northern Australia: Implications for trans-boundary fisheries management in the shared Arafura and Timor Seas. *Fish. Res.* **2017**, *188*, 23–37. [CrossRef]
21. Kim, S.-G.; Lee, W.-I.; Yuseok, M. The estimation of derelict fishing gear in the coastal waters of South Korea: Trap and gill-net fisheries. *Mar. Policy* **2014**, *46*, 119–122. [CrossRef]
22. Treble, M.A.; Stewart, R.E.A. *Impacts and Risks Associated with a Greenland Halibut (Reinhardtius hippoglossoides) Gillnet Fishery in Inshore Areas of NAFO Subarea 0*; Canadian Science Advisory Secretariat: Ottawa, ON, Canada, 2010.
23. Revill, A.S.; Dunlin, G. The fishing capacity of gillnets lost on wrecks and on open ground in UK coastal waters. *Fish. Res.* **2003**, *64*, 107–113. [CrossRef]
24. Ayaz, A.; Acarli, D.; Altinagac, U.; Ozekinci, U.; Kara, A.; Ozen, O. Ghost fishing by monofilament and multifilament gillnets in Izmir Bay, Turkey. *Fish. Res.* **2006**, *79*, 267–271. [CrossRef]
25. Stelfox, M.; Hudgins, J.; Sweet, M. A review of ghost gear entanglement amongst marine mammals, reptiles and elasmobranchs. *Mar. Pollut. Bull.* **2016**, *111*, 6–17. [CrossRef] [PubMed]
26. Sharif, N.F.H.; Mon, S.Z.K. A review on the strength of fishing net: The effect of material, yarn diameter and mesh size progress. *Prog. Eng. Appl. Technol.* **2021**, *2*, 1030–1036.
27. PerkinElmer. *Nylon 6—Influence of Water on Mechanical Properties and Tg*; PerkinElmer Inc.: Waltham, MA, USA, 2007.
28. NoFir, Life Cycle Assessment of EUfir System. A European System for Collecting and Recycling Discarded Equipment from the Fishing and Fish Farming Industry, Life Cycle Engineering. 18 January 2016. Available online: https://www.lcengineering.eu/portfolio_page/lca-of-eufir/ (accessed on 6 June 2022).
29. Al-Oufi, H.; McLean, E.; Kumar, A.S.; Claereboudt, M.; Al-Habsi, M. The effects of solar radiation upon breaking strength and elongation of fishing nets. *Fish. Res.* **2004**, *66*, 115–119. [CrossRef]
30. Welden, N.A.; Cowie, P.R. Degradation of common polymer ropes in a sublittoral marine environment. *Mar. Pollut. Bull.* **2017**, *118*, 248–253. [CrossRef]
31. Kershaw, P.J.; Rochman, C.M. (Eds.) *Sources, Fate and Effects of Microplastics in the Marine Environment: Part 2 of a Global Assessment*; Reports and Studies—IMO/FAO/UNESCO-IOC/UNIDO/WMO/IAEA/UN/UNEP/UNDP Joint Group of Experts on the Scientific Aspects of Marine Environmental Protection (GESAMP) Eng No. 93; International Maritime Organization: London, UK, 2016; p. 220.
32. Greene, K.L.; Tonjes, D.J. Degradable plastics and their potential for affecting solid waste systems. *WIT Trans. Ecol. Environ.* **2014**, *180*, 91–102. [CrossRef]
33. Norwegian Environment Agency. Bio-Based and Biodegradable Plastics: An Assessment of the Value Chain for Bio-Based and Biodegradable Plastics in Norway. 2018. Available online: https://www.miljodirektoratet.no/globalassets/publikasjoner/m1206/m1206.pdf (accessed on 6 June 2022).
34. Emadian, S.M.; Onay, T.T.; Demirel, B. Biodegradation of bioplastics in natural environments. *Waste Manag.* **2017**, *59*, 526–536. [CrossRef] [PubMed]
35. Basel Convention. Baseline Report on Plastic Waste. UNEP/CHW/PWPWG.1/INF/4. 2020. Available online: http://www.basel.int/Implementation/Plasticwaste/PlasticWastePartnership/Consultationsandmeetings/PWPWG1/tabid/8305/Default.aspx (accessed on 6 June 2022).
36. Alaerts, L.; Augustinus, M.; Van Acker, K. Impact of Bio-Based Plastics on Current Recycling of Plastics. *Sustainability* **2018**, *10*, 1487. [CrossRef]
37. Green, D.S.; Boots, B.; Blockley, D.J.; Rocha, C.; Thompson, R. Impacts of Discarded Plastic Bags on Marine Assemblages and Ecosystem Functioning. *Environ. Sci. Technol.* **2015**, *49*, 5380–5389. [CrossRef]
38. Vendries, J.; Sauer, B.; Hawkins, T.R.; Allaway, D.; Canepa, P.; Rivin, J.; Mistry, M. The Significance of Environmental Attributes as Indicators of the Life Cycle Environmental Impacts of Packaging and Food Service Ware. *Environ. Sci. Technol.* **2020**, *54*, 5356–5364. [CrossRef]

39. Al-Salem, S.M. Study of the Degradation Behaviour of Virgin and Biodegradable Plastic Films in Marine Environment Using ASTM D 6691. *J. Polym. Environ.* **2021**, *30*, 2329–2340. [CrossRef]
40. Halim, N.A.; Mon, S.Z.K. A review on impact of environmental conditions on fishing net mechanical properties. *Prog. Eng. Appl. Technol.* **2021**, *2*, 957–964.
41. Isangedighi, I.A.; David, G.S.; Obot, O.I. Plastic Waste in the Aquatic Environment: Impacts and Management. *Environment* **2018**, *2*, 15–43. [CrossRef]
42. Gall, S.; Thompson, R. The impact of debris on marine life. *Mar. Pollut. Bull.* **2015**, *92*, 170–179. [CrossRef]
43. IUCN. The IUCN Red List of Threatened Species. Version 2018–2019. Available online: http://www.iucnredlist.org (accessed on 6 June 2022).
44. Wilcox, C.; Mallos, N.J.; Leonard, G.H.; Rodriguez, A.; Hardesty, B.D. Using expert elicitation to estimate the impacts of plastic pollution on marine wildlife. *Mar. Policy* **2016**, *65*, 107–114. [CrossRef]
45. Matsuoka, T.; Nakashima, T.; Nagasawa, N. A review of ghost fishing: Scientific approaches to evaluation and solutions. *Fish. Sci.* **2005**, *71*, 691–702. [CrossRef]
46. Good, T.P.; June, J.A.; Etnier, M.A.; Broadhurst, G. Derelict fishing nets in Puget Sound and the Northwest Straits: Patterns and threats to marine fauna. *Mar. Pollut. Bull.* **2010**, *60*, 39–50. [CrossRef] [PubMed]
47. Hardesty, B.D.; Good, T.P.; Wilcox, C. Novel methods, new results and science-based solutions to tackle marine debris impacts on wildlife. *Ocean Coast. Manag.* **2015**, *115*, 4–9. [CrossRef]
48. Page, B.; Welling, A.; Chambellant, M.; Goldsworthy, S.D.; Dorr, T.; van Veen, R. Population status and breeding season chronology of Heard Island fur seals. *Polar Biol.* **2003**, *26*, 219–224. [CrossRef]
49. Kuzin, A.E.; Trukhin, A.M. Entanglement of Steller sea lions (Eumetopias jubatus) in man-made marine debris on Tyuleniy Island, Sea of Okhotsk. *Mar. Pollut. Bull.* **2022**, *177*, 113521. [CrossRef]
50. Fowler, C.W. Marine debris and northern fur seals: A case study. *Mar. Pollut. Bull.* **1987**, *18*, 326–335. [CrossRef]
51. Butterworth, A.; Sayer, S. The Welfare Impact on Pinnipeds of Marine Debris and Fisheries. *Anim. Welf.* **2017**, 215–239. [CrossRef]
52. Vauk, G.J.; Schrey, E. Litter pollution from ships in the German Bight. *Mar. Pollut. Bull.* **1987**, *18*, 316–319. [CrossRef]
53. Williams, A.T.; Tudor, D.T. Litter Burial and Exhumation: Spatial and Temporal Distribution on a Cobble Pocket Beach. *Mar. Pollut. Bull.* **2001**, *42*, 1031–1039. [CrossRef]
54. Gregory, M.R. Environmental implications of plastic debris in marine settings entanglement, ingestion, smothering, hangers-on, hitch-hiking and alien invasions. *Philos. Trans. R. Soc. Lond. B Biol. Sci.* **2009**, *364*, 2013–2025. [CrossRef]
55. Amaral-Zettler, L.A.; Zettler, E.R.; Mincer, T.J. Ecology of the plastisphere. *Nat. Rev. Microbiol.* **2020**, *18*, 139–151. [CrossRef]
56. Kiessling, T.; Gutow, L.; Thiel, M. *Marine Anthropogenic Litter*; Bergmann, M., Gutow, L., Klages, M., Eds.; Springer Open: Berlin/Heidelberg, Germany, 2015; pp. 141–181.
57. Beaumont, N.J.; Aanesen, M.; Austen, M.C.; Börger, T.; Clark, J.R.; Cole, M.; Hooper, T.; Lindeque, P.K.; Pascoe, C.; Wyles, K.J. Global ecological, social and economic impacts of marine plastic. *Mar. Pollut. Bull.* **2019**, *142*, 189–195. [CrossRef] [PubMed]
58. Peng, X.; Dasgupta, S.; Zhong, G.; Du, M.; Xu, H.; Chen, M.; Chen, S.; Ta, K.; Li, J. Large debris dumps in the northern South China Sea. *Mar. Pollut. Bull.* **2019**, *142*, 164–168. [CrossRef] [PubMed]
59. Landrigan, P.J.; Stegeman, J.J.; Fleming, L.E.; Allemand, D.; Anderson, D.M.; Backer, L.C.; Brucker-Davis, F.; Chevalier, N.; Corra, L.; Czerucka, D.; et al. Human Health and Ocean Pollution. *Ann. Glob. Health* **2020**, *86*, 151. [CrossRef]
60. KIMO. Economic Impacts of Marine Litter. 2010. Available online: https://www.kimointernational.org/wp/wp-content/uploads/2017/09/KIMO_Economic-Impacts-of-Marine-Litter.pdf (accessed on 6 June 2022).
61. Antonelis, K.; Huppert, D.; Velasquez, D.; June, J. Dungeness Crab Mortality Due to Lost Traps and a Cost–Benefit Analysis of Trap Removal in Washington State Waters of the Salish Sea. *N. Am. J. Fish. Manag.* **2011**, *31*, 880–893. [CrossRef]
62. Scheld, A.M.; Bilkovic, D.M.; Havens, K.J. The Dilemma of Derelict Gear. *Sci. Rep.* **2016**, *6*, 19671. [CrossRef] [PubMed]
63. McIlgorm, A.; Raubenheimer, K.; McIlgorm, D.E. *Update of 2009 APEC Report on Economic Costs of Marine litter to APEC Economies*; A report to the APEC Ocean and Fisheries Working Group by the Australian National Centre for Ocean Resources and Security (ANCORS); University of Wollongong: Wollongong, Australia, 2020.
64. Vlachogiann, T. Understanding the Socio-Economic Implications of Marine Litter in the Adriatic-Ionian Microregion. IPA-AdriaticDeFishGear Project and MIO-ECSDE. 2017. Available online: https://mio-ecsde.org/project/understanding-the-socio-economic-implications-of-marine-litter-in-the-adriatic-ionian-macroregion-ipa-adriatic-defishgear-project-and-mio-ecsde-2017/ (accessed on 6 June 2022).
65. UNEP. Marine Litter Socio-Economic Study. Nairobi. 2017. Available online: https://wedocs.unep.org/20.500.11822/26014 (accessed on 6 June 2022).
66. Qiang, M.; Shen, M.; Xie, H. Loss of tourism revenue induced by coastal environmental pollution: A length-of-stay perspective. *J. Sustain. Tour.* **2019**, *28*, 550–567. [CrossRef]
67. NOAA (United Sates National Oceanic and Atmospheric Administration). *Assessing the Economic Benefits of Reductions in Marine Litter: A Pilot Study of Beach Recreation in Orange County, California*; National Oceanic and Atmospheric Administration: Silver Spring, MD, USA, 2014.
68. UNEP. Marine Plastic Debris and Microplastics: Global Lessons and Research to Inspire and Guide Policy Change. Nairobi. 2016. Available online: https://wedocs.unep.org/handle/20.500.11822/7720 (accessed on 6 June 2022).

69. Jeffrey, C.F.G.; Havens, K.J.; Slacum, H.W., Jr.; Bilkovic, D.M.; Zaveta, D.; Scheld, A.M.; Willard, S.; Evans, J.D. *Assessing Ecological and Economic Effects of Derelict Fishing Gear: A Guiding Framework*; Virginia Institute of Marine Science, Collage of William and Mary: Williamsburg, VA, USA, 2016. [CrossRef]
70. Deloitte. The Price Tag of Plastic Pollution: An Economic Assessment of River Plastic. 2019. Available online: https://www2.deloitte.com/content/dam/Deloitte/nl/Documents/strategy-analytics-and-ma/deloitte-nl-strategy-analytics-and-ma-the-price-tag-of-plastic-pollution.pdf (accessed on 6 June 2022).
71. Burt, A.; Raguain, J.; Sanchez, C.; Brice, J.; Fleischer-Dogley, F.; Goldberg, R.; Talma, S.; Syposz, M.; Mahony, J.; Letori, J.; et al. The costs of removing the unsanctioned import of marine plastic litter to small island states. *Sci. Rep.* **2020**, *10*, 14458. [CrossRef]
72. Woo-Rack, S. Progress in Addressing Marine Litter in Korea: Recent Policies and Efforts to Protect the Marine Environment from Marine Litter. In Proceedings of the NOWPAP-TEMM Joint Workshop on Marine Litter Management, Bali, Indonesia, 4–5 June 2018.
73. Ministry of Oceans and Fisheries. Efforts to Combat Marine Litter in Japan. Economy Reports and Presentations. In *Capacity Building for Marine Litter Prevention and Management in the APEC Region Phase II*; Appendix E; APEC—Ocean and Fisheries Working Group (OFWG): Busan, Korea, 2018; pp. 286–290.
74. Skirtun, M.; Sandra, M.; Strietman, W.J.; Burg, S.W.V.D.; De Raedemaecker, F.; Devriese, L.I. Plastic pollution pathways from marine aquaculture practices and potential solutions for the North-East Atlantic region. *Mar. Pollut. Bull.* **2021**, *174*, 113178. [CrossRef]
75. Ragaert, K.; Delva, L.; Van Geem, K. Mechanical and chemical recycling of solid plastic waste. *Waste Manag.* **2017**, *69*, 24–58. [CrossRef]
76. Garcia, J.M.; Robertson, M.L. The future of plastics recycling. *Science* **2017**, *358*, 870–872. Available online: https://science.sciencemag.org/content/358/6365/870 (accessed on 6 June 2022). [CrossRef]
77. Thiounn, T.; Smith, R.C. Advances and approaches for chemical recycling of plastic waste. *J. Appl. Polym. Sci.* **2020**, *58*, 1347–1364. [CrossRef]
78. European Commission. Communication from the Commission to the European Parliament, the Council, the European Economic and Social Committee and the Committee of the Regions A European Strategy for Plastics in a Circular Economy. 2018. Available online: https://eur-lex.europa.eu/legal-content/EN/TXT/?uri=COM:2018:28:FIN (accessed on 6 June 2022).
79. Stolte, A.; Schneider, F. Recycling Options for Derelict Fishing Gear. 2018. Available online: www.marelittbaltic.eu (accessed on 6 June 2022).
80. Richardson, K.; Hardesty, B.D.; Wilcox, C. Estimates of fishing gear loss rates at a global scale: A literature review and meta-analysis. *Fish Fish.* **2019**, *20*, 1218–1231. [CrossRef]
81. Shamsuyeva, M.; Endres, H.-J. Plastics in the context of the circular economy and sustainable plastics recycling: Comprehensive review on research development, standardization and market. *Compos. Part C Open Access* **2021**, *6*, 100168. [CrossRef]
82. Skvorčinskienė, R.; Striūgas, N.; Navakas, R.; Paulauskas, R.; Zakarauskas, K.; Vorotinskienė, L. Thermal Analysis of Waste Fishing Nets for Polymer Recovery. *Waste Biomass Valorization* **2019**, *10*, 3735–3744. [CrossRef]
83. Law, K.L.; Narayan, R. Reducing environmental plastic pollution by designing polymer materials for managed end-of-life. *Nat. Rev. Mater.* **2002**, *7*, 104–116. [CrossRef]
84. Knott, B.C.; Erickson, E.; Allen, M.D.; Gado, J.E.; Graham, R.; Kearns, F.L.; Pardo, I.; Topuzlu, E.; Anderson, J.J.; Austin, H.P.; et al. Characterization and engineering of a two-enzyme system for plastics depolymerization. *Proc. Natl. Acad. Sci. USA* **2020**, *117*, 25476–25485. [CrossRef]
85. Juan, R.; Domínguez, C.; Robledo, N.; Paredes, B.; Galera, S.; García-Muñoz, R.A. Challenges and Opportunities for Recycled Polyethylene Fishing Nets: Towards a Circular Economy. *Polymers* **2021**, *13*, 3155. [CrossRef]
86. Spadea, S.; Farina, I.; Carrafiello, A.; Fraternali, F. Recycled nylon fibers as cement mortar reinforcement. *Constr. Build. Mater.* **2015**, *80*, 200–209. [CrossRef]
87. Srimahachota, T.; Yokota, H.; Akira, Y. Recycled Nylon Fiber from Waste Fishing Nets as Reinforcement in Polymer Cement Mortar for the Repair of Corroded RC Beams. *Materials* **2020**, *13*, 4276. [CrossRef]
88. Sivathanu, N.; Anantham, N.V.; Peer, M.S. An experimental investigation on waste fishing net as an alternate fuel source for diesel engine. *Environ. Sci. Pollut. Res.* **2019**, *26*, 20530–20537. [CrossRef]
89. Monteiro, D.; Rangel, B.; Alves, J.L. Design as a vehicle for using waste of fishing nets and ropes to create new products. *Eng. Soc.* 2016, p. 67. Available online: https://www.esat.kuleuven.be/stadius/engineering4society/files/E4S2016_Proceedings.pdf#page=67 (accessed on 6 June 2022).
90. Gabriella Now! Ocean Collection. Available online: https://www.gabriella.pl/gabriella-now.html (accessed on 6 June 2022).
91. Orska Design. Available online: https://orska.pl/blog/blog/jak-sie-tworzy-nowy-material-kilka-slow-o-kamieniu-mare (accessed on 6 June 2022).
92. Boomplastic. Available online: https://boomplastic.com/circula-lawka-z-recyklingu/ (accessed on 6 June 2022).
93. Noizz Fashion. Available online: https://noizz.pl/fashion/te-okulary-powstaja-z-sieci-rybackich-ktore-stanowia-46-proc-plastiku-w-morzach/qynfssy (accessed on 6 June 2022).
94. Yv, Y. 2022. Available online: https://design-milk.com/this-kelp-chair-is-3d-printed-using-recycled-fishing-nets-wood-fiber/?utm_source=rss&utm_medium=rss&utm_campaign=this-kelp-chair-is-3d-printed-using-recycled-fishing-nets-wood-fiber (accessed on 6 June 2022).

Article

# ZnO/CQDs Nanocomposites for Visible Light Photodegradation of Organic Pollutants

Elena E. Toma, Giuseppe Stoian, Bogdan Cojocaru, Vasile I. Parvulescu and Simona M. Coman *

Department of Organic Chemistry, Biochemistry and Catalysis, Faculty of Chemistry, University of Bucharest, Regina Elisabeta Blvd., No. 4-12, 030016 Bucharest, Romania
* Correspondence: simona.coman@chimie.unibuc.ro

**Abstract:** Currently, carbon quantum dots (CQDs) have been widely investigated as an enhancing photocatalytic component of various nanocomposites. In this study, hetero-structures containing carbon quantum dots (CQDs) associated to zinc oxide were prepared following two one-pot procedures: (i) a hydrothermal approach in which commercial ZnO was used as carrier for CQDs; and (ii) an approach in which the ZnO/CQDs samples were produced in situ by adding zinc acetate to an aqueous suspension of CQDs. CQDs were prepared in advance by a low-temperature hydrothermal (LHT) treatment of useless humins wastes produced by the glucose dehydration in an acidic medium. These samples were characterized by several techniques such asadsorption-desorption isotherms of liquid nitrogen at 77K, X-ray diffraction (XRD), infrared diffuse reflectance with Fourier transform (DRIFT) and UV-vis spectroscopy. The photocatalytic behavior of these materials was investigated in the degradation of methylene blue (MB). The obtained results revealed electronic interactions between CQDs and ZnO which have as an effect an enhancement of the charge separation and diminution of the charge recombination. In accordance, a correlation between the photocatalytic activity and the intrinsic properties of ZnO/CQDs has been evidenced. The highest photocatalytic activity corresponded to the heterostructure containing highly dispersed narrow sized CQDs onto ZnO. Under visible light irradiation and after 180 min of irradiation, MB was degraded by as much as 97.6%.

**Keywords:** zinc oxide; carbon quantum dots; nanocomposites; photodegradation; organic dyes; methylene blue

Citation: Toma, E.E.; Stoian, G.; Cojocaru, B.; Parvulescu, V.I.; Coman, S.M. ZnO/CQDs Nanocomposites for Visible Light Photodegradation of Organic Pollutants. *Catalysts* 2022, 12, 952. https://doi.org/10.3390/catal12090952

Academic Editor: Didier Robert

Received: 3 August 2022
Accepted: 24 August 2022
Published: 26 August 2022

**Publisher's Note:** MDPI stays neutral with regard to jurisdictional claims in published maps and institutional affiliations.

**Copyright:** © 2022 by the authors. Licensee MDPI, Basel, Switzerland. This article is an open access article distributed under the terms and conditions of the Creative Commons Attribution (CC BY) license (https://creativecommons.org/licenses/by/4.0/).

## 1. Introduction

Photocatalysis is providing a promising alternative to the highly energetic and pollutant chemical transformations [1,2]. As a photocatalyst, zinc oxide (ZnO) attracted a certain interest due to its quite efficient behavior in environmental applications such as the cleaning of wastewaters or the polluted air [3]. Also, its high photosensitivity correlates well to other important features such as its non-toxicity and low cost [3]. However, its large bandgap (3.37 eV), massive charge carrier recombination, and photoinduced corrosion–dissolution under extreme pH conditions, leading to inert $Zn(OH)_2$, are all barriers to its extensive applicability [4].

Then, the photocatalytic efficiency of ZnO is highly dependent on the surface morphology and particles size [5–8]. Therefore, in order to inhibit the surface-bulk charge carrier recombination, intensive research has been carried out to improve its performances by tailoring the surface-bulk structure and altering its photogenerated charge transfer pathways [4]. This also correlates well with the requirement for new photocatalytic materials able to shift the adsorption from UV towards the visible region [9].

The luminescence of ZnO in both the visible and UV regions is assigned to the intrinsic defects created during synthesis. Accordingly, its broad visible luminescence is associated to its morphology and crystalline structure [4]. Further, engineering O and Zn vacancies of this oxide may even tune its photoluminescence emission [10]. In this way, its photocatalytic behavior under visible irradiation may be tuned by coupling with another semiconductor, with carbon nanoparticles, and by doping with transition metals [11–13] or non-metal elements (i.e., C or N) [14,15]. However, most of these hybrid ZnO nanocomposites still suffer from an inefficient use of sunlight (<5%) and an unstable nature.

Recently, due to their unique properties, carbon quantum dots (CQDs) became a focal point of interest as a new class of carbon nanomaterials [16]. Their synthesis mainly focuses easy routes of preparation, surface passivation and functionalization [17], small particle sizes, biocompatibility, photoluminescence (PL) properties, a high-temperature stability, a chemically inert structure, and a low toxicity. Correlated to these, CQDs already received interest in many fields, such as photocatalysis [18], sensors [19] and biomedical applications [20,21].

For the particular case of photocatalysis, CQDs became potential competitors to conventional semiconductor quantum dots (QDs) synthesized from the elements of the groups III–VI or toxic heavy metals [22]. Moreover, their hybridization with different materials such as $TiO_2$, ZnO, $SiO_2$, $Fe_2O_3$, $Ag_3PO_4$, or $Cu_2O$ has been reported as a new opportunity to improve the charge separation and the reduction of the charge recombination, leading thus to the enhancement of the photocatalytic efficiency [23–27].

Very recently Widiyandari et al. [25]. reviewed the applications of ZnO/CQDs nanocomposites for the degradation of pollutants focusing the progress in their development. However, despite the numerous reports on the CQDs exploiting for different applications, the synthesis of ZnO/CQDs nanocomposites, with the aim to promote ZnO photocatalytic activity under visible light irradiation, was investigated only in a few reports [27–30].

Based on this state of the art, the challenge of the present work was to combine, in an innovative way, photocatalysis, biomass wastes valorization and environmental degradation with the aim to rich a high degree of sustainability. To reach this scope, the synthesis of carbon quantum dots (CQDs) has been carried out from a worthless humins by-product, generated in the dehydration of the glucose. This was treated through an environmentally hydrothermal friendly (LTH) process at low-temperature. Subsequently, for the photocatalytic degradation of the organic dyes, there were synthesized ZnO/CQDs nanocomposites. The synthesis of these followed two hydrothermal routes: (i) in which a commercial ZnO was used as carrier for the CQDs; and (ii) an one-pot process in which a $Zn(CH_3COO)_2 \cdot 2H_2O$ precursor salt was directly added to an aqueous solution of CQDs. After gentle drying, these materials were investigated in the photooxidative degradation of methylene blue (MB), a representative of a class of dye-stuff resistants to biodegradation, under both UV and Vis lights irradiation.

## 2. Results and Discussion

*2.1. ZnO/CQDs Nanocomposites Characterization*

Carbon quantum dots (CQDs) obviously present a core-shell morphology with nanocrystalline or amorphous cores [31] and shells with attached polar groups (e.g., hydroxyls, carboxyls and carbonyls) derived from starting materials [32]. However, when the CQDs are produced via a low-temperature hydrothermal (LTH) approach their size distribution and chemical composition highly depend on two key factors, namely, the reaction time and the synthesis temperature.

In this work the CQDs were synthesized from humins wastes produced by acid dehydration of D-glucose. The LTH approach was applied for the CQDs synthesized at 160–200 °C, for 4–12 h, respectively. Emission spectra at $\lambda_{ex}$ = 310 nm reveal the effect of the hydrothermal parameters on the fluorescence characteristics of CQDs. Both the temperature and time affect the CQDs formation progress and the fluorescence properties. The highest PL intensities were registered for either CQDs synthesized at higher temperatures and shorter synthesis times (i.e., $CQD_{200-4}$) or at lower temperatures and longer synthesis times (e.g., $CQD_{180-12}$) (Figure 1a). At 200 °C, longer reaction times led to the aggregation of the small CQDs exhibiting a lower intensity of fluorescence (not shown in Figure 1a) while, at 180 °C, the intensity of the fluorescence depended on the reaction time, thus suggesting a progress of the LTH process. Therefore, at 180 °C and a short reaction time, the fluorescence is weak. This implies that the reaction is just in the initial stage and the concentration of the quantum dots formed is low. The fluorescence intensity gradually increases with the reaction time increases, which suggests an increase in the carbon dot concentration. The solution with the brightest luminescent is obtained after 12 h (Figure 1b).

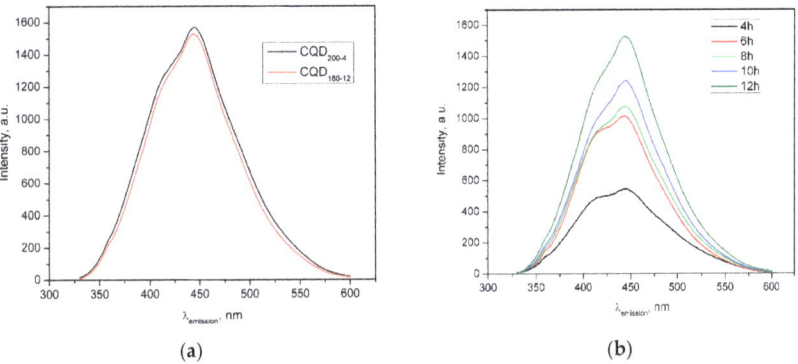

**Figure 1.** (**a**) Emission spectra ($\lambda_{ex}$ = 310 nm) of the synthesized CQDs at 200 °C and 4 h and at 180 °C and 12 h; (**b**) emission spectra ($\lambda_{ex}$ = 310 nm) of the synthesized CQDs at 180 °C as a function of the reaction time.

ATR-FTIR measurements confirmed the influence of the synthesis parameters upon the nature of the CQDs functionalities. As ATR-IR spectra show (Figure 2), the bands at 1220–1100 cm$^{-1}$ correspond to the C-O-C stretching, at 1370 cm$^{-1}$ to the O-H in-plane deformation vibration, at 1640 cm$^{-1}$ to the C=C stretching, at 1730 cm$^{-1}$ to the C=O stretching and at 3300 cm$^{-1}$ to the –OH stretching vibration, most probably attached to the surface and at the CQDs edge. The intense and broad band centered at 3300 cm$^{-1}$ proves the high concentration of the hydroxyl groups, irrespective of the LTH parameters. Also, their presence justify a very good dispersion and high stability of these solid nanoparticles in water [33]. However, the intensity of the absorption band at 1730 cm$^{-1}$ (characteristic to the C=O stretching) diminished with the increase of the hydrothermal temperature from 180 to 200 °C that may indicate a decreased population of the carboxylic group in the shell composition, also in accordance to the literature data [34].

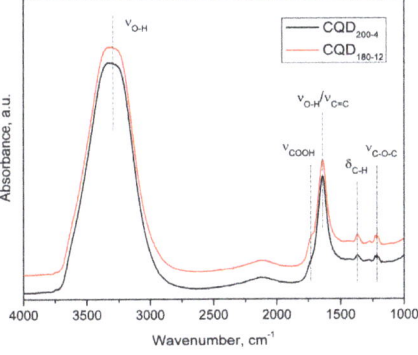

**Figure 2.** ATR-FTIR spectra in the 1000–4000 cm$^{-1}$ range of the CQDs synthesized at 200 °C and 4 h and at 180 °C and 12 h.

The structural properties of the CQDs were as well investigated by UV-Vis (Figure 3a) and XRD patterns (Figure 3b). The inset in Figure 3a shows the images of the CQD$_{200\text{-}4}$ solutions collected in visible and UV light (365 nm). The faint yellow color under visible light and the distinct blue emission under the UV light at 365 nm correspond to a red-shift.

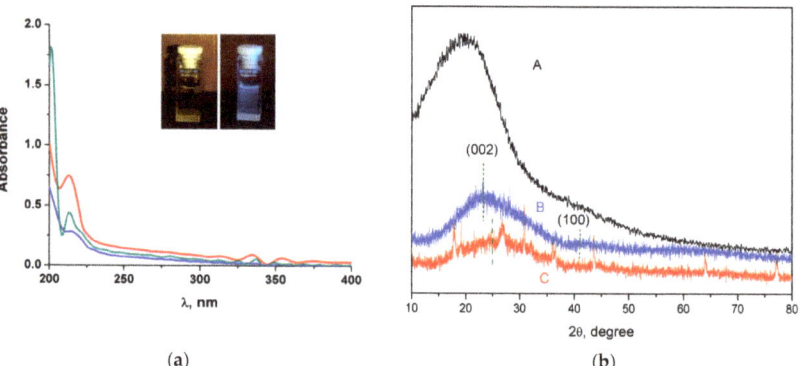

**Figure 3.** (a) The UV-vis absorption spectra for CQD$_{180\text{-}4}$ (blue), CQD$_{180\text{-}12}$ (green) and CQD$_{200\text{-}4}$ (red) samples; (b) the XRD pattern of humins (A), CQD$_{200\text{-}4}$ (B) and CQD$_{180\text{-}12}$ (C).

The UV absorption band at 220 nm is attributed to the π-π* transition of the C=C bonds with a tail extending into the visible range (Figure 3a). The relatively weak absorptions at 335 and 354 nm may originate from the n–π* transition of the C=C-C=O groups (335 nm) and from the overtone transition of the oxygen bridged bond (R-O-R') (354 nm) in the sp$^3$ hybrid region. The XRD pattern (Figure 3b) of CQDs revealed two diffraction lines centered at 22 and 42°, indicating a graphite-like carbon structure in their core [35]. For the CQD$_{180\text{-}12}$, the lower diffraction line intensity along with a relatively high full width half maximum suggest the presence of particles with a smaller size [34].

The PL emission spectra of the CQDs excited at $\lambda_{ex}$ = 250–450 nm (Figure 4a exemplifies the PL spectra of CQD$_{200\text{-}4}$) consist of two overlapping spectral bands [36], with a double PL peaks at 410 and 445 nm, respectively. In accordance to Yuan et al. [37] this is a consequence of the electron transition between the edge states functional groups and carbon states, such as conjugate π states (or sp$^2$ area). The different origins of the PL emission are also confirmed by the decrease of the relative intensity of the band at 410 nm with the increase of the relative intensity of the band at 445 nm (namely with the increase of the excitation wavelength).

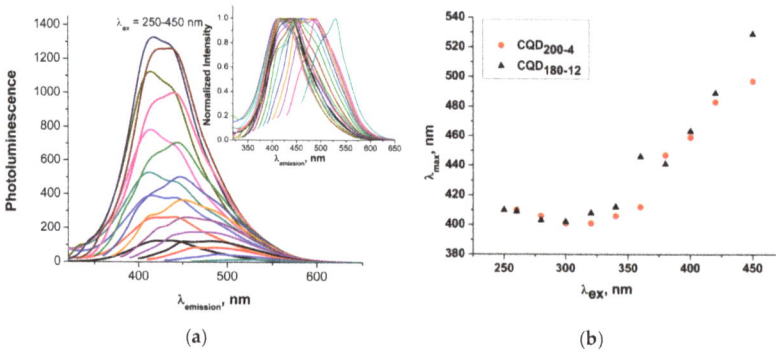

**Figure 4.** (a) PL spectra of the CQD$_{200\text{-}4}$ aqueous solution excited in a range of $\lambda_{ex}$ = 250–450 nm; (b) the maximum of the wavelength PL emission ($\lambda_{max}$) as a function of the excitation wavelength ($\lambda_{ex}$).

For the excitation in the range of 250–340 nm no perceptible excitation-emission dependence ($\lambda_{em}$ = 410–445 nm) has been observed while a distinct red-shift from $\lambda_{em}$ = 445 to $\lambda_{em}$ = 530 nm was clearly evidenced when $\lambda_{ex}$ shifted from 360 to 450 nm (Figure 4a). This excitation-independent PL behavior is derived from the CQDs homogeneous particle size [38] while the high polarity of the nanosized and the $sp^2$-carbon networks of CQDs is usually claimed as being responsible for the excitation-dependent PL behavior [39]. In accordance, it can be assumed that the excitation-independent PL behavior of the shorter wavelength part of the emission is due to the predominant homogeneous size of CQDs, while the excitation-dependent behavior of the longer wavelength part is attributed to the presence of a low population of CQDs with larger sizes [40]. An additional argument that comes to support the assumption of the existence of predominate homogeneous CQDs size is given by the wavelength corresponding to the maximum emission ($\lambda_{max}$). This remained approximately constant for the excitation ($\lambda_{ex}$) in the 240–340 nm range while for the range of 360–450 nm the position of the maximum emission shifted with the increase of the excitation wavelength (Figure 4b).

The quantum yield (QY) ($\lambda_{ex}$ = 366 nm) of the CQDs varied in the: $CQD_{200-4}$ (22.6%) > $CQD_{180-12}$ (21%) > $CQD_{180-4}$ (8%), order, namely, with values comparable to most of the reported QYs in the literature [38]. In accordance with Zhu et al. [41], the higher QY of $CQD_{200-4}$ can be attributed to its edge slightly smaller concentration of the non-radiative recombination carboxyl (–COOH) sites (see also Figure 2). The smaller size of the CQDs obtained at 180 °C and after 12 h could justifies the increased QY value compared to that of CQDs produced at the same temperature but after only 4 h.

ZnO/CQD nanocomposites were subsequently produced using the above synthesized CQDs through either the hydrothermal (ZnO(C)/CQD) or the one-pot hydrothermal (ZnO(OP)/CQD) routes. The corresponding XRD patterns are given in Figure 5.

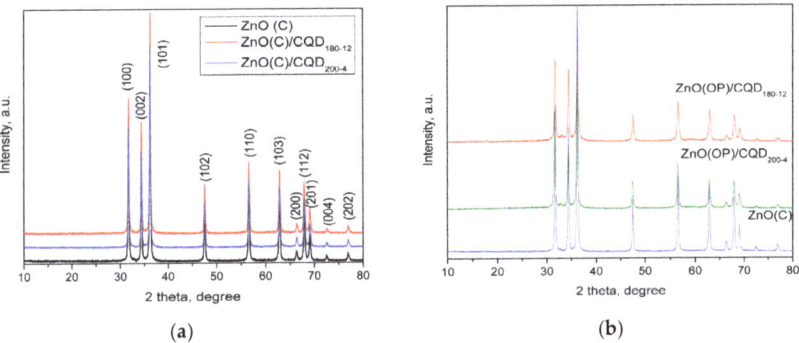

**Figure 5.** (a) The XRD patterns of the ZnO(C)/CQD nanocomposites; (b) the XRD patterns of the ZnO(OP)/CQD nanocomposites.

The diffraction lines at 2θ = 31.68°, 34.35°, 36.09°, 47.36°, 56.48°, 62.70°, 66.23°, 67.87°, 68.99°, 72.55°, and 76.77° were assigned to (100), (002), (101), (102), (110), (103), (200), (112), (201), (004) and (202) crystalline planes of ZnO (wurtzite, with hexagonal phase structure (JCPD36-1451)) [42]. The patterns of ZnO(C)/CQDs were identical with those of the ZnO(C) carrier (Figure 5a) that demonstrates the crystal structure of ZnO has not been modified by CQDs. Also, the XRD patterns of the ZnO(C)/CQD (Figure 5a) do not evidenced diffraction lines characteristic to CQDs indicating a high dispersion of these on the external surface of ZnO. This may also correspond to a small loading, poor crystallinity and high dispersion of the CQDs into the ZnO(C)/CQD composites.

XRD patterns of the samples prepared via the one-pot hydrothermal approach (OP) also showed lines assigned to the ZnO (wurtzite, hexagonal phase structure (JCPD36-1451)) (Figure 5b).

The crystallites average size was determined from the Debye-Scherrer equation taking the reflection (101) of hexagonal ZnO wurtzite (1) [43]:

$$d = \frac{k\lambda}{\beta \cos\theta} \quad (1)$$

where: $d$ is the crystallite size in nm; $k = 0.94$; $\lambda$ is the wavelength of the X-ray (1.54178 Å); $\theta$ is the half-diffraction angle and $\beta$ is the full width at half-maximum (FWHM) in radians for the $2\theta$ value.

As Table 1 shows, the average crystallite sizes of the ZnO(C)-based samples varied in a narrow range (33–35 nm), while for those prepared via the one-pot hydrothermal approach (i.e., ZnO(OP)-based samples), the average crystallite sizes were even smaller (cca 28 nm). The broadening of the (101) diffraction line also supports the smaller sizes of all of the prepared samples, irrespective of the applied approach.

**Table 1.** The crystallite size of the ZnO/CQD nanocomposites.

| Sample | $\beta$, Degree | (101) Plane, $2\theta$ | $\theta$, Degree | Crystallite Size, nm |
|---|---|---|---|---|
| ZnO(C) | 0.2599 | 36.2158 | 18.1079 | 33.61 |
| ZnO(C)/CQD$_{180-12}$ | 0.2490 | 36.2637 | 18.1318 | 35.09 |
| ZnO(C)/CQD$_{200-4}$ | 0.2469 | 36.2322 | 18.1161 | 35.39 |
| ZnO(OP)/CQD$_{180-12}$ | 0.3053 | 36.1265 | 18.0632 | 28.61 |
| ZnO(OP)/CQD$_{200-4}$ | 0.3139 | 36.1327 | 18.0663 | 27.83 |

Similar to ZnO–graphene composites [44], the XRD patterns of ZnO(OP)/CQD show a shifting of the main diffraction lines (i.e., (100), (002), (101) planes) to a lower degree (Figure 6 and Table 1, column 3) compared to the pure ZnO(C), indicating both a change in the ZnO lattice constants and a stronger interaction with CQDs [45,46].

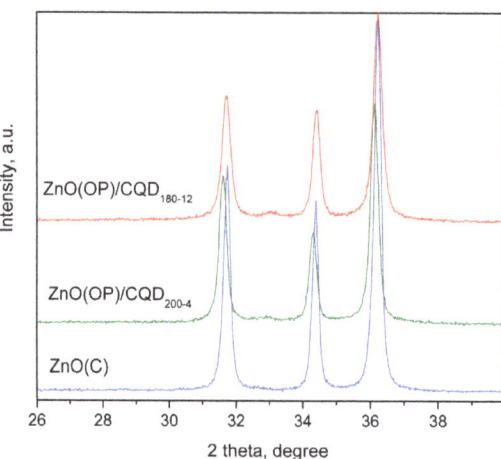

**Figure 6.** XRD patterns of on $2\theta = 30\text{--}38°$ of ZnO(OP)/CQD samples.

The lattice parameters ($a$, $b$ and $c$) were calculated using Equation (2), while the plane d-spacing ($d$) was calculated using the Braggs' Equation (3) [47]:

$$a = b = \frac{\lambda}{\sqrt{3}\sin\theta_{100}} \text{ and } c = \frac{\lambda}{\sin\theta_{002}} \quad (2)$$

$$d = \frac{\lambda}{2\sin\theta_{101}} \quad (3)$$

where: $\theta_{100}$ and $\theta_{002}$ are the diffraction lines angles of the (100) and (002) planes, respectively.

As Table 2 shows, for ZnO(OP)/CQD the lattice parameters (*a, b* and *c*) were higher [48]. Obviously, such a modification is due to the presence of CQDs and the enlarged d-spacing may suggest a relatively high density of the ZnO stabilized CQDs layers.

**Table 2.** The lattice parameters (a–c) and the inter-planer d-spacing of the ZnO/CQD samples.

| Sample | 2θ, Degree | Lattice Parameters, Å | | c/a | d-Spacing, BRAGG |
|---|---|---|---|---|---|
| | | c | a = b | | |
| ZnO(C) | 36.2158 | 5.2157 | 3.2561 | 1.6018 | 2.4803 |
| ZnO(C)/CQD$_{180-12}$ | 36.2637 | 5.2140 | 3.2533 | 1.6027 | 2.4771 |
| ZnO(C)/CQD$_{200-4}$ | 36.2322 | 5.2140 | 3.2547 | 1.6020 | 2.4795 |
| ZnO(OP)/CQD$_{180-12}$ | 36.1265 | 5.2246 | 3.2678 | 1.5988 | 2.4867 |
| ZnO(OP)/CQD$_{200-4}$ | 36.1327 | 5.2246 | 3.2713 | 1.5971 | 2.4859 |
| JCPDS 36-1451 | 36.2150 | 5.2150 | 3.2560 | 1.6016 | - |

For both ZnO and ZnO/CQD the adsorption-desorption isotherms of nitrogen (Figure 7) correspond to a type IV in the IUPAC classification. These indicate the presence of a small percentage of micropores filled with nitrogen at extremely low relative pressures and of a multilayer evidenced at high relative pressures. Finally, a capillary condensation prevails as the pressure gets higher. The H3-type hysteresis provides an indication of slit-shaped pores generated through the particles' aggregation.

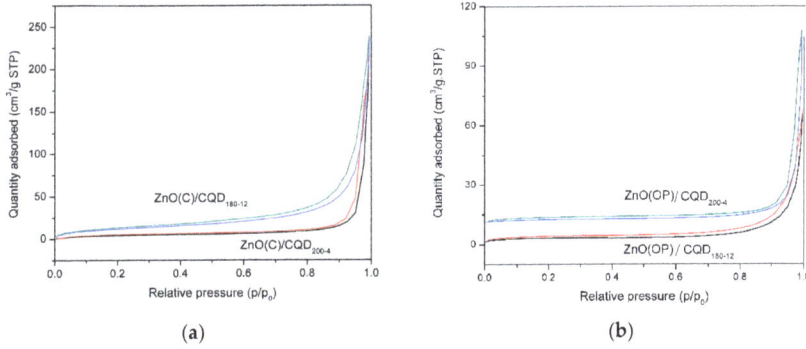

**Figure 7.** Nitrogen adsorption-desorption isotherms of the ZnO(C)/CQDs (**a**) and the ZnO(OP)/CQDs (**b**) samples.

The BET surface areas, pore volumes ($V_p$), average pore diameter ($D_p$) and particles size ($d_{BET}$) are given in Table 3.

**Table 3.** BET analysis surface area and porosity of ZnO/CQD samples.

| Sample | BET (m$^2$/g) | $V_p$ (cm$^3$/g) | Average $D_p$ (nm) | $d_{BET}$ (nm) |
|---|---|---|---|---|
| ZnO(C) | 12 | 0.033 | 26.5 | 89.1 |
| ZnO(C)/CQD$_{200-4}$ | 15 | 0.136 | 52.0 | 71.3 |
| ZnO(C)/CQD$_{180-12}$ | 18 | 0.062 | 25.8 | 59.4 |
| ZnO(OP)/CQD$_{200-4}$ | 7 | 0.038 | 25.7 | 152.8 |
| ZnO(OP)/CQD$_{180-12}$ | 11 | 0.045 | 37.2 | 97.2 |

The average diameter of the particles ($d_{BET}$) was calculated from $S_{BET}$ using the formula:

$$d_{BET} \text{ (nm)} = \frac{6000}{S_{BET}(m^2/g) \times \rho(g/cm^3)} \quad (4)$$

where: $\rho$ is the ZnO density (5.61 g/cm$^3$).

As Table 3 shows, except sample ZnO(C)/CQD$_{200-4}$ sample with an average Dp (nm) of 52 nm, all other samples possess average pores in the mesoporous size range, which coincides with the type IV adsorption isotherm. BET surface areas were in the range of 7–18 m$^2$/g while the pore volumes between 0.038 and 0.136 cm$^3$/g. The diminution of the surface and pore volume for ZnO(OP)-samples may be assigned to the intercalation of the CQDs between the ZnO layers but also as a consequence of the agglomeration of the nanoparticles. Indeed, for the samples prepared via the one-pot hydrothermal approach (denoted with (OP)) the increased particle size (d$_{BET}$) corresponds to a higher agglomeration of the smaller crystallite ZnO. These results are also supported by the XRD measurements.

The electronic interaction between ZnO and CQDs was further confirmed by the DRIFT spectra. The strong bands at 642 and 872 cm$^{-1}$ are attributed to the Zn-O stretching vibrations [49] while the bands at around 1635 and 3464 cm$^{-1}$ are attributed to the O-H stretching vibrations of chemisorbed water molecules [50,51] (Figure 8). However, the band at 1635 cm$^{-1}$ may also be assigned to the C=C stretching vibration of the numerous sp$^2$ species that enriched the aromatic structure of the CQDs [52]. The band at 1569 cm$^{-1}$ is assigned to the C=C stretching of polycyclic aromatic structures, while that at 1418 cm$^{-1}$ to C-O vibrations of the oxygen containing groups [46].

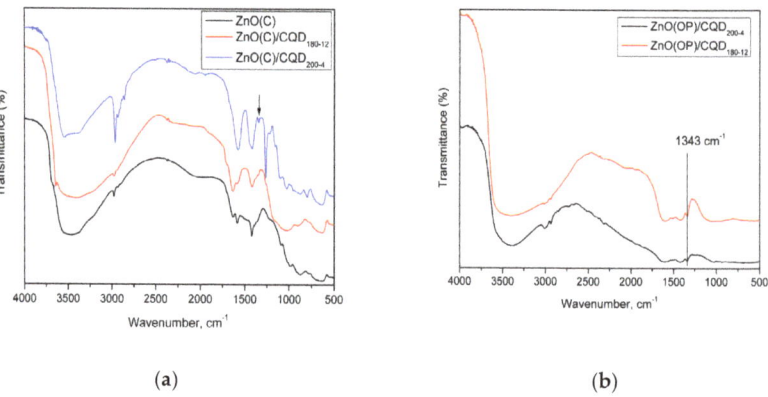

**Figure 8.** DRIFT spectra of (**a**) ZnO(C)/CQDs and (**b**) ZnO(OP)/CQDs samples.

The absorption band at 1260 cm$^{-1}$ is ascribed to the C-O-C stretching of the furan ring deformation while the new band at 1338–1345 cm$^{-1}$ for ZnO/CQDs to a new Zn-COO$^-$ group formed as a result of the CQDs interaction with the ZnO carrier (Figure 9).

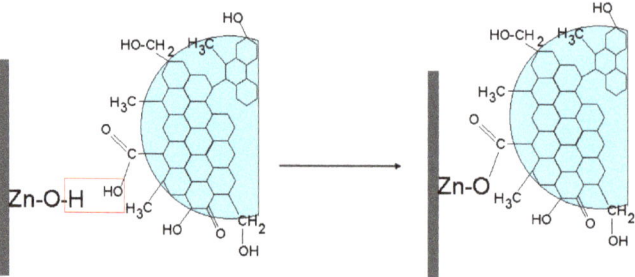

**Figure 9.** Schematically representation of the esterification reaction between the carboxyl groups from CQDs and the hydroxyls groups from ZnO.

Photo-absorption is one of the key factors affecting the photocatalytic performance of the photocatalysts. As shown in Figure 10, ZnO has almost no absorption above 370 nm that is in accordance to its band gap. However, the ZnO/CQDs composite showed an absorption in the visible light region, ranging from 450 to 650 nm, likely derived from the interaction between the CQDs and ZnO. As consequence, the increased intensity of this absorption band may serve as an indication of the contribution of CQDs to the enhanced visible light absorption of the composite [53].

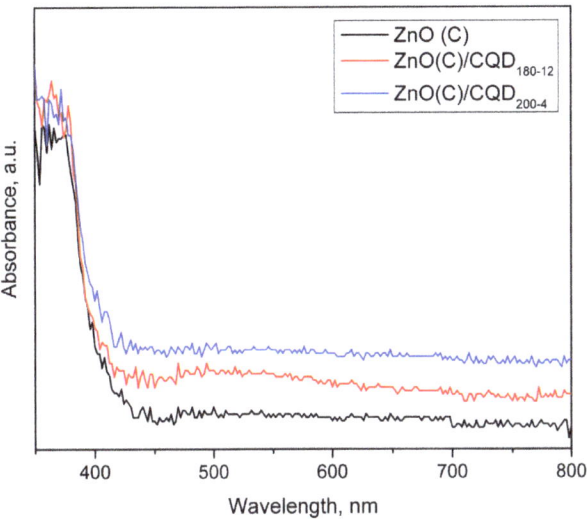

**Figure 10.** UV-Vis diffuse reflectance spectra of ZnO(C)/CQDs samples.

*2.2. Methylene Blue (MB) Photodegradation*

The pH solution influences the ionization of the MB and the catalyst surface [54,55], neutral and basic environments leading to better efficiency than acidic environments. The $pH_{PZC}$ of ZnO which lies between 8 and 9. Therefore, as the pH of the solution increases $pH_{PZC}$ of ZnO becomes zero or less positive hence, the percentage of MB adsorption increases with increase in pH, leading to an enhanced degradation. The low degradation efficiency in acidic pH is related to the presence $H^+$ ions which compete with cationic pollutant molecules (i.e., MB) for the adsorption onto the adsorbent surface [56].

Therefore, the photocatalytic degradation of the methylene blue has been carried out in both the absence (i.e., pH = 6 for the used MB solution) and the presence of the $Na_2CO_3/NaHCO_3$ buffer (pH = 9.2), and under UV and Vis light irradiation. Under the UV irradiation zinc oxide may suffer photocorrosion reactions. However, very recently it has been showed that due to the fast charge separation, the presence of CQDs can protect the ZnO semiconductor from photocorrosion and, therefore, enhance its structural stability [53].

MB is a chromogenic agent whose N-S heterocycle group attached to benzene includes lone pair electron in which thiol group is the main chromophoric group [57,58]. The UV-Vis bands at 612 and 666 nm are characteristic for the large conjugated system of the N-S heterocycle group while that at 292 nm characterizes the phenothiazine structure [58]. The collected UV-Vis spectra during the MB photo-degradation in the presence of ZnO(C) and under UV irradiation, in the absence or the presence of the $Na_2CO_3/NaHCO_3$ buffer are shown in Figure 11.

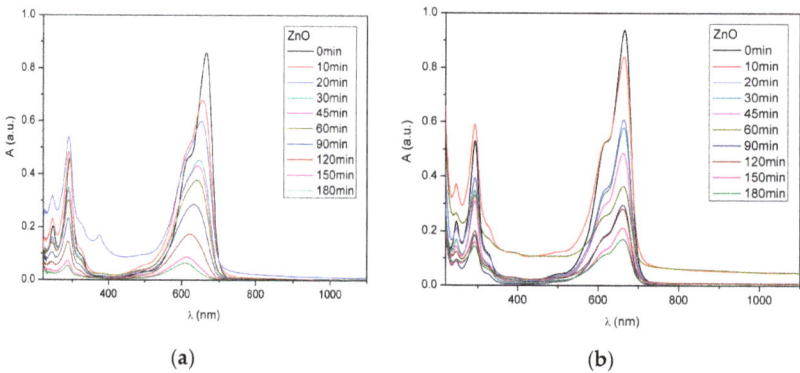

**Figure 11.** UV–Vis spectra collected during the MB degradation versus reaction time on the ZnO(C) catalyst, under UV irradiation and (**a**) in the absence of the $Na_2CO_3/NaHCO_3$ buffer and (**b**) in the presence of the $Na_2CO_3/NaHCO_3$ buffer.

During the photocatalytic reaction the intensity of the bands located in the visible region suffered an important decrease (Figure 11a,b) that is also an indication of a destruction of the heterocycle conjugated structure, both at pH = 6 and pH = 9.2. This effect was synchronized with a decoloration process, supporting the recorded blue shifts (i.e., hypsochromic effects) associated to N-demethylation of the dimethylamino group of MB [57,58]. The diminution of the phenothiazine structure occurred through opening of the ring via the attack of hydroxyl radicals.

However, as Figure 11 shows the MB degradation takes place with a conversion of 98.2%, at pH = 6 and after 180 min of irradiation with UV (Figure 11a), while in the presence of the $Na_2CO_3/NaHCO_3$ buffer (Figure 11b) the degradation occurred in a lower extend (i.e., 85% conversion after 180 min). This behavior can be due to the $Na_2CO_3/NaHCO_3$ buffer which acts as a $OH^\bullet$ scavenger, in a similar way as reported for processes taking place in the presence of the $Na_2CO_3$ base [59]. Therefore, MB degradation tests in the presence of the ZnO(C)/CQDs and ZnO(OP)/CQDs semiconductors were subsequently carried out in the absence of the buffer $Na_2CO_3/NaHCO_3$. The obtained results under UV irradiation are shown in Figure 12.

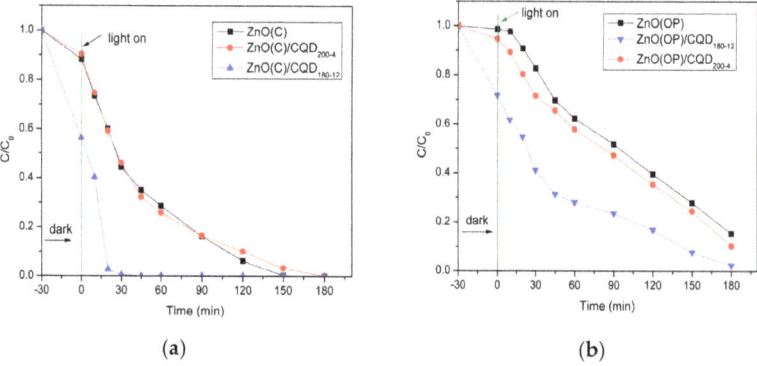

**Figure 12.** The degradation of MB (665 nm band) in time in the presence of (**a**) ZnO(C)/CQD and (**b**) ZnO(OP)/CQD under UV irradiation.

The promoting effects of the CQDs system may differ according to the utilized irradiation [53]. Under the UV exposure, CQDs act as a reservoir trapping photogenerated electrons from ZnO and promoting the separation of the photogenerated electron-hole pairs.

However, the rate of MB decomposition is different as a function of the ZnO/CQDs features (Figure 12). A faster decomposition takes place in the presence of the $CQD_{180-12}$-based nanocomposites irrespective of the ZnO carrier nature. The conversion of MB reached 98.8% in the presence of $ZnO(C)/CQD_{180-12}$ in only 30 min (Figure 12a), while, in the presence of the $ZnO(OP)/CQD_{180-12}$, it decreased to 89.1% even after a longer time (i.e., 180 min) (Figure 12b). The higher photoactivity of the $CQD_{180-12}$—based samples can be associated to the presence of a higher loading the -COOH acidic groups onto the CQDs surface, making it easier the chemisorption of the basic MB groups. Moreover, the steric hindrance of the $CQD_{180-12}$ is smaller owing a small size promotion of an effective photon and hydroxyl radicals transfer [38].

As XRD patterns showed (Figure 5b) the ZnO(OP) posses larger particle sizes generated by a higher agglomeration of the smaller ZnO crystallites during the synthesis. ZnO(OP)/CQDs are also characterized by a reduced surface area and pore volume, most probably due to the intercalation and agglomeration of the CQDs between the ZnO layers. Thus, a lower conversion in the MB degradation might be attributed to a decreased active surface.

A mechanism for the UV photocatalytic degradation of dyes, also fitting the results of this work, was not long ago proposed by Zou and co-workers [46]. In accordance with this, the UV irradiation of the ZnO/CQD heterostructure at energies higher or equal to the band-gap of ZnO excites the electrons ($e^-$) from the valence band into the conduction band. These move freely to the surface of CQDs layer through a photo-induced charge transfer process, while the holes ($h^+$) leave in the valence band of ZnO (step 1). Then, holes ($h^+$) migrate to the surface and react with $H_2O$ or $OH^-$ to produce $OH^\bullet$ radicals (step 3). The photo-generated electrons can further react with the dissolved oxygen on the surface of CQDs generating $O_2^{\bullet-}$ radicals (step 2) which after protonation generate hydroperoxy $HO_2^\bullet$ radicals (step 4), producing hydroxyl radical $OH^\bullet$ and hydrogen peroxide (step 5). Hydrogen peroxide is a supplementary source for the hydroxyl radical $OH^\bullet$ production (step 6). These species represent strong oxidizing agents able to degrade MB and, finally, to mineralize them towards $CO_2$ and $H_2O$ (step 7).

1. $CQD@ZnO + h\nu \rightarrow e^- (CQD) + h^+ (ZnO)$
2. $e^- + O_2 \rightarrow O_2^{\bullet-}$
3. $h^+ + OH^- \rightarrow OH^\bullet$
4. $O_2^{\bullet-} + H_2O \rightarrow HO_2^\bullet + OH^-$
5. $HO_2^\bullet + H_2O \rightarrow H_2O_2 + OH^\bullet$
6. $H_2O_2 \rightarrow 2OH^\bullet$
7. $OH^\bullet + dye \rightarrow CO_2 + H_2O$

Under visible light irradiation the MB degradation takes place at a lower level by comparison with the UV irradiation, irrespective of the ZnO/CQDs nature (Figure 13).

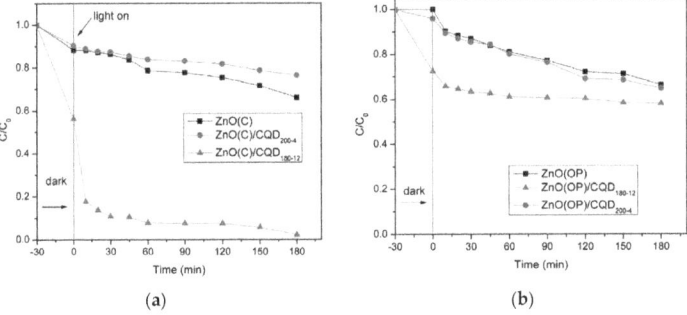

**Figure 13.** The degradation of MB (665 nm band) in time in the presence of (**a**) ZnO(C)/CQD and (**b**) ZnO(OP)/CQD under Vis light irradiation.

According to Heng et al. [53], under irradiation with visible light, the CQDs act as a photosensitizer enlarging the absorption range of metal oxide into the visible light region through electronic coupling between π states of CQDs and conduction band of the metal oxide. This corresponds to a transfer of excited electrons from CQDs to the conduction band of the ZnO. The best results in MB degradation were obtained in the presence of ZnO(C)/CQD$_{180-12}$ sample and are well correlated with the UV-vis absorption measurements results (Figure 10). In the presence of this sample, the MB conversion reached a level of 97.6%, after 180 min, while, in the presence of the pristine ZnO(C) the MB conversion was four times lower (i.e., 25%). The same level of MB conversion (26%) was registered in the presence of the ZnO(OP) sample, while onto ZnO(OP)/CQD$_{180-12}$ the decomposition of the MB reached a higher level (32%) but much lower compared to the one obtained in the presence of the ZnO(C)/CQD$_{180-12}$. Once again, the photocatalytic efficiency seems to be governed by the same factors: the CQDs and ZnO features. In the case of the ZnO(OP)/CQD samples, the self-aggregation of CQDs on the ZnO surface reduces fewer sites for activation and absorption of visible light by carbon [60].

## 3. Methods and Materials

### 3.1. Synthesis of ZnO/CQDs Nanocomposites

Carbon quantum dots (CQDs) were synthesized following a low temperature hydrothermal (LTH) methodology in which the humins, generated from the decomposition of glucose, were used as raw material. Briefly, 30 mg of humins were dispersed in 30 mL of water. To this suspension, 0.15 mL of glacial $CH_3COOH$ were added. Then, the solution was placed in a Teflon-lined stainless-steel autoclave and heated at 160–200 °C for 4–12 h. After the hydrothermal treatment, the aqueous mixture was transferred into a 250 mL separating funnel and 40 mL n-butanol was added. The formed layers were vigorously mixed thoroughly for some time, the aqueous layer, containing CQDs, was separated and further analyzed.

Obtained CQDs were subsequently used for the preparation of the ZnO/CQDs nanocomposites by using two methods as described below.

Hydrothermal approach: In a typical procedure, 0.4 g ZnO was slowly added under stirring to a solution of 20 mL of $H_2O$, 6 mL $C_2H_5OH$, and 2 mL solution of CQDs and the mixture was kept at room temperature for 4 h. The resulting suspension was then placed in a Teflon-lined stainless-steel autoclave and kept at a constant temperature of 140 °C for 4 h. After hydrothermal treatment, the ZnO/CQDs product was collected by centrifugation, washed with deionized water several times to remove the impurities and finally dried in air for 12 h at 80 °C. The obtained samples were denoted ZnO(C)/CQD$_{200-4}$ and ZnO(C)/CQD$_{180-12}$, were (C) indicates a commercial ZnO.

One-pot hydrothermal approach: The ZnO/CQDs hybrid nanostructures were synthesized by a hydrothermal method at a low temperature [26]. Typically, 0.274 g (125 mM) of $Zn(CH_3COO)_2 \cdot 2H_2O$ were dissolved in the previously prepared CQDs solution at room temperature. The pH of the solution was adjusted to 10 through dropwise addition of 8 M NaOH aqueous solution and stirring vigorously for 15 min at room temperature. Then, the solution was transferred into a Teflon-beaker and heated at 80 °C for 3 h. The final product was collected by centrifugation, washed with deionized water several times to remove the impurities and finally dried in air for 12 h at 80 °C. The obtained samples were denoted ZnO(OP)/CQD$_{200-4}$ and ZnO(OP)/CQD$_{180-12}$ where (OP) indicates the "one-pot" approach.

### 3.2. Characterization Techniques

ZnO/CQDs materials were characterized by techniques as: adsorption-desorption isotherms of liquid nitrogen at 77K, X-ray diffraction (XRD), IR diffuse reflectance with Fourier transform (DRIFT) and UV-vis spectroscopy.

Textural characteristics (surface area, pore volume and pore diameter) were determined from the adsorption-desorption isotherms of nitrogen at 77 K using a Micromeritics ASAP 2020 Surface Area and Porosity Analyzer.

Powder X-ray Diffraction patterns were collected at room temperature using a Shimadzu XRD-7000 apparatus with the Cu Kα monochromatic radiation of 1.5406 Å, 40 kV, 40 mA at a scanning rate of 1.0 2θ min$^{-1}$, in the 2θ range of 10–90°.

DRIFT spectra were recorded with a Thermo spectrometer 4700 (400 scans with a resolution of 4 cm$^{-1}$) in the range of 600–4000 cm$^{-1}$.

UV-vis spectra were recorded with a SPECORD 250-222P108 in a range of 304–1100 nm with a scan rate of 50 nm per second.

### 3.3. Methylene Blue Photooxidative Degradation

MB photooxidative degradation tests were performed in a LZC-4b photoreactor (Luzchem Research Inc., Gloucester, ON, Canada) provided with LED lamps (112 W, 445–465 nm) or in a home-made system provided with UV lamps (VilberLorumat, 240 W, 365 nm). Both photoreactors are provided as well as an exhaust/ventilation system. Experiments were conducted in a quartz cylinder. The distance between the lamp and the reaction cylinder was fixed at 20 cm. In each experiment 30 mg of catalyst (3.0 g/L) was added to 10 mL of a MB aqueous solution (30 mg/L; 9.4 × 10$^{-5}$ mol/L) at its natural pH of 6.0. Some experiments were also carried out at a constant pH of 9.2 maintained by a $Na_2CO_3/NaHCO_3$ buffer. The suspension of the photocatalyst and the homogeneity of the reacting mixture were ensured by magnetic stirring. Before switching on the lamp, the suspension was kept in darkness for 30 min and under stirring to reach the thermodynamic equilibrium. During the runs, samples of reacting suspension (500 µL) were withdrawn at 10, 20, 30, 45, 60, 90, 120, 150, and 180 min, filtered through a 0.45 µm hydrophilic membrane (HA, Millipore) and mixed with 2.5 mL of $H_2O$ before being analyzed. The spectrum was registered on a SPECORD 250-222P108 in a range from 190 nm to 1100 nm with a scan rate of 10 nm per second, a delta lambda of 1 nm and a split of 4 nm. Cuvettes of quartz were used with a path length of 1 cm.

### 3.4. Reaction Products Analysis

The concentration of MB after illumination was determined by a UV-Vis spectrometer (SPECORD 250-222P108). The intensity of the main absorption peak (666 nm) of the MB dye was referred to as a measure of the residual MB dye concentration. Control experiments with pristine ZnO were also performed.

For the calibration curve stock solution of 500 mL with a concentration of 30 mg/L was prepared (15 mg MB dissolved in distilled water in a volumetric flask). The stock solution was used to prepare other five different MB concentrations as: 5, 10, 15, 20, and 25 mg/L. For each concentration, the UV-vis spectra was registered (Figure 14a) and the corresponding absorbance value for the 665 nm band was used to build the calibration curve (Figure 14b). The calibration curve presents a linear relation and the relationship coefficient was 0.9990 (Figure 14b).

The MB conversion was defined as follows:

$$Conversion\ (\%) = \frac{C_0 - C}{C_0} \times 100 \qquad (5)$$

where: $C_0$ is the initial concentration of MB and $C$ is the concentration of the MB at a certain reaction time, respectively.

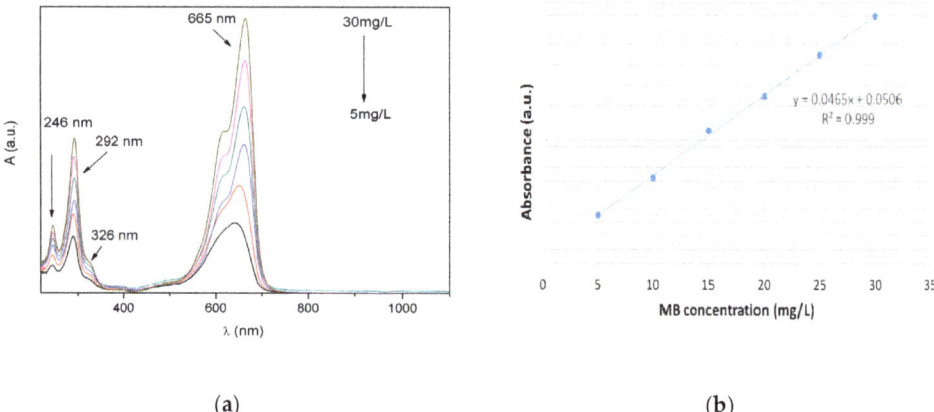

**Figure 14.** (**a**) UV-vis spectra of MB at different concentrations; (**b**) the calibration curve.

## 4. Conclusions

In summary, ZnO/CQDs were synthesized by two methods, namely a hydrothermal approach, in which CQDs were deposited onto the surface of commercial ZnO, and a one-pot hydrothermal approach, in which CQDs were added during the synthesis of the ZnO from a zinc acetate precursor. For the first time, the CQDs used for the production of these nanocomposites were obtained from worthless humins by-product by an environmentally friendly low-temperature hydrothermal (LTH) process. ZnO display a wurtzite (hexagonal phase) structure and adding it to CQDs generates hierarchical heterophases formed by ZnO and CQDs. The characterization techniques indicate a series of differences between these materials in terms of surface area, particle size, dispersion, and location of the CQDs. For ZnO(C)/CQD the CQDs are preponderantly docked on the external surface of the ZnO, with a high dispersion, while for ZnO(OP)/CQDs the agglomerated CQDs are preponderantly stacked between the ZnO layers, generating a sandwich-like structure.

Under UV irradiation the CQDs act as electron reservoirs that resist the recombination of electron-hole pairs. However, an increased photocatalytic activity of ZnO/CQDs can be induced by the presence of small sized CQDs (i.e., $CQD_{180-12}$). The larger size of CQDs (i.e., $CQD_{200-4}$) caused decreased photocatalytic performance of ZnO.

As these results shows, the presence of CQDs on the surface of ZnO can also promote a large spectrum of absorption, from UV into visible one, and the excellent photocatalytic activity of the ZnO(C)/$CQD_{180-12}$ nanocomposite under Vis light irradiation could be attributed to visible light sensitization of ZnO.

Among the investigated catalysts ZnO(C)/$CQD_{180-12}$ was able to degrade MB at a conversion of 99.3% under UV irradiation for 30 min and over 97.6% under Vis light irradiation for 180 min.

**Author Contributions:** Conceptualization, S.M.C.; methodology, E.E.T. and G.S.; validation, B.C.; formal analysis, E.E.T., G.S. and B.C.; investigation, E.E.T., G.S. and B.C.; writing—original draft preparation, S.M.C.; writing review and editing, S.M.C. and V.I.P.; visualization, S.M.C. and V.I.P.; supervision, S.M.C.; funding acquisition, B.C. All authors have read and agreed to the published version of the manuscript.

**Funding:** This study was funded by the Government of Romania, Ministry of Research and Innovation, project PNIII-P4-ID-PCE2020-2207 nr. 235/2021.

**Conflicts of Interest:** The authors declare no conflict of interest.

## References

1. Hoffmann, M.R.; Martin, S.T.; Choi, W.; Bahnemann, D.W. Environmental Applications of Semiconductor Photocatalysis. *Chem. Rev.* **1995**, *95*, 69–96.
2. Yang, X.; Wang, D. Photocatalysis: From Fundamental Principles to Materials and Applications. *ACS Appl. Energy Mater.* **2018**, *1*, 6657–6693.
3. Djurišić, A.B.; Chen, X.; Leung, Y.H.; Ng, A.M.C. ZnO Nanostructures: Growth, Properties and Applications. *J. Mater. Chem.* **2012**, *22*, 6526–6535.
4. Kumar, S.G.; Rao, K.S.R.K. Zinc Oxide Based Photocatalysis: Tailoring Surface-Bulk Structure and Related Interfacial Charge Carrier Dynamics for Better Environmental Applications. *RSC Adv.* **2015**, *5*, 3306–3351.
5. Li, D.; Li, D.K.; Wu, H.Z.; Liang, F.; Xie, W.; Zou, C.W.; Shao, L.X. Defects Related Room Temperature Ferromagnetism in Cu-Implanted ZnO Nanorod Arrays. *J. Alloys Compd.* **2014**, *591*, 80–84.
6. Gupta, M.K.; Lee, J.H.; Lee, K.Y.; Kim, S.W. Two-Dimensional Vanadium-Doped ZnO Nanosheet-Based Flexible Direct Current Nanogenerator. *ACS Nano* **2013**, *7*, 8932–8939.
7. Liu, X.; Hu, M.; Chu, X.; Yan, Q. Synthesis and Field Emission Properties of Highly Ordered Ti-Doped ZnO Nanoarray Structure. *J. Mater. Sci. Mater. Electron.* **2013**, *24*, 2839–2845.
8. Qi, L.; Li, H.; Dong, L. Simple Synthesis of Flower-like ZnO by a Dextran Assisted Solution Route and Their Photocatalytic Degradation Property. *Mater. Lett.* **2013**, *107*, 354–356.
9. Xiang, Q.; Yu, J.; Jaroniec, M. Graphene-Based Semiconductor Photocatalysts. *Chem. Soc. Rev.* **2012**, *41*, 782–796.
10. Rauwel, E.; Galeckas, A.; Rauwel, P.; Sunding, M.F.; Fjellvaåg, H. Precursor-Dependent Blue-Green Photoluminescence Emission of ZnO Nanoparticles. *J. Phys. Chem. C* **2011**, *115*, 25227–25233.
11. Wang, Z.L. Splendid One-Dimensional Nanostructures of Zinc Oxide: A New Nanomaterial Family for Nanotechnology. *ACS Nano* **2008**, *2*, 1987–1992.
12. Polyakov, A.Y.; Nesterov, A.V.; Goldt, A.E.; Zubyuk, V.; Dolgova, T.; Yadgarov, L.; Visic, B.; Fedyanin, A.A.; Tenne, R.; Goodilin, E.A. Optical Properties of Multilayer Films of Nanocomposites Based on WS 2 Nanotubes Decorated with Gold Nanoparticles. *J. Phys. Conf. Ser.* **2015**, *643*, 012046.
13. Donkova, B.; Vasileva, P.; Nihtianova, D.; Velichkova, N.; Stefanov, P.; Mehandjiev, D. Synthesis, Characterization, and Catalytic Application of Au/ZnO Nanocomposites Prepared by Coprecipitation. *J. Mater. Sci.* **2011**, *46*, 7134–7143.
14. Kurtz, M.; Strunk, J.; Hinrichsen, O.; Muhler, M.; Fink, K.; Meyer, B.; Wöll, C. Active Sites on Oxide Surfaces: ZnO-Catalyzed Synthesis of Methanol from CO and $H_2$. *Angew. Chem. Int. Ed.* **2005**, *44*, 2790–2794.
15. Yuan, J.; Choo, E.S.G.; Tang, X.; Sheng, Y.; Ding, J.; Xue, J. Synthesis of ZnO–Pt Nanoflowers and Their Photocatalytic Applications. *Nanotechnology* **2010**, *21*, 185606.
16. Wang, Y.; Hu, A. Carbon Quantum Dots: Synthesis, Properties and Applications. *J. Mater. Chem. C* **2014**, *2*, 6921–6939.
17. Pan, M.; Xie, X.; Liu, K.; Yang, J.; Hong, L.; Wang, S. Fluorescent Carbon Quantum Dots-Synthesis, Functionalization and Sensing Application in Food Analysis. *Nanomaterials* **2020**, *10*, 930.
18. Zhang, Z.; Zheng, T.; Li, X.; Xu, J.; Zeng, H. Progress of Carbon Quantum Dots in Photocatalysis Applications. *Part. Part. Syst. Charact.* **2016**, *33*, 457–472.
19. Molaei, M.J. A Review on Nanostructured Carbon Quantum Dots and Their Applications in Biotechnology, Sensors, and Chemiluminescence. *Talanta* **2019**, *196*, 456–478.
20. Luo, P.G.; Yang, F.; Yang, S.T.; Sonkar, S.K.; Yang, L.; Broglie, J.J.U.; Liu, Y.; Sun, Y.P. Carbon-Based Quantum Dots for Fluorescence Imaging of Cells and Tissues. *RSC Adv.* **2014**, *4*, 10791–10807.
21. Singh, I.; Arora, R.; Dhiman, H.; Pahwa, R. Carbon Quantum Dots: Synthesis, Characterization and Biomedical Applications. *Turkish J. Pharm. Sci.* **2018**, *15*, 219–230.
22. Sumanth Kumar, D.; Jai Kumar, B.; Mahesh, H.M. Quantum Nanostructures (QDs): An Overview. In *Synthesis of Inorganic Nanomaterials*; Elsevier: Amsterdam, The Netherlands, 2018; pp. 59–88.
23. Muthulingam, S.; Lee, I.H.; Uthirakumar, P. Highly Efficient Degradation of Dyes by Carbon Quantum Dots/N-Doped Zinc Oxide (CQD/N-ZnO) Photocatalyst and Its Compatibility on Three Different Commercial Dyes under Daylight. *J. Colloid Interface Sci.* **2015**, *455*, 101–109.
24. Yu, H.; Zhao, Y.; Zhou, C.; Shang, L.; Peng, Y.; Cao, Y.; Wu, L.Z.; Tung, C.H.; Zhang, T. Carbon Quantum Dots/$TiO_2$ Composites for Efficient Photocatalytic Hydrogen Evolution. *J. Mater. Chem. A* **2014**, *2*, 3344–3351.
25. Al Ja'farawy, M.S.; Kusumandari; Purwanto, A.; Widiyandari, H. Carbon Quantum Dots Supported Zinc Oxide (ZnO/CQDs) Efficient Photocatalyst for Organic Pollutant Degradation—A Systematic Review. *Environ. Nanotechnol. Monit. Manag.* **2022**, *18*, 100681.
26. Bozetine, H.; Wang, Q.; Barras, A.; Li, M.; Hadjersi, T.; Szunerits, S.; Boukherroub, R. Green Chemistry Approach for the Synthesis of ZnO-Carbon Dots Nanocomposites with Good Photocatalytic Properties under Visible Light. *J. Colloid Interface Sci.* **2016**, *465*, 286–294.
27. Yu, H.; Zhang, H.; Huang, H.; Liu, Y.; Li, H.; Ming, H.; Kang, Z. ZnO/Carbon Quantum Dots Nanocomposites: One-Step Fabrication and Superior Photocatalytic Ability for Toxic Gas Degradation under Visible Light at Room Temperature. *New J. Chem.* **2012**, *36*, 1031–1035.

28. Zhang, X.; Pan, J.; Zhu, C.; Sheng, Y.; Yan, Z.; Wang, Y.; Feng, B. The Visible Light Catalytic Properties of Carbon Quantum Dots/ZnO Nanoflowers Composites. *J. Mater. Sci. Mater. Electron.* **2015**, *26*, 2861–2866.
29. Zhang, X.-Y.; Liu, J.-K.; Wang, J.-D.; Yang, X.-H. Mass Production, Enhanced Visible Light Photocatalytic Efficiency, and Application of Modified ZnO Nanocrystals by Carbon Dots. *Ind. Eng. Chem. Res.* **2015**, *54*, 1766–1772.
30. Han, C.; Yang, M.-Q.; Weng, B.; Xu, Y.-J. Improving the Photocatalytic Activity and Anti-Photocorrosion of Semiconductor ZnO by Coupling with Versatile Carbon. *Phys. Chem. Chem. Phys.* **2014**, *16*, 16891–16903.
31. Liu, H.; Ye, T.; Mao, C. Fluorescent Carbon Nanoparticles Derived from Candle Soot. *Angew. Chem. Int. Ed.* **2007**, *46*, 6473–6475.
32. Dekaliuk, M.O.; Viagin, O.; Malyukin, Y.V.; Demchenko, A.P. Fluorescent Carbon Nanomaterials: "Quantum Dots" or Nanoclusters? *Phys. Chem. Chem. Phys.* **2014**, *16*, 16075–16084.
33. Li, H.; He, X.; Kang, Z.; Huang, H.; Liu, Y.; Liu, J.; Lian, S.; Tsang, C.H.A.; Yang, X.; Lee, S.T. Water-Soluble Fluorescent Carbon Quantum Dots and Photocatalyst Design. *Angew. Chem. Int. Ed.* **2010**, *49*, 4430–4434.
34. Chen, X.; Zhang, W.; Wang, Q.; Fan, J. C8-Structured Carbon Quantum Dots: Synthesis, Blue and Green Double Luminescence, and Origins of Surface Defects. *Carbon N. Y.* **2014**, *79*, 165–173.
35. Liu, Y.; Zhou, L.; Li, Y.; Deng, R.; Zhang, H. Highly fluorescent nitrogen-doped carbon dots with excellent thermal and photo stability applied as invisible ink for loading important information and anti-counterfaiting. *Nanoscale* **2017**, *9*, 491–496.
36. Dong, Y.; Shao, J.; Chen, C.; Li, H.; Wang, R.; Chi, Y.; Lin, X.; Chen, G. Blue Luminescent Graphene Quantum Dots and Graphene Oxide Prepared by Tuning the Carbonization Degree of Citric Acid. *Carbon N. Y.* **2012**, *50*, 4738–4743.
37. Yuan, F.; Yuan, T.; Sui, L.; Wang, Z.; Xi, Z.; Li, Y.; Li, X.; Fan, L.; Tan, Z.; Chen, A.; et al. Engineering Triangular Carbon Quantum Dots with Unprecedented Narrow Bandwidth Emission for Multicolored LEDs. *Nat. Commun.* **2018**, *9*, 2249–2260.
38. Wu, P.; Wu, X.; Li, W.; Liu, Y.; Chen, Z.; Liu, S. Ultra-Small Amorphous Carbon Dots: Preparation, Photoluminescence Properties, and Their Application as TiO2 Photosensitizers. *J. Mater. Sci.* **2019**, *54*, 5280–5293.
39. Liu, H.; Zhao, X.; Wang, F.; Wang, Y.; Guo, L.; Mei, J.; Tian, C.; Yang, X.; Zhao, D. High-Efficient Excitation-Independent Blue Luminescent Carbon Dots. *Nanoscale Res. Lett.* **2017**, *12*, 399–405.
40. Song, Y.; Zhu, S.; Xiang, S.; Zhao, X.; Zhang, J.; Zhang, H.; Fu, Y.; Yang, B. Investigation into the Fluorescence Quenching Behaviors and Applications of Carbon Dots. *Nanoscale* **2014**, *6*, 4676–4683.
41. Zhu, S.; Song, Y.; Zhao, X.; Shao, J.; Zhang, J.; Yang, B. The Photoluminescence Mechanism in Carbon Dots (Graphene Quantum Dots, Carbon Nanodots, and Polymer Dots): Current State and Future Perspective. *Nano Res.* **2015**, *8*, 355–381.
42. Dulub, O.; Boatner, L.A.; Diebold, U. STM Study of the Geometric and Electronic Structure of ZnO(0 0 0 1)-Zn, (0 0 0 $\bar{1}$)-O, (1 0 $\bar{1}$ 0), and (1 1 $\bar{2}$ 0) Surfaces. *Surf. Sci.* **2002**, *519*, 201–217.
43. Zhang, Q. Effects of Calcination on the Photocatalytic Properties of Nanosized TiO2 Powders Prepared by TiCl4 Hydrolysis. *Appl. Catal. B Environ.* **2000**, *26*, 207–215.
44. Fu, Y.; Han, G.; Chang, Y.; Dong, J. The Synthesis and Properties of ZnO-Graphene Nano Hybrid for Photodegradation of Organic Pollutant in Water. *Mater. Chem. Phys.* **2012**, *132*, 673–681.
45. Song, S.; Wu, K.; Wu, H.; Guo, J.; Zhang, L. Multi-shelled ZnO decorated with nitrogen and phosphorus co-doped carbon quantum dots: Synthesis and enhanced photodegradation activity of methylene blue in aqueous solutions. *RSC Adv.* **2019**, *9*, 7362–7374.
46. Li, Y.; Zhang, B.P.; Zhao, J.X.; Ge, Z.H.; Zhao, X.K.; Zou, L. ZnO/Carbon Quantum Dots Heterostructure with Enhanced Photocatalytic Properties. *Appl. Surf. Sci.* **2013**, *279*, 367–373.
47. Pandiyarajan, T.; Karthikeyan, B. Cr Doping Induced Structural, Phonon and Excitonic Properties of ZnO Nanoparticles. *J. Nanoparticle Res.* **2012**, *14*, 647.
48. Chung, F.H. Quantitative Interpretation of X-Ray Diffraction Patterns of Mixtures. I. Matrix-Flushing Method for Quantitative Multicomponent Analysis. *J. Appl. Crystallogr.* **1974**, *7*, 519–525.
49. Sali, S.; Boumaour, M.; Kechouane, M.; Kermadi, S.; Aitamar, F. Nanocrystalline ZnO Film Deposited by Ultrasonic Spray on Textured Silicon Substrate as an Anti-Reflection Coating Layer. *Phys. B Condens. Matter* **2012**, *407*, 2626–2631.
50. Li, Y.; Zhao, Y.; Cheng, H.; Hu, Y.; Shi, G.; Dai, L.; Qu, L. Nitrogen-Doped Graphene Quantum Dots with Oxygen-Rich Functional Groups. *J. Am. Chem. Soc.* **2012**, *134*, 15–18.
51. Xu, T.; Zhang, L.; Cheng, H.; Zhu, Y. Significantly Enhanced Photocatalytic Performance of ZnO via Graphene Hybridization and the Mechanism Study. *Appl. Catal. B Environ.* **2011**, *101*, 382–387.
52. Van Zandvoort, I.; Wang, Y.; Rasrendra, C.B.; Van Eck, E.R.H.; Bruijnincx, P.C.A.; Heeres, H.J.; Weckhuysen, B.M. Formation, Molecular Structure, and Morphology of Humins in Biomass Conversion: Influence of Feedstock and Processing Conditions. *ChemSusChem* **2013**, *6*, 1745–1758.
53. Heng, Z.W.; Chong, W.C.; Pang, Y.L.; Koo, C.H. An Overview of the Recent Advances of Carbon Quantum Dots/Metal Oxides in the Application of Heterogeneous Photocatalysis in Photodegradation of Pollutants towards Visible-Light and Solar Energy Exploitation. *J. Environ. Chem. Eng.* **2021**, *9*, 105199.
54. Moalem-Banhangi, M.; Ghaeni, N.; Ghasemi, S. Saffron derived carbon quantum dot/N-doped ZnO/fulvic acid nanocomposite for sonocatalytic degradation of methylene blue. *Synth. Met.* **2021**, *271*, 116626.
55. Velumani, A.; Sengodan, P.; Arumugam, P.; Rajendran, R.; Santhanam, S.; Palanisamy, M. Carbon quantum dots supported ZnO sphere based photocatalyst for dye degradation application. *Curr. Appl. Phys.* **2020**, *20*, 1176–1184.
56. Soga, T. *Nanostructured Materials for Solar Energy Conversion*; Elsevier: Oxford, UK, 2006.

57. Zhang, T.; Oyama, T.; Aoshima, A.; Hidaka, H.; Zhao, J.; Serpone, N. Photooxidative N-Demethylation of Methylene Blue in Aqueous $TiO_2$ Dispersions under UV Irradiation. *J. Photochem. Photobiol. A Chem.* **2001**, *140*, 163–172.
58. Houas, A.; Lachheb, H.; Ksibi, M.; Elaloui, E.; Guillard, C.; Herrmann, J.M. Photocatalytic Degradation Pathway of Methylene Blue in Water. *Appl. Catal. B Environ.* **2001**, *31*, 145–157.
59. Zhou, B.; Song, J.; Zhang, Z.; Jiang, Z.; Zhang, P.; Han, B. Highly Selective Photocatalytic Oxidation of Biomass-Derived Chemicals to Carboxyl Compounds over Au/TiO2. *Green Chem.* **2017**, *19*, 1075–1081.
60. Behnood, R.; Sodeifian, G. Synthesis of N Doped-CQDs/Ni Doped-ZnO Nanocomposites for Visible Light Photodegradation of Organic Pollutants. *J. Environ. Chem. Eng.* **2020**, *8*, 103821.

Article

# Methane Hydrate Formation in Hollow ZIF-8 Nanoparticles for Improved Methane Storage Capacity

Chong Chen [1,2], Yun Li [1,2,*] and Jilin Cao [1]

[1] Engineering Research Center of Seawater Utilization Technology of Ministry of Education, School of Chemical Engineering and Technology, Hebei University of Technology, Tianjin 300130, China; jiningzccc@163.com (C.C.); caojilin@hebut.edu.cn (J.C.)

[2] Key Laboratory of Gas Hydrate, Guangzhou Institute of Energy Conversion, Chinese Academy of Sciences, Guangzhou 510640, China

\* Correspondence: liyun@hebut.edu.cn

**Abstract:** Methane hydrate has been extensively studied as a potential medium for natural gas storage and transportation. Due to their high specific surface area, tunable porous structure, and surface chemistry, metal–organic frameworks are ideal materials to exhibit the catalytic effect for the formation process of gas hydrate. In this paper, hollow ZIF-8 nanoparticles are synthesized using the hard template method. The synthesized hollow ZIF-8 nanoparticles are used in the adsorption and methane hydrate formation process. The effect of pre-adsorbed water mass in hollow ZIF-8 nanoparticles on methane storage capacity and the hydrate formation rate is investigated. The storage capacity of methane on wet, hollow ZIF-8 is augmented with an increase in the mass ratio of pre-adsorbed water and dry, hollow ZIF-8 ($R_W$), and the maximum adsorption capacity of methane on hollow ZIF-8 with a $R_W$ of 1.2 can reach 20.72 mmol/g at 275 K and 8.57 MPa. With the decrease in $R_W$, the wet, hollow ZIF-8 exhibits a shortened induction time and an accelerated growth rate. The formation of methane hydrate on hollow ZIF-8 is further demonstrated with the enthalpy of the generation reaction. This work provides a promising alternative material for methane storage.

**Keywords:** hollow ZIF-8; methane hydrate; methane storage

Citation: Chen, C.; Li, Y.; Cao, J. Methane Hydrate Formation in Hollow ZIF-8 Nanoparticles for Improved Methane Storage Capacity. *Catalysts* **2022**, *12*, 485. https://doi.org/10.3390/catal12050485

Academic Editors: Simona M. Coman, Madalina Tudorache and Elisabeth Egholm Jacobsen

Received: 10 March 2022
Accepted: 23 April 2022
Published: 26 April 2022

**Publisher's Note:** MDPI stays neutral with regard to jurisdictional claims in published maps and institutional affiliations.

**Copyright:** © 2022 by the authors. Licensee MDPI, Basel, Switzerland. This article is an open access article distributed under the terms and conditions of the Creative Commons Attribution (CC BY) license (https://creativecommons.org/licenses/by/4.0/).

## 1. Introduction

Natural gas is a clean energy source, mainly composed of methane, which has a higher energy density per unit mass and emits less carbon dioxide during combustion than fossil fuels [1,2]. These favorable characteristics make natural gas an ideal alternative energy source. Therefore, it is crucial to improve natural gas storage and transportation methods to maximize the use of this energy source. Compressed natural gas (CNG) is a storage method in which natural gas is stored at a high pressure (15–25 MPa), which requires high-pressure vessels. In contrast, liquefied natural gas (LNG) is more expensive to operate on account of its high critical pressure and the need to maintain a low-temperature (−162 °C) environment during storage and transportation. Both CNG and LNG methods are very unfavorable from the point of view of safety and energy efficiency [3].

Due to its safety and low cost, nature gas hydrate has emerged as a new and promising method to store and transport methane [4,5]. Natural gas hydrates are clathrate compounds in which methane-dominated gas molecules are contained by a cage lattice formed by water molecules with the aid of hydrogen bonds. The most common structure of natural gas hydrates is the sI-type hydrate structure, in which the crystal cell consists of two small pentagonal dodecahedral ($5^{12}$) cages and six large tetrakaidecahedron ($5^{12}6^2$) cages, capable of distributing up to eight methane molecules. If each cage is filled with methane, 1 m$^3$ of methane hydrate can be decomposed into 180 m$^3$ of methane gas under standard conditions, with a significant increase in energy density [6]. However, the long induction time and slow growth kinetics of methane hydrate limit the application of the hydrate method in gas storage and transportation technology.

It has been proved that porous materials as a wet storage medium show great advantages in the storage of methane due to their porous structure and large surface area [7,8]. On the one hand, methane physisorption can occur in porous materials; on the other hand, porous materials can be used as nanoreactors to promote methane hydrate formation by providing abundant gas–liquid contact sites for water and methane, and the catalytic effect results from their surface chemistry [9,10]. Among porous materials, metal–organic frameworks (MOFs) are porous crystalline materials with a large specific surface area, highly tunable pore size, and flexible structure [11–13], thereby showing excellent application potential in the fields of gas storage, sensing [14], catalysis [15], medicine, etc. At present, the methane hydrate formation has been explored in various MOF materials, including MIL-100 [16], ZIF-8 [7], MIL-101 [17], HKUST-1 [18], ZIF-67 [19], etc. Compared with solid MOFs with the same composition and size, hollow-structured MOF materials have the advantages of low bulk density, large surface area, and high porosity [20,21]. The large surface provides numerous contact sites between water and methane to promote the formation of methane hydrate, and high porosity allows for a high storage capacity for methane. In addition, the low bulk density is conducive to achieving high gravimetric and volumetric yield. On the basis of these properties, we envisage that hollow-structured MOFs would be a good wet storage medium for methane. Currently, however, there are a lack of studies on the effect of hollow-structured MOFs in methane hydrate formation. In this work, hollow ZIF-8 nanoparticles, taken as an example of hollow-structured MOFs, are synthesized using the hard template method and used for the wet storage of methane for the first time. The effect of water content on the storage capacity of hollow ZIF-8 nanoparticles for methane is investigated, and the formation kinetics and enthalpy of the methane hydrate in hollow ZIF-8 nanoparticles are evaluated. The satisfactory storage performance of hollow ZIF-8 nanoparticles will lead to their actual application in methane storage and transportation.

## 2. Results and Discussion

### 2.1. Characterization of Hollow ZIF-8 Materials

The hard template method was employed to synthesize hollow ZIF-8 materials. As shown in Figure 1a, polystyrene (PS) microspheres, as a template prepared by surfactant-free emulsion polymerization, are monodisperse, with a perfectly smooth surface, and they exhibit an average size of around 310 nm. After solvothermal synthesis, the surface of PS was significantly roughened to form a shell layer of ZIF-8 crystal. The synthesis conditions, including mainly the mixing order of raw materials and the synthesis time, had an influence on the compactness, continuity, and thickness of the ZIF-8 shell on the surface of PS microspheres. As shown in Figure S1, 30 min is the suitable synthesis time to form an intact shell on the surface of PS microspheres. The reduced synthesis time (10 min) resulted in an incomplete coverage of the ZIF-8 shell on PS micropheres. When $Zn(NO_3)_2 \cdot 6H_2O$ and 2-methylimidazole were pre-mixed, followed by the addition of PS microspheres, and the mixture reacted for 30 min, PS@ZIF-8 microspheres with a size of ca. 322 nm were obtained. The synthesized shell thickness was ca. 6 nm. After removal of the template, the spherical shape of the synthesized hollow particles collapsed, which resulted from a thin shell with a weak mechanical strength, which was insufficient to maintain the intact hollow spheres (Figure S2). Using this mixing order of raw materials, the homogeneous nucleation of ZIF-8 crystals in the bulk solution led to a reduction in the nuleation density on the surface of the PS microspheres [22], which is not conducive to accessing a continuous and dense shell. In contrast, when $Zn(NO_3)_2 \cdot 6H_2O$ was initially mixed with PS microspheres in methanol, followed by the addition of 2-methylimidazole, the particle size of the synthesized PS@ZIF-8 composite particle was about 350 nm (Figure 1b). From the broken, hollow ZIF-8 particles, it was seen that the PS templates were successfully removed by DMF solvent, and the prepared hollow ZIF-8 microspheres had a complete structure, maintaining a good spherical morphology and uniform size (Figure 1c). As shown in the high-magnification TEM image

(Figure 1d), the contrast between the obvious inner bright area and the dark area on the edge clearly confirms the formation of a hollow structure, and the thickness of the ZIF-8 shell is about 20 nm. The priority of mixing $Zn(NO_3)_2 \cdot 6H_2O$ and PS microspheres could promote the preferential adsorption of $Zn^{2+}$ on PS microspheres, which is conducive to the heterogenous nucleation and growth of ZIF-8 crystals on the surface of PS microspheres. As a result, the thick, continuous ZIF-8 shell obtained was compared with that obtained by pre-mixing $Zn(NO_3)_2 \cdot 6H_2O$ and 2-methylimidazole. Therefore, a ZIF-8 shell with a thickness of about 20 nm was synthesized after one cycle of solvothermal synthesis, and it was sufficient to obtain hollow and spherical particles with a cavity size of about 310 nm. This was different from the result obtained by the ZIF-8 shell with a thickness of ca. 15 nm after one cycle of the solvothermal method, which did not maintain a perfectly spherical shape after the removal of a PS template with a size of 870 nm, as reported in the literature [23]. This result could be attributed to the thickened and continuous shell and reduced cavity size [24,25].

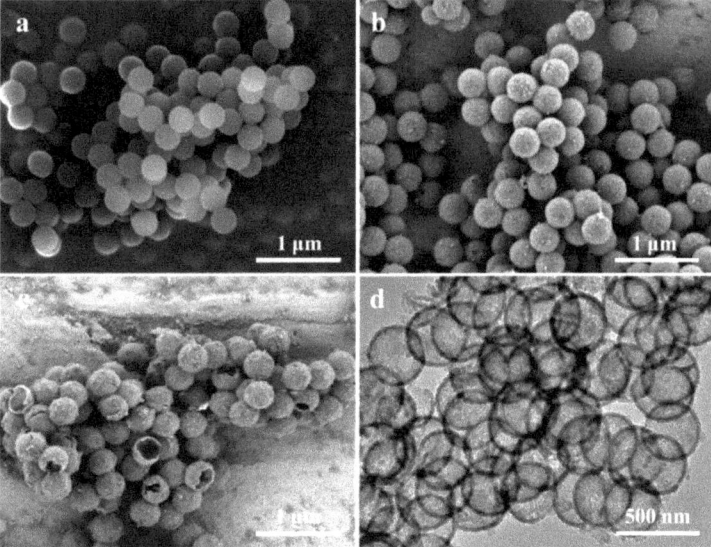

**Figure 1.** SEM images of PS (**a**) and PS@ZIF-8 (**b**); SEM (**c**) and TEM(**d**) images of hollow ZIF-8 particles.

Figure 2 shows XRD patterns of PS@ZIF-8 core–shell nanoparticles and hollow ZIF-8 nanoparticles. The diffraction peaks of PS@ZIF-8 core–shell nanoparticles were in good agreement with the simulated XRD pattern of ZIF-8, confirming the formation of a ZIF-8 shell layer on PS microspheres. After the template removal, the XRD pattern of PS@ZIF-8 and hollow ZIF-8 nanoparticles presented the same diffraction peaks with their intensities, indicating that the shell structure of ZIF-8 was still well-preserved after removal of the PS microspheres.

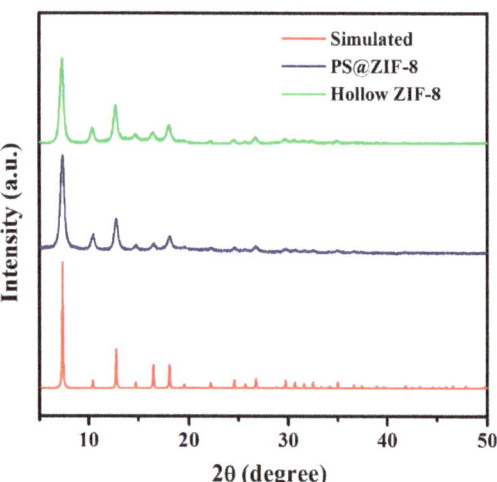

**Figure 2.** XRD patterns of ZIF-8, PS@ZIF-8, and hollow ZIF-8 particles with the simulated XRD pattern from the crystallographic data of ZIF-8.

To further demonstrate the complete removal of the PS template, Figure 3 gives FTIR spectra of PS microspheres, ZIF-8, PS@ZIF-8, and hollow ZIF-8 nanoparticles. The FTIR spectrum of PS microspheres shows the adsorption peaks at 1495, 1600, 1650–2000, 2800–3000, and 3000–3100 cm$^{-1}$. The peaks at 1495 and 1600 cm$^{-1}$ could be attributed to the stretching vibration of C=C from the benzene ring, and a generalized frequency peak of the benzene ring appeared at 1650–2000 cm$^{-1}$. The bands at 3000–3100 cm$^{-1}$ could be assigned to the stretching vibrations of C–H on the benzene ring. The synthesized PS@ZIF-8 nanoparticles exhibited new peaks. The peaks at 900–1350 cm$^{-1}$ corresponded to the bending vibration of the imidazole ring, and the peak at 420 cm$^{-1}$ was attributed to the stretching vibration of Zn–N. The appearance of these peaks indicated the successful synthesis of the ZIF-8 shell on the surface of the PS microsphere. After an etching template with DMF solvent, the bands assigned to PS microspheres diminished, and the synthesized hollow ZIF-8 presented only peaks corresponding to the ZIF-8 crystals, which demonstrates the complete removal of the PS template and the formation of hollow ZIF-8 nanoparticles.

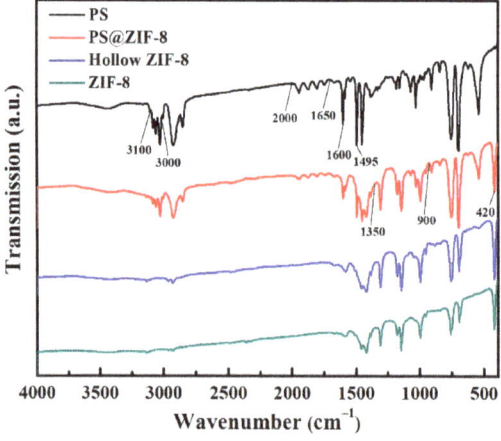

**Figure 3.** FTIR spectra of PS microspheres, ZIF-8, PS@ZIF-8, and hollow ZIF-8 particles.

N$_2$ adsorption–desorption isotherms of PS@ZIF-8 and hollow ZIF-8 nanoparticles at 77 K with a corresponding pore size distribution are shown in Figure 4. PS@ZIF-8 nanoparticles exhibited an I-type adsorption curve, indicating that the synthesized PS@ZIF-8 nanoparticles had a microporous structure. The uptake at high relative pressure of the synthesized PS@ZIF-8 nanoparticles could be attributed to the porosity formed by the accumulation of nanoparticles [26]. After removal of the PS template, the hollow ZIF-8 nanoparticles showed an IV-type adsorption–desorption isotherm with a hysteresis loop containing an H2 type in the pressure range of 0.6–0.9 and an H3 type in the pressure range of 0.9–1.0, suggesting a presence of mesopores in addition to micropores in the hollow ZIF-8 nanoparticles. This phenomenon might be caused by the hollow cavity and inevitable voids between adjacent ZIF-8 nanocrystals after the etching template, as well as the intergranular pores formed by the packing of nanoparticles [27,28]. The Brunauer–Emmett–Teller (BET) surface area of the PS@ZIF-8 nanoparticles was calculated to be 518.8239 m$^2$/g, and the BET surface area of hollow ZIF-8 nanoparticles showed a remarkable improvement, with a value of 1260.9272 m$^2$/g. The corresponding pore volume also increased from 0.3187 cm$^3$/g to 0.8250 cm$^3$/g (Table 1).

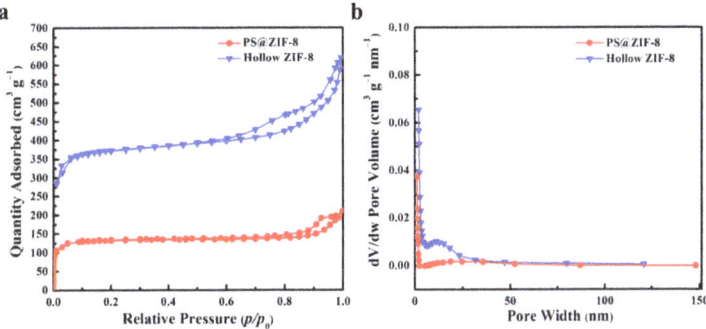

**Figure 4.** N$_2$ adsorption–desorption isotherms of PS@ZIF-8 and hollow ZIF-8 particles (**a**) with the corresponding pore size distribution (**b**).

**Table 1.** Physical characterizations of PS@ZIF-8 and hollow ZIF-8 nanoparticles.

| Sample | $S_{BET}$ (m$^2$/g) | V (cm$^3$/g) |
| --- | --- | --- |
| PS@ZIF-8 | 518.8239 | 0.3187 |
| hollow ZIF-8 | 1260.9272 | 0.8250 |

*2.2. Methane Hydrate Formation in Hollow ZIF-8 Nanoparticles*

The methane adsorption isotherms measured at 275 K for adsorbents obtained by pre-adsorbing different masses of water ($R_W$ = 0, 0.2, 0.8, and 1.2) on dry, hollow ZIF-8 are shown in Figure 5. $R_W$ stands for the mass ratio of pre-adsorbed water and dry hollow ZIF-8 nanoparticles. The adsorption capacity of methane (n) is represented as the amount of methane adsorbed per unit mass of dry, hollow ZIF-8 nanoparticles. According to the IUPAC classification, the adsorption of the dry, hollow ZIF-8 nanoparticles exhibited a type I isotherm, and the adsorption amount gradually increased until saturation at 8–9 MPa, with a maximum adsorption amount of 13.32 mmol/g at 8.58 MPa. The reason for such high adsorption capacity, in addition to the promotion of methane adsorption by the methyl group in the organic ligand [29], may be that the hollow structure provides high pore volume to increase the adsorption capacity of methane. Notably, the methane adsorption isotherms for adsorbents with pre-adsorbed water no longer had the characteristics of type I isotherms, and there was a jump at about 4–5 MPa. When the pressure was less than the jumping pressure, the methane adsorption isotherm of the pre-adsorbed water sample almost coincided with the adsorption isotherm of the dry sample. This phenomenon is caused by the hydrophobic surface properties of ZIF-8, where water is completely repelled by the internal hydrophobic cavity, and the pores are not blocked, making the entire pore fully available for adsorption of methane molecules. At the jumping point, the amount of methane gas adsorbed on the sample rose sharply, corresponding to the generation of a large amount of methane gas hydrate. With $R_W$ of 0.8, the storage capacity of methane was 17.74 mmol/g. The storage capacity of methane gradually increased with the increase in pre-adsorbed water mass, and the maximum storage capacity of methane was 20.72 mmol/g when $R_W$ was 1.2, which raised the methane storage capacity by 55.6% more than the dry, hollow ZIF-8 nanoparticles, indicating that higher pre-sorption water mass could greatly increase the storage capacity of methane on the hollow ZIF-8 nanoparticles. The above results demonstrate that hollow ZIF-8 nanoparticles are conducive to achieving high methane storage capacity.

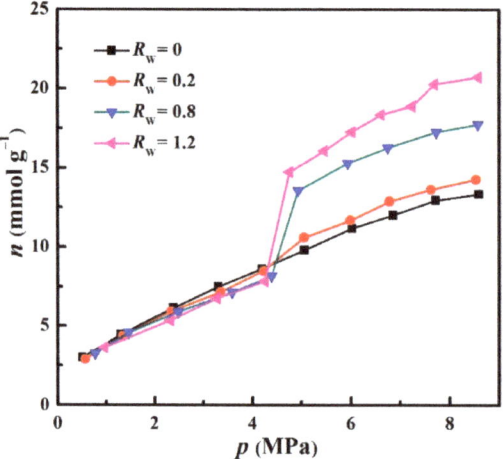

**Figure 5.** Methane adsorption isotherms of hollow ZIF-8 nanoparticles with different amounts of pre-adsorbed water at 275 K.

To study the generation rate at the jump point of the methane adsorption isotherm on the hollow ZIF-8 nanoparticles, the methane pressure values with time were recorded at 275 K, when $R_W$ was 0.8 and 1.2, respectively. The corresponding curves were made as shown in Figure 6. When $R_W$ was 0.8, the induction time of the methane hydrate formation at 275 K was 3.5 h. With the increase in $R_W$ to 1.2, the induction time of the methane hydrate formation further increased to 4.75 h. In addition, it could be observed that the methane growth rate on hollow ZIF-8 nanoparticles was faster at $R_W$ of 0.8 than that at $R_W$ of 1.2. The process of hydrate generation took 5 h to reach equilibrium when $R_W$ was 0.8. With $R_W$ of 1.2, although the equilibrium adsorption capacity of the methane was remarkably improved (Figure 5), the time to reach the equilibrium pressure increased significantly and reached about 7 h. This may be because the increase in water mass made a considerable thickness of the water layer accumulate on the surface of the hollow ZIF-8 nanoparticles, which is very unfavorable for the diffusion of gas from the gas-solid interface to the porous surface. As a result, methane could not quickly reach the pores favorable for hydrate generation, and the rate of hydrate generation was greatly reduced. Hollow ZIF-8 nanoparticles with $R_W$ of 1.2 could obtain a slightly higher storage capacity for methane and a slower nucleation and growth kinetics than that at $R_W$ of 0.8. Taking these two factors into account, $R_W$ for subsequent methane adsorption experiments on hollow ZIF-8 nanoparticles with pre-adsorbed water was set to 0.8. The methane adsorption experiment was conducted in pure water, but in the absence of hollow ZIF-8 nanoparticles. There was no occurrence of methane adsorption (>12 h), demonstrating the promotion effect of the hollow ZIF-8 nanoparticles on methane hydrate formation. In addition to a large number of contact sites between water and methane provided by the hollow ZIF-8 nanoparticles, the hydrophobic surface of ZIF-8, which promotes the formation of hydrogen, bonded to facilitate the growth of methane hydrate. Furthermore, a large number of the adsorbed methane molecules in the pores of the hollow ZIF-8 were partially moved to the outside surface of the hollow ZIF-8 to expedite the hydrate growth [30]. Therefore, hollow ZIF-8 nanoparticles play a catalytic role to promote methane hydrate formation.

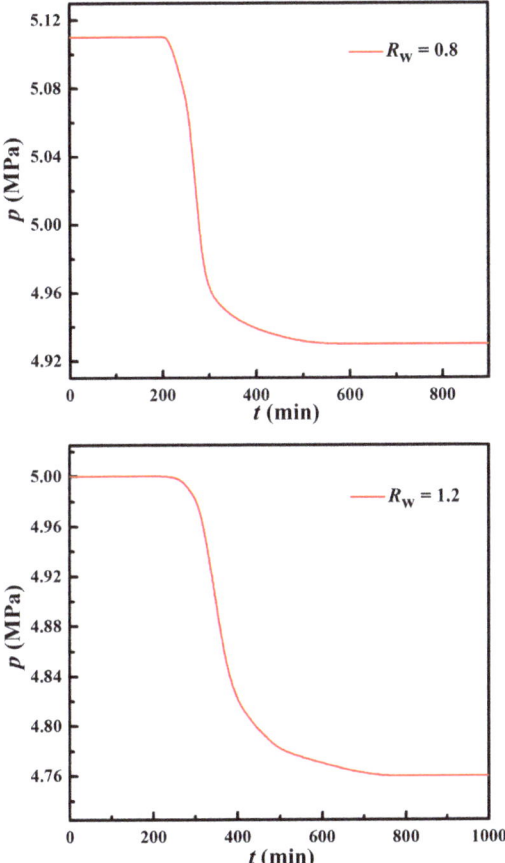

**Figure 6.** Dynamic curves of methane adsorption on the wet, hollow ZIF-8 nanoparticles with different amounts of pre-adsorbed water.

### 2.3. Formation Enthalpy of Methane Hydrate in Hollow ZIF-8 Nanoparticles

To further demonstrate that the increase in methane storage in the wet, hollow ZIF-8 nanoparticles results from the generation of methane hydrate, the adsorption isotherms of methane in wet, hollow ZIF-8 ($R_W$ = 0.8) at four temperatures were performed. As shown in Figure 7, as the temperature increased, the pressure at the jump point gradually increased. Before jumping in pressure, the methane adsorption capacity at different temperatures was almost the same. After the jumping pressure, the methane storage capacity showed a decreasing trend along with the temperature. The above results suggest that the temperature has a significant influence on methane hydrate formation pressure and the equilibrium adsorption capacity of methane. The increased temperature is not conducive to methane storage, which is consistent with the properties of methane hydrate.

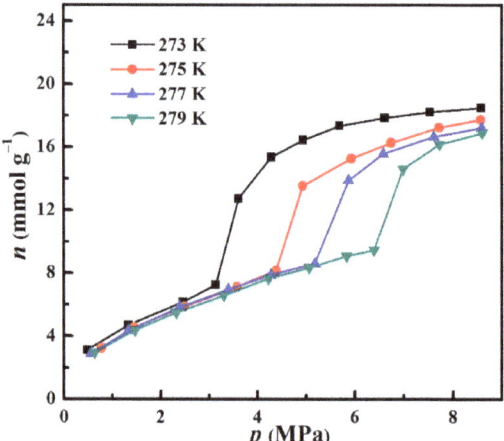

**Figure 7.** Methane adsorption isotherms of hollow ZIF-8 nanoparticles ($R_W$ = 0.8) at different temperatures.

According to the Clausius–Clapeyron equation (1), the enthalpy of methane hydrate generation ($\Delta H^{form}$) can be calculated from the jump pressures corresponding to the different temperatures based on the methane adsorption isotherms at 273 K, 275 K, 277 K, and 279 K in the wet, hollow ZIF-8 material ($R_W$ = 0.8):

$$\frac{dp}{dT} = \frac{\Delta H}{T \Delta V} \tag{1}$$

where $\Delta H$ is the molar enthalpy of phase change, $\Delta V$ is the molar volume change caused by the phase change, and R is the gas constant 8.314 J·mol$^{-1}$ K$^{-1}$. For the actual gas, the pressure ($p$) is replaced by the fugacity ($f$), which is related to $p$ and temperature ($T$). Equation (1) is transformed as follows:

$$\frac{df}{dT} = \frac{\Delta H^{form}}{T \Delta V} \tag{2}$$

Because the experimental temperatures are above 273 K, and water exists in the liquid state before hydrate generation, the generation of methane hydrate can be expressed by Equation (3):

$$CH_4(gas) + nH_2O(liquid) \Leftrightarrow CH_4 \cdot nH_2O(solid) \tag{3}$$

From the equation, it can be seen that the methane hydrate generation and decomposition reaction is a two-component and three-phase system. The change in the molar volume in the hydrate generation process is

$$\Delta V = V_{CH_4 \cdot nH_2O(solid)} - V_{CH_4(gas)} - V_{H_2O(liquid)} \approx -V_{CH_4(gas)} = -\frac{RT}{f} \tag{4}$$

where $\Delta H$ is assumed to be a constant that does not vary with $T$. Equation (4) taken into Equation (2) can be deformed into the Equation (5):

$$\ln f = \pm \frac{\Delta H}{RT} + C_1 (\text{constant}) \tag{5}$$

The formula (5) for calculating the enthalpy change from the formation reaction is the sign (+); the formula (5) for calculating the enthalpy change from the decomposition reaction is the sign (−).

According to the definition of fugacity, there is

$$f = p \times \phi \tag{6}$$

The relationship between the fugacity factor $\phi$ and the compression factor $Z$ is

$$\ln \phi = \int_0^p \frac{(Z-1)}{p} dp \tag{7}$$

According to the methane hydrate generation pressure corresponding to the methane adsorption isotherm at different temperatures, $\phi$ is calculated from Equation (7), and then $f$ is calculated from Equation (6). The results of methane hydrate generation $p$, $\phi$, and $f$ at different temperatures are shown in Table 2.

**Table 2.** State data of the jump pressure at different temperatures.

| $T$ (K) | $p$ (MPa) | $\phi$ | ln$f$ |
|---|---|---|---|
| 273 | 3.61 | 0.9303 | 1.2115 |
| 275 | 4.93 | 0.9007 | 1.4907 |
| 277 | 5.87 | 0.8803 | 1.6424 |
| 279 | 6.98 | 0.8583 | 1.7902 |

The fitting plot of ln$f$ versus reciprocal temperature is shown in Figure 8. From Equation (5), the slope of the fitting line in Figure 8 is $\Delta H^{form}/R$. The calculated enthalpy of the generation reaction on the wet, hollow ZIF-8 nanoparticles is $-59.22$ kJ/mol, and the negative sign represents the exothermal process. This is essentially the same as the enthalpy of the formation of methane hydrate in pure water ($-59.5$ kJ/mol). Therefore, it is very reasonable to infer that methane hydrate is formed on the wet, hollow ZIF-8 nanoparticles.

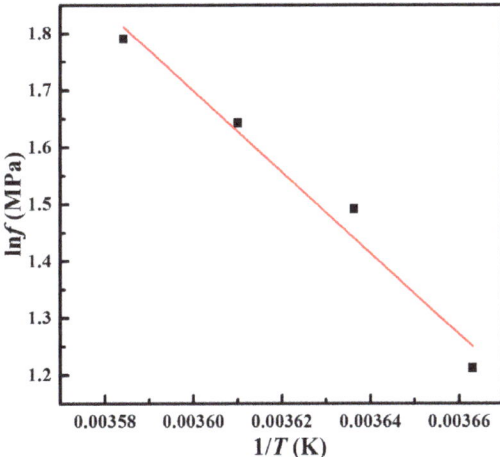

**Figure 8.** Linear fit of ln$f$ with $1/T$.

### 2.4. Stability of Hollow ZIF-8 Nanoparticles

To check the stability of the hollow ZIF-8 nanoparticles, the wet, hollow ZIF-8 nanoparticles with $R_W$ of 0.8 were used as an absorbent to conduct repeated methane adsorption experiments at 275 K. As shown in Figure 9, the jump point in the adsorption isotherm in the second cycle was essentially the same as in the first cycle, and the storage capacity of methane on the wet, hollow ZIF-8 nanoparticles was almost unchanged. The above results demonstrate that hollow ZIF-8 nanoparticles could be recyclable, which might be ascribed to the hydrophobicity and water stability of ZIF-8.

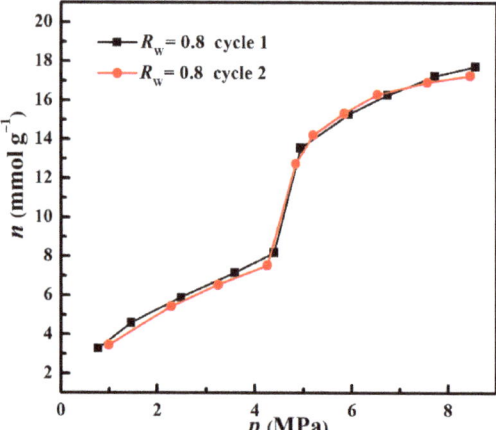

**Figure 9.** Methane adsorption isotherms at 275 K in the wet, hollow ZIF-8 with $R_W$ of 0.8 after different cycles.

## 3. Materials and Methods

### 3.1. Materials

Styrene (99%) and 2-methylimidazole (98%) were supplied by Aladdin. Sodium styrene sulfonate (98%), sodium bicarbonate (99.99%), and potassium persulfate (99.5%) were purchased from Macklin. Methanol anhydrous (99.8%), zinc nitrate hexahydrate (99%), and N, N-dimethylformamide (99.5%) were purchased from Kermel. Styrene of chemical reagent grade was distilled under vacuum before use.

### 3.2. Preparation of PS Microspheres

Homogeneous PS colloidal microspheres were synthesized by the surfactant-free emulsion polymerization method [31]. The specific process is as follows: First, 0.0217 g of sodium styrene sulfonate and 0.1082 g of sodium bicarbonate were dissolved in 200 mL of distilled water in a round bottom flask at 80 °C under stirring for 10 min. Then, 21.6450 g of styrene was added. After stirring for 1 h, 0.1082 g of potassium persulfate was added. The reaction was allowed to proceed for 18 h under a nitrogen atmosphere with magnetic stirring at 350 rpm.

### 3.3. Preparation of Hollow ZIF-8 Nanoparticles

PS@ZIF-8 core–shell nanoparticles were prepared by the hard template method. First, 0.0600 g of $Zn(NO_3)_2 \cdot 6H_2O$ and 0.0300 g of PS microspheres were mixed in 16 mL of methanol, and then 0.1660 g of 2-methylimidazole was added, followed by sonication for 1 min. The resulting mixture was heated in a water bath at 70 °C for 30 min. Subsequently, the reaction mixture was cooled naturally to room temperature, and the products were collected by centrifugation, washed, and dried. Due to the mass difference, the pure nanosized ZIF-8 nanoparticles were separated from the core–shell nanoparticles during the preparation process by adjusting the centrifugation speed. The PS@ZIF-8 core–shell nanoparticles were immersed in DMF solvent to remove the PS template, and the obtained hollow ZIF-8 nanoparticles were dried at 70 °C for 12 h.

### 3.4. Characterizations

A Bruker D8 Focus Powder X-ray diffractometer (XRD) from Germany was used to characterize the crystalline phase of the products. A Nova Nano SEM450 field emission scanning electron microscope (SEM) and a Talos F200S field emission transmission electron microscope (TEM) from the USA were used to observe the morphology of the products. The specific surface area, pore structure, and pore size of the samples were determined using an ASAP 2020 specific surface and porosity analyzer produced by Micromeritics in the USA. Based on the 77 K nitrogen adsorption equilibrium data, the specific surface area was calculated by the Brunauer–Emmett–Teller (BET) method, and the pore size distribution was calculated by the Barret–Joyner–Halenda (BJH) method. The removal of PS was analyzed using a V80X FTIR spectrometer manufactured by Bruker, Germany, and pressed into tablets using a tablet press for determination.

*3.5. Methane Adsorption Measurements*

The methane adsorption test was carried out by the volumetric method, and the corresponding apparatus is shown in Figure 10 [32,33]. The temperature fluctuation range of the thermostatic bath was ±0.1 °C. The pressure sensor had a pressure range of 0–20 MPa and an accuracy of 0.05%. The pressure range of the test was 0–10 MPa. Before measuring the adsorption by volumetric method, the sample was heated to 100 °C, and then degassed under vacuum at 100 °C for 12 h. The entire system was vacuumed, and the adsorption tank containing the wet sample needed to be cooled at −10 °C for 1 h before each vacuuming to reduce the moisture loss of wet samples during the vacuuming. The calculation of $n$ is described in Ref [32].

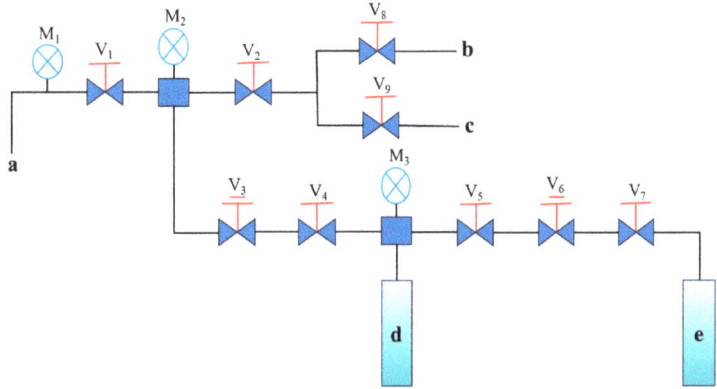

**Figure 10.** Experimental apparatus of methane adsorption: a—methane cylinder; b—vacuum pump; c—vent; d—reference tank in a thermostatic bath; e—adsorption tank in a low-temperature bath; $V_1$, $V_2$, $V_4$, $V_5$, $V_7\sim_9$—cutoff valve; $V_3$, $V_6$—metering valve; $M_1$, $M_2$—pressure gauge; $M_3$—pressure sensor.

## 4. Conclusions

Hollow-structured ZIF-8 with a high specific surface area was prepared by the hard template method. The synthesized hollow-ZIF-8 nanoparticles could promote the formation of methane hydrate. The pre-adsorbed water mass has a significant influence on the methane storage capacity. With an increase in $R_W$, the methane storage capacity on the hollow ZIF-8 nanoparticles shows a gradually increasing tendency. The maximum storage capacity of methane is 20.72 mmol/g on the wet, hollow ZIF-8 nanoparticles with $R_W$ of 1.2 at 275 K and 8.57 MPa, which increases the methane storage capacity by 55.6% more than the dry, hollow ZIF-8 nanoparticles under the same condition. Compared with the methane formation process on the hollow ZIF-8 nanoparticles with $R_W$ of 1.2, the hollow ZIF-8 nanoparticles with $R_W$ of 0.8 possess a shortened induction time of 3.5 h and an accelerated growth process. The calculated enthalpy of the generation reaction (−59.22 kJ/mol) on the wet, hollow ZIF-8 nanoparticles with $R_W$ of 0.8 is close to the formation enthalpy of the methane hydrate in pure water, demonstrating the formation of the methane hydrate in hollow ZIF-8 nanoparticles. Due to the hydrophobicity and water stability of ZIF-8, the hollow ZIF-8 nanoparticles could be recyclable. This work demonstrates that hollow ZIF-8 nanoparticles are a promising material for the wet storage of methane, which would be conducive to realizing the practical application of methane storage and transportation.

**Supplementary Materials:** The following supporting information can be downloaded at: https://www.mdpi.com/article/10.3390/catal12050485/s1, Figure S1: SEM images of PS@ZIF-8 nanoparticles synthesized for 10 min (a) and 30 min (b).; Figure S2: SEM images of PS@ZIF-8 nanoparticles synthesized by mixing the precursors first and then adding PS (a) and after removal of PS (b).

**Author Contributions:** Conceptualization, Y.L. and J.C.; methodology, C.C. and Y.L.; software, C.C.; validation, C.C., Y.L. and J.C.; formal analysis, Y.L.; investigation, C.C.; data curation, Y.L. and J.C.; writing—original draft preparation, C.C.; writing—review and editing, Y.L. and J.C.; visualization, C.C.; supervision, Y.L. and J.C.; project administration, Y.L.; funding acquisition, Y.L. All authors have read and agreed to the published version of the manuscript.

**Funding:** This work was funded by the Key Laboratory of Gas Hydrate, Guangzhou Institute of Energy Conversion, Chinese Academy of Sciences, China (No. E029kf1601), the National Natural Science Foundation of China (No. 22008049), and the Natural Science Foundation of Hebei Province, China (No. B2020202081).

**Conflicts of Interest:** The authors declare no conflict of interest.

## References

1. Huang, K.; Miller, J.B.; Huber, G.W.; Dumesic, J.A.; Maravelias, C.T. A General Framework for the Evaluation of Direct Nonoxidative Methane Conversion Strategies. *Joule* **2018**, *2*, 349–365. [CrossRef]
2. Chong, Z.R.; Yang, S.H.B.; Babu, P.; Linga, P.; Li, X.-S. Review of Natural Gas Hydrates as an Energy Resource: Prospects and Challenges. *Appl. Energy* **2016**, *162*, 1633–1652. [CrossRef]
3. Casco, M.E.; Silvestre-Albero, J.; Ramírez-Cuesta, A.J.; Rey, F.; Jordá, J.L.; Bansode, A.; Urakawa, A.; Peral, I.; Martínez-Escandell, M.; Kaneko, K.; et al. Methane Hydrate Formation in Confined Nanospace Can Surpass Nature. *Nat. Commun.* **2015**, *6*, 6432. [CrossRef] [PubMed]
4. Bhattacharjee, G.; Goh, M.N.; Arumuganainar, S.E.K.; Zhang, Y.; Linga, P. Ultra-Rapid Uptake and the Highly Stable Storage of Methane as Combustible Ice. *Energy Environ. Sci.* **2020**, *13*, 4946–4961. [CrossRef]
5. Veluswamy, H.P.; Kumar, A.; Kumar, R.; Linga, P. An Innovative Approach to Enhance Methane Hydrate Formation Kinetics with Leucine for Energy Storage Application. *Appl. Energy* **2017**, *188*, 190–199. [CrossRef]
6. Sloan, E.D. Fundamental Principles and Applications of Natural Gas Hydrates. *Nature* **2003**, *426*, 353–359. [CrossRef] [PubMed]
7. Mu, L.; Liu, B.; Liu, H.; Yang, Y.; Sun, C.; Chen, G. A Novel Method to Improve the Gas Storage Capacity of ZIF-8. *J. Mater. Chem.* **2012**, *22*, 12246–12252. [CrossRef]
8. Cuadrado-Collados, C.; Mouchaham, G.; Daemen, L.; Cheng, Y.; Ramirez-Cuesta, A.; Aggarwal, H.; Missyul, A.; Eddaoudi, M.; Belmabkhout, Y.; Silvestre-Albero, J. Quest for an Optimal Methane Hydrate Formation in the Pores of Hydrolytically Stable Metal-organic Frameworks. *J. Am. Chem. Soc.* **2020**, *142*, 13391–13397. [CrossRef]
9. Casco, M.E.; Zhang, E.; Grätz, S.; Krause, S.; Bon, V.; Wallacher, D.; Grimm, N.; Többens, D.M.; Hauß, T.; Borchardt, L. Experimental Evidence of Confined Methane Hydrate in Hydrophilic and Hydrophobic Model Carbons. *J. Phys. Chem. C* **2019**, *123*, 24071–24079. [CrossRef]
10. Borchardt, L.; Casco, M.E.; Silvestre-Albero, J. Methane Hydrate in Confined Spaces: An Alternative Storage System. *ChemPhysChem* **2018**, *19*, 1298–1314. [CrossRef]
11. Peng, Y.; Krungleviciute, V.; Eryazici, I.; Hupp, J.T.; Farha, O.K.; Yildirim, T. Methane Storage in Metal-organic Frameworks: Current Records, Surprise Findings, and Challenges. *J. Am. Chem. Soc.* **2013**, *135*, 11887–11894. [CrossRef] [PubMed]
12. Silva, P.; Vilela, S.M.F.; Tomé, J.P.C.; Almeida Paz, F.A. Multifunctional Metal-organic Frameworks: From Academia to Industrial Applications. *Chem. Soc. Rev.* **2015**, *44*, 6774–6803. [CrossRef]
13. Jimenez, D.F.; Moggach, S.A.; Wharmby, M.T.; Wright, P.A. Opening the Gate: Framework Flexibility in ZIF-8 Explored by Experiments and Simulations. *J. Am. Chem. Soc.* **2011**, *133*, 8900–8902. [CrossRef]
14. Andrés, M.A.; Vijjapu, M.T.; Surya, S.G.; Shekhah, O.; Salama, K.N.; Serre, C.; Eddaoudi, M.; Roubeau, O.; Gascón, I. Methanol and Humidity Capacitive Sensors Based on Thin Films of MOF Nanoparticles. *ACS Appl. Mater. Interfaces* **2020**, *12*, 4155–4162. [CrossRef] [PubMed]
15. Bavykina, A.; Kolobov, N.; Khan, I.S.; Bau, J.A.; Ramirez, A.; Gascon, J. Metal-organic Frameworks in Heterogeneous Catalysis: Recent Progress, New Trends, and Future Perspectives. *Chem. Rev.* **2020**, *120*, 8468–8535. [CrossRef]
16. Casco, M.E.; Rey, F.; Jordá, J.L.; Rudić, S.; Fauth, F.; Martínez-Escandell, M.; Rodríguez-Reinoso, F.; Ramos-Fernández, E.V.; Silvestre-Albero, J. Paving the Way for Methane Hydrate Formation on Metal–Organic Frameworks (MOFs). *Chem. Sci.* **2016**, *7*, 3658–3666. [CrossRef]
17. He, Z.; Zhang, K.; Jiang, J. Formation of $CH_4$ Hydrate in a Mesoporous Metal-organic Framework MIL-101: Mechanistic Insights from Microsecond Molecular Dynamics Simulations. *J. Phys. Chem. Lett.* **2019**, *10*, 7002–7008. [CrossRef] [PubMed]

18. Denning, S.; Majid, A.A.; Lucero, J.M.; Crawford, J.M.; Carreon, M.A.; Koh, C.A. Metal-organic Framework HKUST-1 Promotes Methane Hydrate Formation for Improved Gas Storage Capacity. *ACS Appl. Mater. Interfaces* **2020**, *12*, 53510–53518. [CrossRef]
19. Denning, S.; Majid, A.A.; Lucero, J.M.; Crawford, J.M.; Carreon, M.A.; Koh, C.A. Methane Hydrate Growth Promoted by Microporous Zeolitic Imidazolate Frameworks ZIF-8 and ZIF-67 for Enhanced Methane Storage. *ACS Sustain. Chem. Eng.* **2021**, *9*, 9001–9010. [CrossRef]
20. Lai, X.; Halpert, J.E.; Wang, D. Recent Advances in Micro-/Nano-Structured Hollow Spheres for Energy Applications: From Simple to Complex Systems. *Energy Env. Sci.* **2012**, *5*, 5604–5618. [CrossRef]
21. Chen, M.; Ye, C.; Zhou, S.; Wu, L. Recent Advances in Applications and Performance of Inorganic Hollow Spheres in Devices. *Adv. Mater.* **2013**, *25*, 5343–5351. [CrossRef]
22. Zhang, X.F.; Liu, Y.G.; Li, S.H.; Kong, L.Y.; Liu, H.O.; Li, Y.S.; Han, W.; Yeung, K.L.; Zhu, W.D.; Yang, W.S.; et al. New Membrane Architecture with High Performance: ZIF-8 Membrane Supported on Vertically Aligned ZnO Nanorods for Gas Permeation and Separation. *Chem. Mater.* **2014**, *26*, 1975–1981. [CrossRef]
23. Lee, H.J.; Cho, W.; Oh, M. Advanced Fabrication of Metal-organic Frameworks: Template-Directed Formation of Polystyrene@ZIF-8 Core-shell and Hollow ZIF-8 Microspheres. *Chem. Commun.* **2012**, *48*, 221–223. [CrossRef] [PubMed]
24. Hutchinson, J.W. Imperfection Sensitivity of Externally Pressurized Spherical Shells. *J. Appl. Mech.* **1967**, *34*, 49–55. [CrossRef]
25. Shan, A.X.; Chen, Z.C.; Li, B.Q.; Chen, C.P.; Wang, R.M. Monodispersed, Ultrathin NiPt Hollow Nanospheres with Tunable Diameter and Composition via a Green Chemical Synthesis. *J. Mater. Chem. A* **2015**, *3*, 1031–1036. [CrossRef]
26. Pan, Y.C.; Liu, Y.Y.; Zeng, G.F.; Zhao, L.; Lai, Z.P. Rapid Synthesis of Zeolitic Imidazolate Framework-8 (ZIF-8) Nanocrystals in an Aqueous System. *Chem. Commun.* **2011**, *47*, 2071–2073. [CrossRef]
27. Yang, Y.F.; Wang, F.W.; Yang, Q.H.; Hu, Y.L.; Yan, H.; Chen, Y.Z.; Liu, H.R.; Zhang, G.Q.; Lu, J.L.; Jiang, H.L.; et al. Hollow Metal–Organic Framework Nanospheres via Emulsion-Based Interfacial Synthesis and Their Application in Size-Selective Catalysis. *ACS Appl. Mater. Interfaces* **2014**, *6*, 18163–18171. [CrossRef] [PubMed]
28. Zhang, J.Q.; Li, L.; Xiao, Z.X.; Liu, D.; Wang, S.; Zhang, J.J.; Hao, Y.T.; Zhang, W.Z. Hollow Sphere $TiO_2$–$ZrO_2$ Prepared by Self-Assembly with Polystyrene Colloidal Template for Both Photocatalytic Degradation and $H_2$ Evolution from Water Splitting. *ACS Sustain. Chem. Eng. A* **2016**, *4*, 2037–2046. [CrossRef]
29. Sumida, K.; Rogow, D.L.; Mason, J.A.; McDonald, T.M.; Bloch, E.D.; Herm, Z.R.; Long, J.R.; Bae, T.-H. Carbon Dioxide Capture in Metal-organic Frameworks. *Chem. Rev.* **2012**, *112*, 724–781. [CrossRef]
30. Wang, Z.; Duan, J.; Chen, S.; Fu, Y.; Zhang, Y.; Wang, D.; Pei, J.; Liu, D. Molecular Insights into Hybrid $CH_4$ Physisorption-Hydrate Growth in Hydrophobic Metal-organic Framework ZIF-8: Implications for $CH_4$ Storage. *Chem. Eng. J.* **2022**, *430*, 132901. [CrossRef]
31. Im, S.H.; Lim, Y.T.; Suh, D.J.; Park, O.O. Three-Dimensional Self-Assembly of Colloids at a Water-Air Interface: A Novel Technique for the Fabrication of Photonic Bandgap Crystals. *Adv. Mater.* **2002**, *14*, 1367–1369. [CrossRef]
32. Zhou, L.; Sun, Y.; Zhou, Y. Enhancement of the Methane Storage on Activated Carbon by Preadsorbed Water. *AIChE J.* **2002**, *48*, 2412–2416. [CrossRef]
33. Zhou, L.; Liu, X.; Sun, Y.; Li, J.; Zhou, Y. Methane Sorption in Ordered Mesoporous Silica SBA-15 in the Presence of Water. *J. Phys. Chem. B* **2005**, *109*, 22710–22714. [CrossRef] [PubMed]

Review

# Recent Progress on Sulfated Nanozirconia as a Solid Acid Catalyst in the Hydrocracking Reaction

Serly Jolanda Sekewael [1], Remi Ayu Pratika [2], Latifah Hauli [2], Amalia Kurnia Amin [2], Maisari Utami [3] and Karna Wijaya [2,*]

1 Department of Chemistry, Faculty of Mathematics and Natural Science, University of Pattimura, Ambon 97235, Indonesia; sjsekewael@yahoo.com
2 Department of Chemistry, Faculty of Mathematics and Natural Science, Universitas Gadjah Mada, Yogyakarta 55281, Indonesia; remi.ayu.pratika@mail.ugm.ac.id (R.A.P.); latifah.hauli@mail.ugm.ac.id (L.H.); amalia.kurnia.a@mail.ugm.ac.id (A.K.A.)
3 Department of Chemistry, Faculty of Mathematics and Natural Sciences, Universitas Islam Indonesia, Yogyakarta 55281, Indonesia; maisariutami@uii.ac.id
* Correspondence: karnawijaya@ugm.ac.id

**Abstract:** Zirconia has advantageous thermal stability and acid–base properties. The acidity character of $ZrO_2$ can be enhanced through the sulfation process forming sulfated zirconia ($ZrO_2$-$SO_4$). An acidity test of the catalyst produced proved that the sulfate loading succeeded in increasing the acidity of $ZrO_2$ as confirmed by the presence of characteristic absorptions of the sulfate group from the FTIR spectra of the catalyst. The $ZrO_2$-$SO_4$ catalyst can be further modified with transition metals, such as Platinum (Pt), Chromium (Cr), and Nickel (Ni) to increase catalytic activity and catalyst stability. It was observed that variations in the concentrations of Pt, Cr, and Ni produced a strong influence on the catalytic activity as the acidity and porosity of the catalyst increased with their addition. The activity, selectivity, and catalytic stability tests of Pt/$ZrO_2$-$SO_4$, Cr/$ZrO_2$-$SO_4$ and Ni/$ZrO_2$-$SO_4$ were carried out with their application in the hydrocracking reaction to produce liquid fuel. The percentage of liquid fractions produced using these catalysts were higher than the fraction produced using pure $ZrO_2$ and $ZrO_2$-$SO_4$ catalyst.

**Keywords:** catalyst; zirconia; sulfated; acidity

## 1. Introduction

Zirconium dioxide, known as zirconia, is a crystalline oxide of zirconium that found form in the mineral baddeleyite. Zirconia is a white material that does not react with water or another solvent, and that has acid–base properties and excellent thermal dan chemical stabilization. $ZrO_2$ materials are of wide interest and development in their application in various fields, such as heterogeneous catalyst, optics, electronics, magnetics, and ceramics owing to the high melting point ($\geq 2700$ °C), low thermal conductivity, corrosion resistance, and good thermal and mechanical strength [1–3].

Modification of zirconia to increase its catalytic activity has been developed. Many studies have showed that modification zirconia such as sulfated process or metal supported was effective during the chemical process [3–5]. Sulfated zirconia catalyst via hydrothermal treatment for hydrocracking of LDPE plastic waste into liquid fuels was examined by Utami et al. [6]. The total acidity of zirconia increased after the sulfation process, thus increasing the amount of liquid yield. However, the catalytic activity of sulfated zirconia catalyst during the hydrocracking reaction at high temperature decreases due to the deactivation catalyst. This process therefore requires an appropriate catalyst to increase the catalytic activity of sulfated zirconia, as well as the acidity and liquid yield [7,8].

Supported noble metals such as Platinum (Pt), Chromium (Cr), and Nickel (Ni) as a promoter have shown good catalytic activity in the hydrocracking reaction. The synthesis of Pt/sulfated zirconia [9], Cr/sulfated zirconia [10], and Ni/sulfated zirconia [11] catalyst in the hydrocracking reaction reported the enhance of acidity of sulfated zirconia and the liquid product of hydrocracking after the addition of promoter Pt, Cr, and Ni, respectively.

## 2. Zirconium Dioxide ($ZrO_2$)

Zirconium dioxide ($ZrO_2$) is a polymorphic material with three crystalline phases, namely the monoclinic, tetragonal, and cubic phases, as shown in Figure 1. The monoclinic and tetragonal phases are stable up to temperatures of 1170 and 2360 °C, respectively, while the cubic phase is stable at temperatures above 2680 °C. $ZrO_2$ as a catalyst is normally used in its monoclinic and metastable tetragonal crystalline phases [12–14]. Transformation of the crystalline phases of $ZrO_2$ is driven by changes in temperature. $ZrO_2$ calcined at <800 °C forms monoclinic and metastable tetragonal phases [15,16].

The transformations of one $ZrO_2$ phase to another are accompanied by changes in lattice parameters. The rate of phase transformation of $ZrO_2$ is also influenced by the particle size of the $ZrO_2$ precursor used. The larger the $ZrO_2$ particle size, the faster the phase transformation occurs. Nano-sized particles have a high surface area, allowing for more of the atoms of the particles to interact and form bonds [15,17,18]. Based on surface properties and polymorphic form, $ZrO_2$ is also often used as a catalyst or carrier material because it has acidic and basic properties [19,20]. In addition, $ZrO_2$ has a structure with vacant sites on its surface that can allow cations to easily enter [21,22].

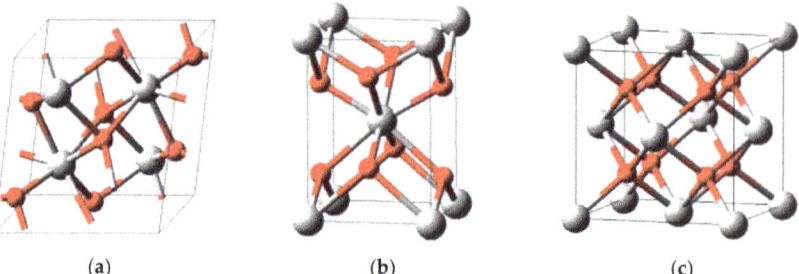

(a) (b) (c)

**Figure 1.** Crystal structures of (**a**) monoclinic, (**b**) tetragonal, and (**c**) cubic $ZrO_2$. Copyright Elsevier, reprinted from Ref. [23].

## 3. $ZrO_2$-$SO_4$ Catalyst

Zirconia is a catalyst containing high Lewis acid sites. Ore [15] mentioned that $ZrO_2$ can be modified with acid or base to achieve, based on the intended application, the appropriate and desired strong acid or base characteristics. Sulfation is a method that can increase the strength of $ZrO_2$ acid through the formation of Brønsted acids on the zirconia substrate that consequently increases its catalytic activity. $ZrO_2$-$SO_4$ can be prepared from sulfate precursors such as $H_2SO_4$, $(NH_4)_2SO_4$, $(NH_4)_2S_2O_3$, and $(NH_4)_2S$ [24–26].

The formation of $ZrO_2$-$SO_4$ occurs through the chelation of the zirconium cations ($Zr^{4+}$) with sulfate ions. The $SO_4^{2-}$ ions acts as ligands that donate their lone pair of electrons from the O atom, thus forming a coordination bond with two $Zr^{4+}$ as the central atoms and causing the acid molecule to release two protons simultaneously. After relaxation, adsorptive complex molecules are produced from the coordination of $SO_4^{2-}$ ions onto the $ZrO_2$ surface through the sharing of two O atoms [27–29]. Saravan et al. [30] illustrate the surface model of $ZrO_2$-$SO_4$ with the Lewis and Brønsted acid sites shown as shown in Figure 2.

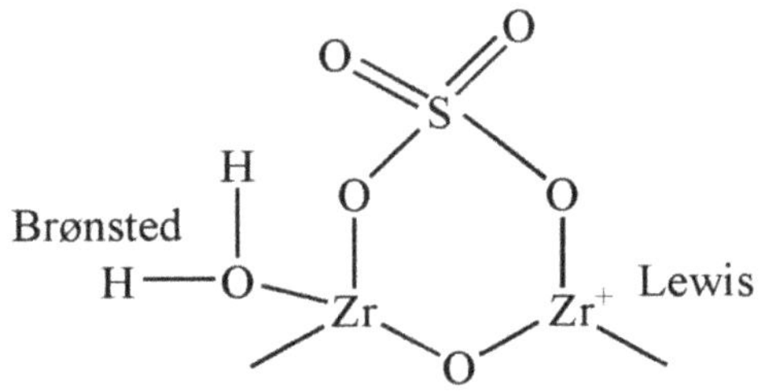

**Figure 2.** Brønsted and Lewis acid sites on $ZrO_2$-$SO_4$. Copyright MDPI, reprinted from Ref. [17].

Modification of zirconia to $ZrO_2$-$SO_4$ produces materials with high Lewis and Brønsted acid strengths. Wang et al. [29] presented an illustration of the $ZrO_2$-$SO_4$ surface as presented in Figure 3. The high acid strength of the Brønsted acid site is associated with the location of $Zr^{4+}$ ions that are adjacent to the S=O bond which attracts electrons from the bisulfate to form the Brønsted acid site.

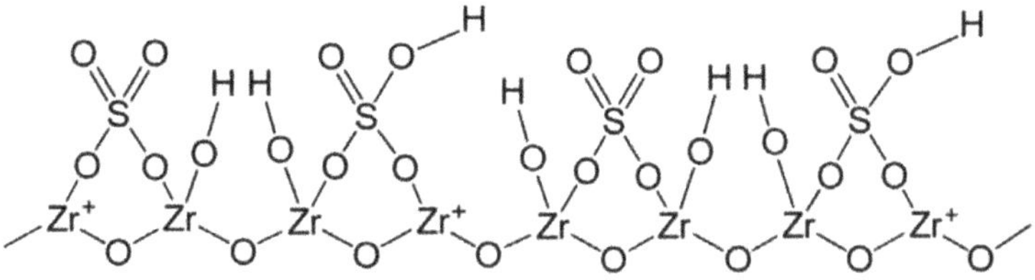

**Figure 3.** $ZrO_2$-$SO_4$ surface model of bidentate chelate type. Reprinted from ref. [29].

### 3.1. Functional Group Characterization for $ZrO_2$-$SO_4$

Research conducted by Utami et al. [31] reported the preparation of sulfated zirconia catalysts with various $H_2SO_4$ concentrations and calcination temperatures. Figure 4 shows the same absorption peak at wave number 424–741 cm$^{-1}$, indicating the Zr–O–Zr bond [32]. Absorptions at 3426–3449 and 1636 cm$^{-1}$ refer to the O-H stretching and bending vibrations of the water molecules adsorbed on the material [33]. In addition, according to Ore et al. [15], the broadband in the absorption region of 3400 cm$^{-1}$ signifies the bridge between the hydroxyl group with two or three Zr atoms.

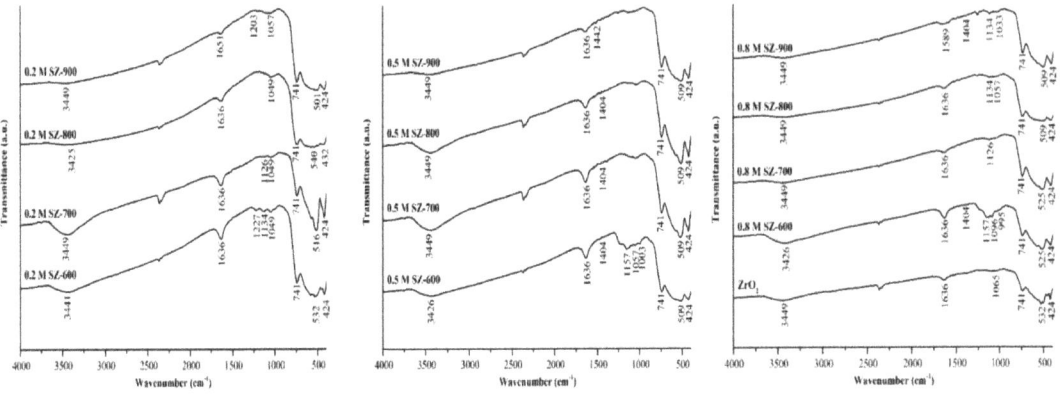

**Figure 4.** FTIR spectra of SZ catalyst at various sulfate concentrations and calcination temperatures. Reprinted with permission from Dr. Utami, Ref. [31]. Copyright 2019 Trans Tech Publication.

The presence of $SO_4^{2-}$ ions on the surface of $ZrO_2$ can be confirmed by the formation of a new peak at 995-1404 cm$^{-1}$ which is typical of the bidentate $SO_4^{2-}$ chelate ion covalently bonded to the $Zr^{4+}$ cation [7,27]. The absorption peaks at 995–1003, 1049–1096, 1134–1157, and 1227 cm$^{-1}$ are S–O symmetric vibrations, S–O asymmetric vibrations, S=O symmetry vibrations, and S=O asymmetric vibrations [34–36]. The peak with low intensity in the area of 1404 cm$^{-1}$ is the stretching vibration of S=O, indicating the formation of $SO_3$ species on the surface of $ZrO_2$ [37]. The presence of characteristic bands of $ZrO_2$-$SO_4$ proves that the impregnation of $H_2SO_4$ on $ZrO_2$ has been successfully carried out.

The use of too high a concentration of $H_2SO_4$ can lead to loss of several absorption bands of the $SO_4^{2-}$ ion which is covalently bonded to $Zr^{4+}$ cations as the $ZrO_2$ structure is degraded [9,37]. Such a case implies that an appropriate or optimum concentration of $H_2SO_4$ is needed for the $ZrO_2$ activation process. The absorption intensity of $ZrO_2$-$SO_4$ increased at a temperature range of 500–600 °C but decreased at temperatures of 700–800 °C. Calcination treatment at 600 °C (SZ-0.8-600) was found to be optimum resulting in the highest $SO_4^{2-}$ dispersion. According to Ore [15], the maximum calcination temperature in the sulfation process is 650 °C. Temperatures above 650 °C cause the decomposition of $SO_4^{2-}$ ions, thereby reducing the acidity and reactivity of the catalyst.

The acidity test of the catalyst was carried out by the gravimetric method based on the amount of $NH_3$ vapor absorbed by the catalyst. Table 1 shows the results of the catalyst acidity test. $ZrO_2$ has a total acidity of 0.18 mmol/g. The acidity of $ZrO_2$ comes from the $Zr^{4+}$ cations that act as Lewis acid sites [7,8]. After modification with $H_2SO_4$, the acidity of $ZrO_2$ increased. SZ-0.8-600 catalyst showed the highest acidity value of 1.06 mmol/g. Ore et al. [15] reported that the number of Brønsted and Lewis acid sites depends on the $SO_4^{2-}$ concentration present on the catalyst surface. Sulfation with optimum sulfate concentration can increase the catalytic activity of the catalyst, while the decrease in catalyst acidity with increasing temperature occurs due to dehydration of protonic sites and loss of $SO_4^{2-}$ groups on the surface of the catalyst [38–40].

**Table 1.** Acidity test for the SZ catalyst at various sulfate concentrations and calcination temperatures [31].

| Catalyst | Acidity (mmol/g) | | | |
| --- | --- | --- | --- | --- |
| | 600 °C | 700 °C | 800 °C | 900 °C |
| $ZrO_2$ | 0.18 | - | - | - |
| SZ-0.2 | 0.39 | 0.32 | 0.29 | 0.26 |
| SZ-0.5 | 0.83 | 0.33 | 0.30 | 0.28 |
| SZ-0.8 | 1.06 | 0.56 | 0.53 | 0.33 |

Qualitatively, the number of Brønsted and Lewis acid sites can be observed absorption spectra intensity that denotes the interaction between the catalyst acid sites and $NH_3$ [41]. Figure 5 presents the FTIR spectra of the catalyst after the acidity test. The absorption peak at 1119-1126 cm$^{-1}$ indicates the presence of $NH_3$ coordinated to the Lewis acid site. The peak at 1404 cm$^{-1}$ confirmed the presence of $NH_4^+$ ions formed by proton transfer from the Brønsted acid site to $NH_3$ [42,43]. The higher the intensity of the absorption band, the higher the number of Brønsted and Lewis acid sites. It was found that the SZ-0.8-600 catalyst had the highest acid site intensities. The absorption intensities of the acid sites decreased with the increase in calcination temperature. Calcination treatment at high temperature decreases the acidity of the catalyst due to the decrease in the number of acid sites on the surface of the catalyst [44,45].

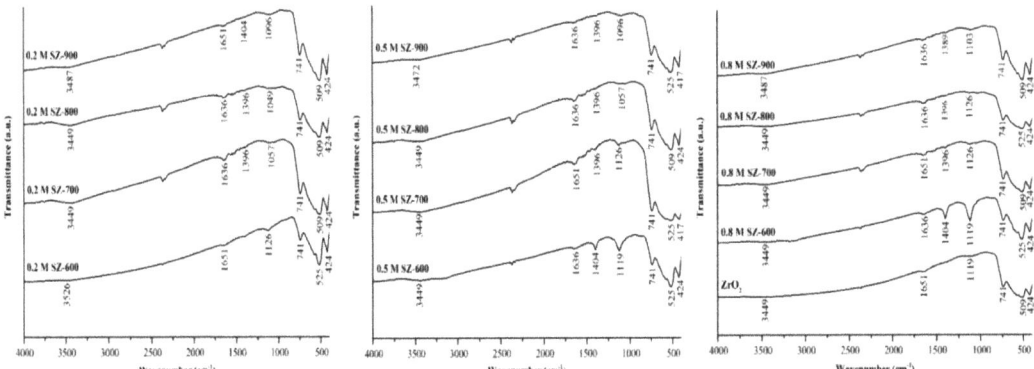

**Figure 5.** FTIR spectra of SZ catalyst at various sulfate concentrations and calcination temperatures after acidity test. Reprinted with permission from Dr. Utami, Ref. [31]. Copyright 2019 Trans Tech Publication.

### 3.2. ZrO$_2$-SO$_4$ Crystal Structure Characterization

Figure 6 presents the diffraction pattern of the SZ catalyst at various sulfate concentrations and calcination temperatures. The main diffraction peaks appear at 2θ = 28.34° (d-111) and 31.64° (d111), referring to the ZrO$_2$ monoclinic crystalline phase [6,15]. In general, the diffraction pattern showed stable crystallinity even after the addition of acid and calcination treatment. However, the intensity of the ZrO$_2$ monoclinic diffraction peak decreased after acid treatment. SZ-0.8-600 catalyst with the highest total acidity showed the lowest monoclinic peak intensity. The addition of a high concentration of H$_2$SO$_4$ causes a large number of SO$_4^{2-}$ ions to cover the surface of ZrO$_2$, decreasing crystallinity [37,38]. The intensities of the monoclinic peaks at temperatures of 800 and 900 °C were higher than those at 600 and 700 °C. This occurred because the high calcination temperature caused SO$_4^{2-}$ ions to decompose from the catalyst surface, increasing the crystallinity of the catalyst.

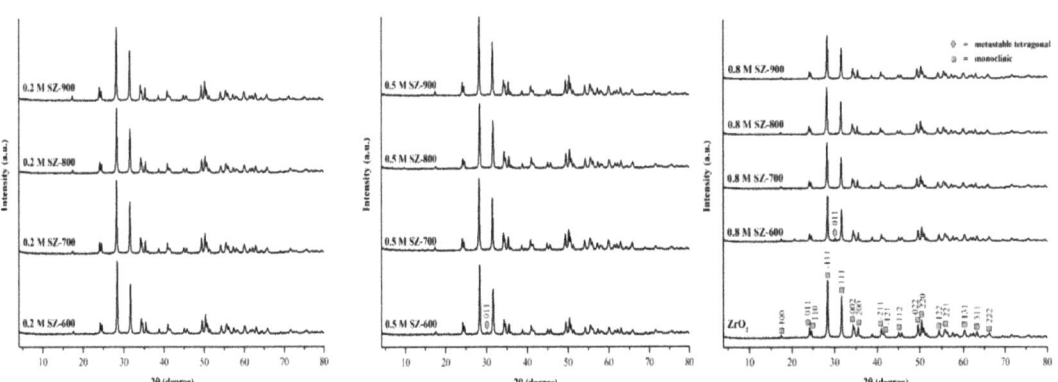

**Figure 6.** Diffraction patterns of SZ catalyst at various sulfate concentrations and calcination temperatures. Reprinted with permission from Dr. Utami, Ref. [31]. Copyright 2019 Trans Tech Publication.

ZrO$_2$-SO$_4$ calcined at a temperature of <800 °C can exist in a metastable tetragonal phase and a monoclinic phase. Similar results were reported by Ore et al. [15] which stated that ZrO$_2$-SO$_4$ calcined at a temperature of 600 °C consists of a mixture of metastable tetragonal and monoclinic phases. The ZrOCl$_2$·8H$_2$O amorphous precursor used made it possible for transformation to a metastable tetragonal structure to occur. However, the $ZrO_2$ used in this study is commercial $ZrO_2$, which contains the monoclinic structure of high stability and crystallinity, hindering it from undergoing phase transformation [46]. The catalyst diffraction pattern in the research by Utami et al. [6] demonstrated the $ZrO_2$ catalyst and its modifications consisting only of monoclinic structure.

## 4. Platinum/Sulfated Zirconia (Pt/ZrO$_2$-SO$_4$) Catalyst

A heterogeneous catalyst is a catalyst material composed of two components, namely the doping and carrier components. The metal catalyst, when used in its pure form, has low thermal stability and tends to sinter that can leading to a decrease in surface area and deactivation [47–49]. Appropriate distribution of metal catalysts on the carrier material having acid–base sites and a large surface area is necessary to avoid sintering [50,51]. The $ZrO_2$-$SO_4$ material has many Brønsted and Lewis acid sites in which, despite its high acidity, this catalyst can be rapidly deactivated. The addition of Pt metal can increase the stability of the catalyst with the simultaneous presence of hydrogen gas ($H_2$) [52–54].

The distribution mechanism of $H_2$ through the Pt surface on the $ZrO_2$-$SO_4$ carrier is illustrated in Figure 7. The $H_2$ molecule dissociates on the surface of the Pt particle homolytically to form two H radicals which then bind to the unpaired electrons in the 5d orbitals. The $H^+$ ions released from Pt are distributed on the $ZrO_2$-$SO_4$ carrier and migrate to the electron-rich O atomic sites, forming Brønsted acid sites [55,56]. Figure 8 shows the physical appearance of $ZrO_2$-$SO_4$ catalyst which a white powder and after the impregnation with Pt metal as a Pt/$ZrO_2$-$SO_4$ catalyst, solid powder which is darker in color is formed due to the presence of the Pt metal that impregnated to the $ZrO_2$-$SO_4$ catalyst [6,9].

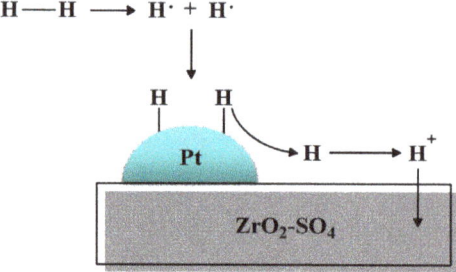

**Figure 7.** Mechanism of $H_2$ distribution through the Pt surface on $ZrO_2$-$SO_4$. Reprinted and modified with permission from Dr. Utami, Ref. [9]. Copyright 2019 The Royal Society of Chemistry.

**Figure 8.** $ZrO_2$-$SO_4$ (**right**) and Pt/$ZrO_2$-$SO_4$ (**left**) catalyst. Reprinted with permission from Dr. Utami, Ref. [42].

*4.1. FTIR and Acidity Characterization of Pt/ZrO$_2$-SO$_4$ Catalyst*

Utami et al. [6] researched the synthesis of Pt-promoted zirconia (Pt/SZ) catalyst and its application in hydrocracking LDPE plastic into liquid fuel. The FTIR spectra of the sulfated zirconia impregnated with Pt metal can be seen in Figure 9. Overall, the Pt1/SZ, Pt2/SZ, and Pt3/SZ spectra showed the same absorption peaks as ZrO$_2$ and SZ. Pt metal impregnated catalysts had characteristic peaks of ZrO$_2$-SO$_4$ at wave number 1065–1126 cm$^{-1}$. The addition of Pt metal caused some of the absorption peaks of ZrO$_2$-SO$_4$ to disappear. This is an early indication that Pt metal had been impregnated on nanoSZ [9].

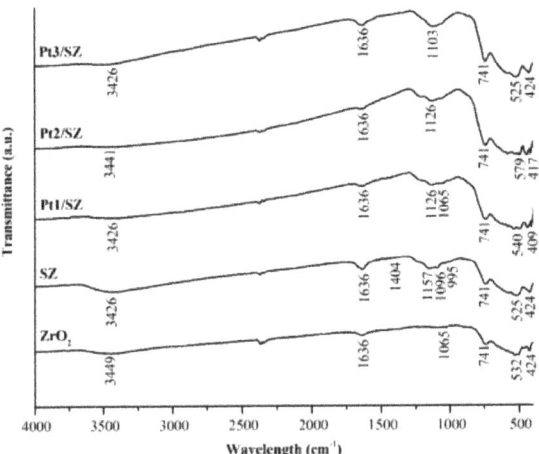

**Figure 9.** FTIR spectra of ZrO$_2$, SZ, Pt1/SZ, Pt2/SZ, and Pt3/SZ. Reprinted with permission from Dr. Utami, Ref. [6]. Copyright 2019 Elsevier.

Table 2 shows the total acidity of the catalysts of the present study. Pt metal impregnation on sulfated zirconia was proven to increase the total acidity of the catalyst significantly (from 1.06 to 10.75 mmol/g). Pt1/SZ, Pt2/SZ, and Pt3/SZ catalysts showed increasing acidity values with increasing concentrations of Pt metal. The increase in the acidity of the catalyst occurs because the Pt metal provides vacant orbitals that can act as electron-pair acceptors (Lewis acid sites) and the presence of unpaired electrons in the d orbitals that form Brønsted acid sites [57,58].

**Table 2.** Acidity test results of ZrO$_2$, SZ, Pt1/SZ, Pt2/SZ, and Pt3/SZ [6].

| Sample | Acidity (mmol/g) |
| --- | --- |
| ZrO$_2$ | 0.18 |
| SZ | 1.06 |
| Pt1/SZ | 10.75 |
| Pt2/SZ | 11.05 |
| Pt3/SZ | 11.14 |

FTIR spectra interpretations of ZrO$_2$, SZ, Pt1/SZ, Pt2/SZ, and Pt3/SZ are shown in Figure 10. Increasing concentration of Pt metal produced increased adsorption of NH$_3$ bound to the Brønsted and Lewis acid sites as shown at wavenumbers of 1396–1404 and 1119 cm$^{-1}$, indicating that the higher the concentration of Pt metal, the higher the number of acid sites contained in the catalyst. The Pt3/SZ catalyst showed the highest intensity of Brønsted and Lewis acid absorptions. Based on the results of the acidity test of the catalyst, the Pt3/SZ catalyst was confirmed to have the highest acidity value.

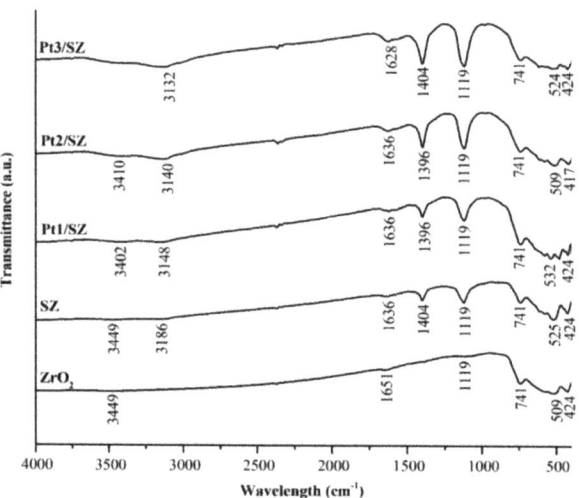

**Figure 10.** FTIR spectra of $ZrO_2$, SZ, Pt1/SZ, Pt2/SZ, and Pt3/SZ after acidity test. Reprinted with permission from Dr. Utami, Ref. [6]. Copyright 2019 Elsevier.

*4.2. XRD and GSA Characterizations of $Pt/ZrO_2$-$SO_4$ Catalyst*

SZ catalysts are shown in Figure 11. Based on crystal identification, all samples showed the presence of the monoclinic phase. According to Ore [15], $SO_4^{2-}$ species can be thermally crystallized through the calcination process and undergo a crystalline phase transformation, further stabilizing the $ZrO_2$ crystalline phase. The addition of Pt metal to SZ would not have caused changes in the crystal structure of the material. A decrease in the intensity of the diffraction peak after the addition of Pt metal was observed. This phenomenon indicated that Pt metal had been successfully impregnated on the SZ surface, where a higher concentration of the impregnated Pt metal would cause the intensity of the monoclinic peak to decrease [6,9].

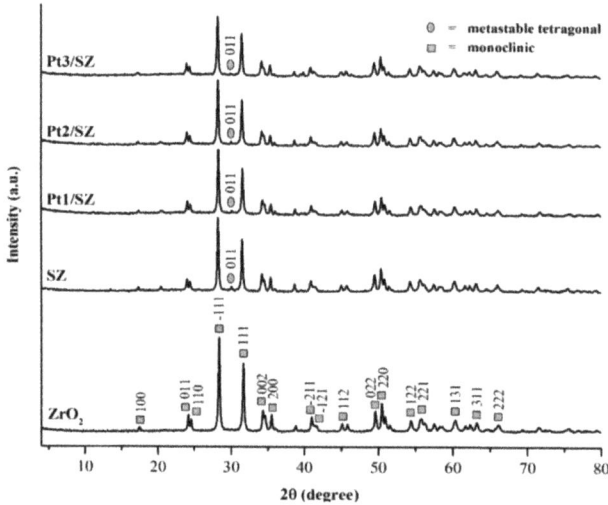

**Figure 11.** Diffraction patterns of $ZrO_2$, SZ, Pt1/SZ, Pt2/SZ, and Pt3/SZ. Reprinted with permission from Dr. Utami, Ref. [6]. Copyright 2019 Elsevier.

Based on the catalyst diffraction pattern, Pt metal characteristic peaks were not detected. This is because the concentration of impregnated Pt metal was relatively low. This is by research conducted by Aboul-Gheit et al. [59] that showed similar results whereby the diffraction peak of Pt metal was not identified after the addition of 0.6% Pt metal to $ZrO_2$-$SO_4$. A relatively low concentration of Pt metal was used in the present study to prevent agglomeration of Pt particles on the $ZrO_2$-$SO_4$ surface, which could cause a decrease in catalytic activity.

Characterization carried out by GSA (Table 3) showed an increase in surface area and pore volume along with the increasing concentration of Pt metal impregnated onto SZ. The Pt1/SZ catalyst had a surface area and pore volume of 13.49 $m^2/g$ and 0.05 $cm^3/g$, respectively, which saw an increase to 20.23 $m^2/g$ and 0.06 $cm^3/g$ for Pt2/SZ and 29.48 $m^2/g$ and 0.08 $cm^3/g$ for Pt3/SZ. Based on the data reported in the study by Utami et al. [9], the increase in the surface area and pore volume of the SZ catalyst can be attributed to the inhibition of the agglomeration process related to the high presence of Pt metal.

**Table 3.** Textural characteristics of $ZrO_2$, SZ, Pt1/SZ, Pt2/SZ, and Pt3/SZ [6].

| Sample | Surface Area ($m^2$/g) | Pore Volume ($cm^3$/g) | Pore Diameter (nm) |
| --- | --- | --- | --- |
| $ZrO_2$ | 22.05 | 0.08 | 3.49 |
| SZ | 12.49 | 0.05 | 3.82 |
| Pt1/SZ | 13.49 | 0.05 | 3.83 |
| Pt2/SZ | 20.23 | 0.06 | 3.84 |
| Pt3/SZ | 29.48 | 0.08 | 3.86 |

The increase in the surface area of the catalyst can also be attributed to the smaller crystal size dimensions. Crystal size results of $ZrO_2$, SZ, Pt1/SZ, Pt2/SZ, and Pt3/SZ (Table 4) presented a decrease in size after metal impregnation of Pt. This indicated that the Pt metal impregnated by the reflux technique was evenly dispersed on the surface and pores of SZ [9]. In contrast, Aboul-Gheit et al. [59] impregnated Pt metal on $ZrO_2$-$SO_4$ through the wet impregnation method and reported an increase in the concentration of Pt metal causing the surface area and pore volume to increase due to the Pt metal being not evenly distributed and covering most of the $ZrO_2$-$SO_4$ pores.

**Table 4.** Crystal sizes of $ZrO_2$, SZ, Pt1/SZ, Pt3/SZ Pt2/SZ and Pt3/SZ [6].

| Sample | Crystal Size (nm) |
| --- | --- |
| $ZrO_2$ | 31.54 |
| SZ | 34.06 |
| Pt1/SZ | 32.80 |
| Pt2/SZ | 31.54 |
| Pt3/SZ | 31.53 |

*4.3. Elemental Composition Characterization Using EDXRF*

Table 5 shows the concentrations of elements contained in the catalysts. The concentrations of Pt metals identified in the Pt1/SZ, Pt2/SZ, and Pt3/SZ samples were 0.35, 0.90, and 1.19%. The metal content of Pt in SZ was strongly influenced by the dispersion ability on the surface and pores of the carrier material. The reduction treatment with $H_2$ gas flow was carried out after the calcination step which aimed to obtain $Pt^0$ particles that would result in dispersion [14].

Table 5. Elemental compositions of ZrO₂, SZ, Pt1/SZ, Pt2/SZ dan Pt3/SZ [6].

| Sample | Elemental Compositions (% w/w) | | | |
|---|---|---|---|---|
| | Zr | Pt | O | S |
| ZrO$_2$ | 64.08 | - | 35.46 | 0.46 |
| SZ | 60.51 | - | 37.42 | 2.07 |
| Pt1/SZ | 75.24 | 0.35 | 24.02 | 0.74 |
| Pt2/SZ | 74.37 | 0.90 | 25.02 | 0.61 |
| Pt3/SZ | 75.34 | 1.19 | 24.40 | 0.26 |

Based on the elemental compositions above, the concentrations of Pt that were observed were lower than the theoretical concentration of Pt metal by the addition of PtCl$_4$ solution used in the impregnation process. This discrepancy occurred due to competition between impregnated Pt metals, causing the formation of multilayer stacking of the active metal in the pore mouth area of the carrier material [60,61]. The active Pt metal that sat at the top position would have weak interaction and would experience easier desorption [62]. Because of this, the amount of active metal in the pore area of the carrier material was observed to be less.

*4.4. Pt metal Composition Identification Using XPS*

Subsequent research related to the synthesis of platinum-loaded sulfated zirconia catalysts using the hydrothermal method was reported by Utami et al. [9]. XPS spectra were used to determine the composition of Pt in the samples based on a comparison of binding energy values. The peaks generated from XPS are not single peaks, so that deconvolution of the peaks was needed to identify the multiple peaks that made up each peak. Figure 12 presents the XPS spectra along with the deconvolution of Pt 4f peaks from Pt/nano ZrO$_2$-SO$_4$ consisting of Pt$^0$ 4f$_{7/2}$, Pt$^0$ 4f$_{5/2}$, and Pt$^{2+}$ 4f$_{7/2}$. The spectra indicated the interaction of electrons between the Pt particles and the nanoZS surface. Table 6 shows the relative area of the deconvoluted peaks of Pt$^0$ 4f$_{7/2}$, Pt$^0$ 4f$_{5/2}$, and Pt$^{2+}$ 4f$_{7/2}$ in the Pt/nano ZrO$_2$-SO$_4$ sample. The data obtained showed that the detected Pt$^0$ composition was 81.82%. The results indicated that the reduction treatment with H$_2$ gas flow at the metal impregnation stage succeeded in forming Pt$^0$ particles.

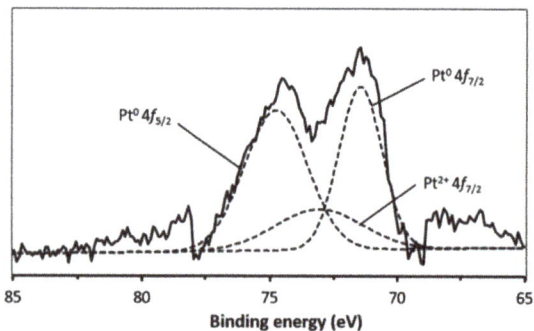
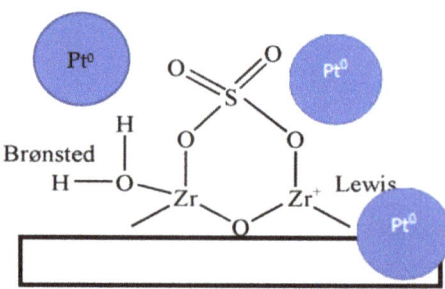

Figure 12. XPS spectra of Pt/nano ZrO$_2$-SO$_4$ in the 4f region and the catalyst model. Reprinted and modified with permission from Dr. Utami, Ref. [9]. Copyright 2019 The Royal Society of Chemistry.

Table 6. Percentage of relative areas of Pt 4f deconvoluted peaks [4].

| Pt 4f Peak | Peak Position (eV) | Relative Area (%) |
|---|---|---|
| Pt$^0$ 4f$_{7/2}$ | 71.45 | 36.36 |
| Pt$^0$ 4f$_{5/2}$ | 74.75 | 45.46 |
| Pt$^{2+}$ 4f$_{7/2}$ | 73.08 | 18.18 |

*4.5. Thermal Stability Characterization with TG/DTA*

The TG/DTA curve provides information about changes in thermal conditions with mass changes in the sample. Figure 13 shows the TGA curves for nano $ZrO_2$, nano $ZrO_2$-$SO_4$, and Pt/nano $ZrO_2$-$SO_4$ samples analyzed at temperatures of 30–900 °C. The TGA curve of nano $ZrO_2$ did not indicate mass decrease indicating that nano Z had good thermal stability [63]. Mass decrease in nano $ZrO_2$-$SO_4$ and Pt/nano $ZrO_2$-$SO_4$ in the range of 50–200 °C by 2.41 and 1.24% were associated with the elimination of water molecules physically adsorbed on the material. At 500–700 °C, decreases in a mass of 5.69 and 2.48%, indicating the decomposition of $SO_4{}^{2-}$ ions bound to the $ZrO_2$ surface. The decomposition of $SO_4{}^{2-}$ ions at 600–1000 °C affected the structural changes of $ZrO_2$-$SO_4$, causing a decrease in catalytic activity [64]. The decomposition of $H_2SO_4$ occurs through a two-stage endothermic process at high temperatures according to Equations (1) and (2).

$$H_2SO_{4\,(aq)} \rightarrow H_2O_{\,(l)} + SO_{3\,(g)} \qquad (1)$$

$$SO_{3\,(g)} \rightarrow SO_{2\,(g)} + \frac{1}{2}O_{2\,(g)} \qquad (2)$$

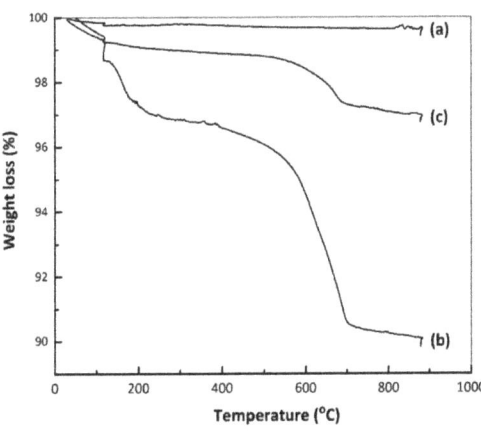

**Figure 13.** TGA curves of (a) nano $ZrO_2$, (b) nano $ZrO_2$-$SO_4$, (c) Pt/nano $ZrO_2$-$SO_4$. Reprinted and modified with permission from Dr. Utami, Ref. [9]. Copyright 2019 The Royal Society of Chemistry.

*4.6. Activity and Selectivity of Pt/ZrO$_2$-SO$_4$ Catalyst in LPDE Hydrocracking Application*

Utami et al. [6] reported the activity and selectivity of the Pt/$ZrO_2$-$SO_4$ catalyst applied in LPDE hydrocracking. The hydrocracking liquid fraction of the LDPE plastic waste thermal cracking is shown in Figure 14. Physically, the hydrocracking liquid fraction has a yellow color and pungent odor, which indicate the success of the hydrocracking process. The percentages of liquid fractions produced using the Pt1/SZ, Pt2/SZ, and Pt3/SZ catalysts were higher than that of the SZ catalyst. The percentage of liquid fraction obtained with the SZ catalyst was 57.92%, while with Pt1/SZ, Pt2/SZ, and Pt3/SZ the liquid fractions were 70.58, 71.15, and 74.60%, respectively. Through MS data, it was found that the hydrocarbon compounds in the gasoline range ($C_5$–$C_{12}$) were more commonly found in the liquid fraction samples that had used Pt3/SZ (catalyst with the highest acidity). Table 7 shows the gasoline fraction percentages from the hydrocracking reaction using Pt1/SZ, Pt2/SZ, and Pt3/SZ catalysts, which were 48.76, 64.22, and 67.51 $w/w$%, respectively. This cased that besides affecting the amount of hydrocracking liquid fraction produced, the addition of Pt metal can also increase selectivity towards the gasoline fraction [65].

**Figure 14.** The physical appearance of the hydrocracking liquid fraction (**a**) without catalyst, and with catalyst (**b**) $ZrO_2$, (**c**) SZ, (**d**) Pt1/SZ, (**e**) Pt2/SZ3, (**f**) Pt3/SZ. Reprinted with permission from Dr. Utami, Ref. [14].

**Table 7.** Distribution of hydrocracking products using different catalysts (T = 250 °C, t = 60 m, catalyst to feed ratio = 1% $w/w$) [6].

| Sample | Hydrocracking Product (% $w/w$) | | | |
|---|---|---|---|---|
| | Liquid | | Solid | Gas |
| | $C_5$–$C_{12}$ | $C_{13}$–$C_{20}$ | | |
| SZ | 42.95 | 14.97 | 0.35 | 41.73 |
| Pt1/SZ | 48.76 | 21.82 | 0.15 | 29.27 |
| Pt2/SZ | 64.22 | 6.93 | 0.16 | 28.69 |
| Pt3/SZ | 67.51 | 7.09 | 0.15 | 25.25 |

The percentage of hydrocarbon compounds in the $C_5$–$C_{12}$ range can be seen to be greater than that of $C_{13}$–$C_{20}$ hydrocarbons. This shows that the hydrocracking liquid fraction of LDPE plastic waste had a higher gasoline fraction than the diesel fraction and that the four types of catalysts used had good selectivity towards the hydrocracking reaction that produces liquid fuel fraction (gasoline fraction).

Figure 15 shows the proportion of hydrocarbon compounds contained in the hydrocracked liquid fraction in the gasoline range, namely olefins, paraffin, isoparaffins, and naphthenes, with a total composition of 56.36 and respective amounts of 20.07, 14.60, and 6.81% $w/w$ with use of the Pt3/SZ catalyst. Aromatic compounds in small amounts were also produced with a composition of 0.70% $w/w$, and only 1.46% $w/w$ was indicated to be non-hydrocarbon compounds. Overall, olefin (unsaturated/double-bonded compound) was dominantly produced from the hydrocracking. This is because LDPE, as the plastic feed used, consists of olefin in which, during the hydrocracking reaction, not all olefins react with the hydride to become paraffin (saturated/single bond compound) [6,9].

*4.7. Stability Test of Pt/ZrO$_2$-SO$_4$*

A stability test of the Pt/nano $ZrO_2$-$SO_4$ catalyst, along with nano $ZrO_2$ and nano $ZrO_2$-$SO_4$ for comparison, was carried out by Utami et al. [9] through a hydrocracking reaction of LDPE plastic waste with a catalyst/feed ratio 1% $w/w$ and a temperature of 250 °C for 60 min. The catalyst stability test was carried out for six cycles with the same reaction conditions. Figure 16 shows that $ZrO_2$, SZ, and Pt3/SZ had good catalytic performances when first used. The hydrocracking reaction with the $ZrO_2$ catalyst showed a significant decrease in the percentage of liquid fraction produced in the fourth cycle, while in SZ this significant reduction occurred in the second cycle.

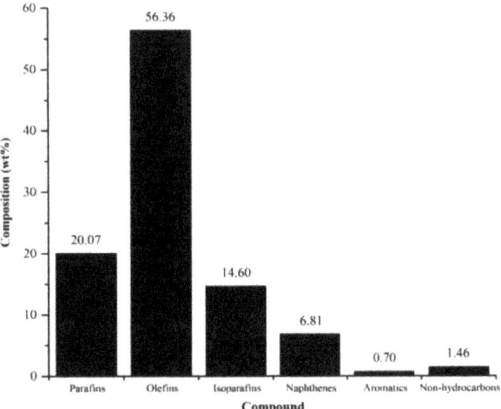

**Figure 15.** The composition of liquid yield in the gasoline fraction from the hydrocracking reaction of LDPE plastic waste using Pt3/SZ at 250 °C. Reprinted with permission from Dr. Utami, Ref. [6]. Copyright 2019 Elsevier.

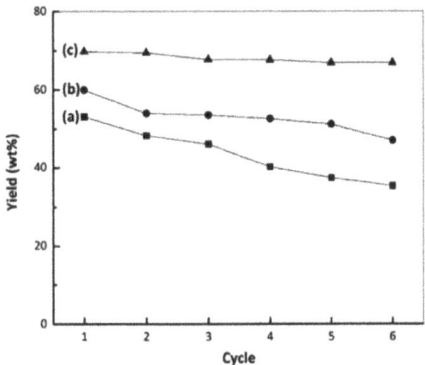

**Figure 16.** Hydrocracking liquid fraction graphs of (a) nano $ZrO_2$, (b) nano $ZrO_2$-$SO_4$, and (c) Pt/nano ZS3 Pt/nanoZS3-600. Reprinted and modified with permission from Dr. Utami, Ref. [9]. Copyright 2019 The Royal Society of Chemistry.

The catalytic activity of Pt3/SZ showed very good stability up to the sixth cycle. Aboul-Gheit et al. [59] reported that the catalytic activity of $ZrO_2$-$SO_4$ in the n-pentane isomerization reaction decreased in the fourth cycle and became inactive in the eighth cycle. In contrast, the Pt/$ZrO_2$-$SO_4$ catalyst showed stable activity until the tenth cycle. The Pt/$ZrO_2$-$SO_4$ catalyst showed high resistance to the deactivation related to the removal of coke from the catalyst surface, thus increasing the stability of the catalyst [36,37].

Figure 16 illustrates the proposed origin of the catalytic stability of $ZrO_2$-$SO_4$ and Pt/nanoZS3-600 nanoscales. Based on the activity and catalytic selectivity data, the conversion of LDPE using nano $ZrO_2$-$SO_4$ produced the highest amount of coke. In addition, this material had a high initial activity but low resistance to the deactivation process as its catalytic properties decreased rapidly in its consequent cycle of use. The formation of a coke can causes the pores and the active sites of the catalyst to be closed and thus reduce activity [44], i.e., the deactivation process that occurred in the $ZrO_2$-$SO_4$ nanocatalyst would have been difficult to control. To restore the activity of nanoZS3-600, catalyst regeneration, i.e., coke removal, is indispensable, especially in large-scale industrial applications. Promisingly, the Pt/nanoZS3-600 catalyst showed good activity, selectivity, and stability even after repeated use.

## 5. Cr/ZrO$_2$-SO$_4$ Catalyst

The metals that are widely used for bifunctional catalysts are usually transition metals with have incomplete orbitals that function as Lewis acid sites. One of the said transition metals is chromium (Cr) [66]. Cr can be doped on a carrier to enhance the catalytic activity of the host [67,68]. The addition of Cr metal to sulfated zirconia can also increase the acidity of the catalyst as the metal would contribute to the presence of Lewis acid sites [69]. The presence of Cr metal on sulfated zirconia also affects increasing the surface area of the catalyst [70,71]. Figure 17 illustrates the reaction mechanism of sulfated zirconia impregnated with Cr metal when interacting with ammonia during an acidity test. Hauli et al. [71] stated that the addition of chromium metal to sulfated zirconia not only increases the surface area of the catalyst but also stabilizes pure zirconia at high calcination temperatures. Increased temperature can thus remove sulfate groups and damage the porosity of the structure, causing the catalyst to deactivate. The physical appearance of the ZrO$_2$-SO$_4$ and Cr/ZrO$_2$-SO$_4$ catalyst (Figure 18) shows solid particle color change after the impregnation of Cr metal to the ZrO$_2$-SO$_4$ catalyst. The gray solid particle is formed due to the presence of Cr metal.

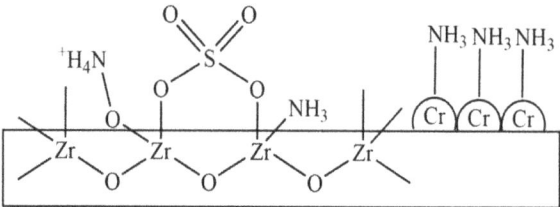

**Figure 17.** Proposed model of ammonia interaction on metal-embedded sulfated zirconia. Reprinted from Ref. [68]. Copyright Elsevier.

**Figure 18.** ZrO$_2$-SO$_4$ (**left**) and Cr/ZrO$_2$-SO$_4$ (**right**) catalyst. Reprinted with permisission from Dr. Hauli, Ref. [71].

### 5.1. FITR and Acidity Characterizations of Cr/ZrO$_2$-SO$_4$ Catalyst

Research on Cr/ZrO$_2$-SO$_4$ catalyst has been reported by Hauli et al. [72]. FTIR results of ZrO$_2$, SZ 0.8-600 (SZ), 0.5% Cr/SZ (Cr1/SZ), 1.0% Cr/SZ (Cr2/SZ), and 1.5% Cr/SZ (Cr3/SZ) are shown in Figure 19. The spectra showed no specific differences between the spectra of Cr-embedded zirconia catalyst and sulfated zirconia. However, the specific absorption peak of the sulfate group at 1053–1224 cm$^{-1}$, decreased with the presence of Cr. This is because the heating process involved in the loading of the metal had allowed sulfates to be released from the ZrO$_2$ surface. The presence of metal produced not sulfate decomposition but rather the release of the sulfate groups from the surface of ZrO$_2$ during the heating process [72]. Hanifah et al. [38] reported similar phenomena, namely that the presence of monometals and bimetals on ZrO$_2$-SO$_4$ release of sulfate group materialized whilst its decomposition did not.

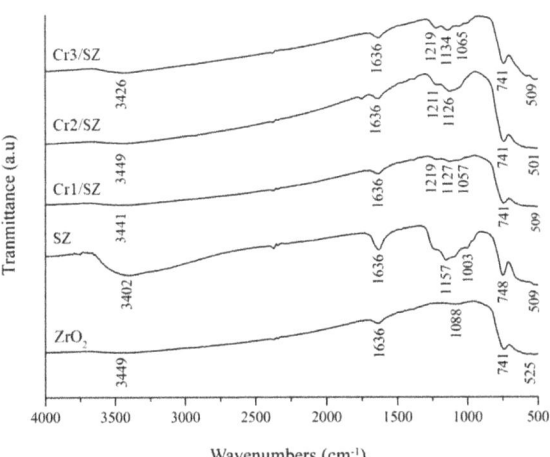

**Figure 19.** FTIR spectra of $ZrO_2$, SZ, Cr1/SZ, Cr2/SZ, Cr3/SZ. Reprinted with permission from Dr. Hauli, Ref. [72]. Copyright 2019 Trans Tech Publication.

The presence of empty orbitals in Cr metal allows electrons from other atoms to take their place in the orbital in what would then contribute to higher acidity of the Cr-containing material. Acidity test results of Cr-containing catalysts are presented in Table 8 confirming this. The acidity of the catalyst increased after the addition of Cr metal. The Cr1/SZ catalyst was the catalyst with the highest acidity value with 8.22 mmol/g. The low acidity value at high Cr-metal concentration was caused by the presence of metal particle aggregates that had covered the active metal sites on the carrier material. The more Cr metal that was embedded, the greater the amount of metal not accommodated in the pores of the carrier material, leading to the formation of aggregates.

**Table 8.** Acidity values of the catalysts [72].

| Sample | Acidity (mmol/g) |
| --- | --- |
| SZ | 3.81 |
| Cr1/SZ | 6.24 |
| Cr2/SZ | 8.22 |
| Cr3/SZ | 6.75 |

The acidity test of the catalyst was carried out to determine the total acidity value of each catalyst. The FTIR spectra after the acidity test results for the $ZrO_2$, SZ, Cr1/SZ, Cr2/SZ, and Cr3/SZ catalysts are shown in Figure 20. The results show the same absorption peaks at wave numbers 1119 and 1404 cm$^{-1}$, confirming the presence of the Lewis and Brønsted acid sites on each catalyst. Cr-embedded sulfated zirconia catalyst showed a higher intensity of Lewis and Brønsted acid sites than before Cr metal was added. The spectra of the Cr/SZ catalyst had the highest acid site intensity, indicating that Cr2/SZ had a high acidity value. The results of the acidity test of the catalysts confirmed that Cr2/SZ had the highest acidity value of 8.22 mmol/g.

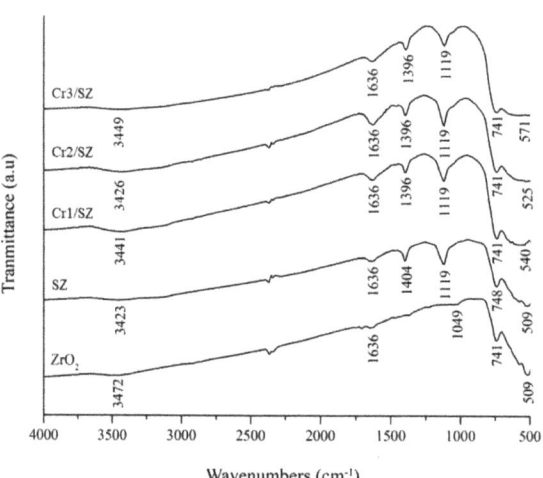

**Figure 20.** FTIR spectra of SZ, Cr1/SZ, Cr2/SZ, and Cr3/SZ after acidity test. Reprinted with permission from Dr. Hauli, Ref. [72]. Copyright 2019 Trans Tech Publication.

### 5.2. XRD Characterization of Cr/ZrO$_2$-SO$_4$ Catalyst

The diffraction patterns of ZrO$_2$, SZ, Cr1/SZ, Cr2/SZ, and Cr3/SZ catalysts are presented in Figure 21. The bearing of Cr metal on sulfated zirconia provided no change in the crystal phase structure of ZrO$_2$. The main peaks were found in the 2θ region around 28°, 31°, and 50° denoting plane distances of d$_{-111}$ (3,2 Å), d$_{111}$ (2,8 Å), d$_{220}$ (1,8 Å) with the highest intensity in the 28° region. Sulfate and Cr impregnation on ZrO$_2$ have a lower intensity peak than ZrO$_2$. This is due to the presence of sulfate on ZrO$_2$ and to Cr metal covering the surface of SZ, inhibiting the growth of ZrO$_2$ crystals and thereby decreasing the crystallinity of the materials [73,74].

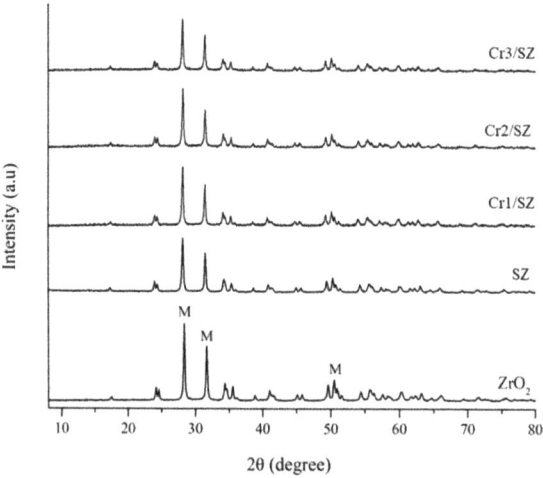

**Figure 21.** Diffraction patterns of ZrO$_2$, SZ, Cr1/SZ, Cr2/SZ, Cr3/SZ. Reprinted with permission from Dr. Hauli, Ref. [72]. Copyright 2019 Trans Tech Publication.

*5.3. SAA Characterization of Cr/ZrO$_2$-SO$_4$ Catalyst*

Characterization of pore characteristics including surface area, pore diameter, and pore volume was carried out by SAA analysis. The results of the SAA measurements of the catalysts are shown in Table 9. As can be seen, the addition of Cr metal to sulfated zirconia increased the surface area, pore diameter, and pore volume of the catalyst. This can be attributed to the uniform distribution of Cr metal on the catalyst surface [69,70]. The surface area of the catalyst, however, decreased upon the addition of higher concentrations of Cr metal, namely at Cr3/ZS, due to the entry of Cr metal into the catalyst pores causing agglomeration of metal atoms that covered the catalyst pores [74].

Table 9. Pore characteristics of the catalysts [72].

| Sample | Surface Area (m$^2$/g) | Pore Diameter (nm) | Pore Volume (cm$^3$/g) |
| --- | --- | --- | --- |
| ZrO$_2$ | 12.27 | 3.70 | 0.07 |
| SZ | 7.79 | 3.68 | 0.08 |
| Cr1/SZ | 12.62 | 3.70 | 0.12 |
| Cr2/SZ | 14.56 | 3.70 | 0.08 |
| Cr3/SZ | 11.91 | 9.04 | 0.04 |

The adsorption and desorption isotherm patterns for ZrO$_2$, SZ, Cr1/SZ, Cr2/SZ, and Cr3/SZ catalysts are shown in Figure 22. Based on the IUPAC classification, all catalysts showed a type IV isotherm pattern, which is characteristic of the isotherm pattern for mesoporous materials with pore diameter sizes of 2–50 nm. The adsorption–desorption isotherm patterns of the Cr1/SZ and Cr2/SZ catalysts demonstrated monolayer absorption of nitrogen gas on the surface when $P/P_0 < 0.2$. At a relative pressure of $0.2 < P/P_0 < 1$, the isotherm curve experienced a sharp increase in volume representing a multilayer arrangement. In the Cr3/SZ catalyst, the absorption of a monolayer of nitrogen gas on the surface occurred when $P/P_0 < 0.4$ and experienced a sharp increase in volume at $0.4 < P/P_0 < 1$. The ZrO$_2$, Cr2/SZ, and Cr3/SZ catalyst had a type H4 hysteresis, while SZ and Cr1/SZ had type H3. Type H3 hysteresis showed no absorption limit at high $P/P_0$. Type H4 hysteresis is associated with narrow slit pores [4,72]. These three types of catalysts exhibit the characteristics of porous materials. This form of porous material is composed of aggregates of particles such as plates that form pore gaps [10].

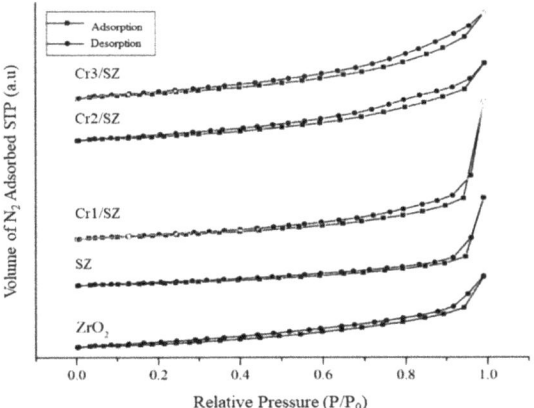

Figure 22. Adsorption desorption isotherm patterns of ZrO$_2$, SZ, Cr1/SZ, Cr2/SZ, and Cr3/SZ. Reprinted with permission from Dr. Hauli, Ref. [72]. Copyright 2019 Trans Tech Publication.

## 5.4. Activity and Selectivity Tests of Cr/ZrO$_2$-SO$_4$ Catalyst

A catalytic activity test of the Cr/ZrO$_2$-SO$_4$ catalyst in the hydrocracking reaction of LDPE into liquid fuel has been carried out by Hauli et al. [72]. The hydrocracking reaction was carried out at a temperature of 250 °C using ZrO$_2$, SZ, Cr1/SZ, Cr2/SZ, and Cr3/SZ catalysts. The percentages of conversion yields of the hydrocracking of each catalyst are presented in Table 10. The presence of sulfate and Cr metal in ZrO$_2$ can increase the yield of liquid products from the hydrocracking reaction. The liquid product increased along with the increase in the amount of Cr metal deposited on ZrO$_2$. This can be attributed to the role of acid sites in the catalyst (from sulfate and metal Cr) which contribute to the active sites of the catalyst, thereby increasing its catalytic activity [50].

**Table 10.** Distribution of LDPE plastic hydrocrack products on various catalysts [72].

| Catalyst | Product Conversion (% b/b) | | |
|---|---|---|---|
| | Liquid | Coke | Gas |
| ZrO$_2$ | 17.39 | 0.36 | 32.23 |
| SZ | 28.72 | 0.34 | 29.51 |
| Cr1/SZ | 33.48 | 0.01 | 25.78 |
| Cr2/SZ | 40.99 | 0.01 | 24.84 |
| Cr3/SZ | 37.51 | 0.01 | 26.77 |

The highest conversion of liquid product was produced using the Cr2/SZ catalyst, which was 40.99%, with a lesser amount of 37.51% with the use of the Cr3/SZ catalyst. Moreover, the Cr2/SZ catalyst had a higher acidity value than Cr3/SZ. In addition, the surface area of the catalyst also affects the catalytic activity of the catalyst in the hydrocracking of LDPE plastics [75,76]. Accordingly, the Cr2/SZ had a larger surface area, resulting in greater conversion. The modification of Cr on sulfated zirconia demonstrated a reduction in coke formation, thereby increasing the feed interaction on the active sites of the catalysts.

The selectivity of the catalysts towards liquid products in the hydrocracking reaction are shown in Figure 23. All the catalysts showed higher selectivity in the gasoline fraction than the diesel fraction. The presence of sulfate and Cr in sulfated zirconia increased the selectivity towards the gasoline fraction (C$_5$–C$_{12}$) and decreased the selectivity towards the diesel fraction (C$_{13}$–C$_{20}$), as expected. The highest selectivity of the gasoline fraction was obtained from the use of the Cr2/SZ catalyst, which was at 93.42%.

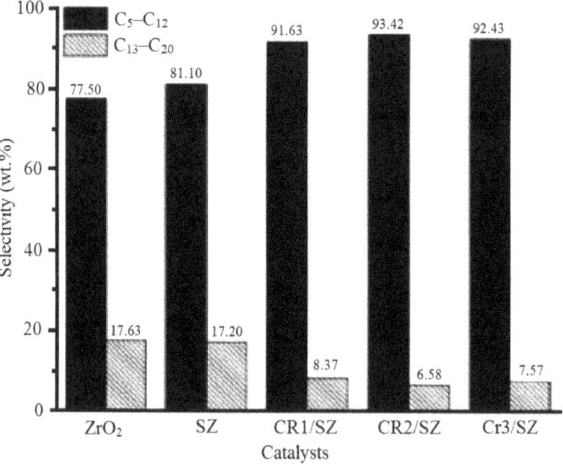

**Figure 23.** Liquid product selectivity of ZrO$_2$, SZ, Cr1/SZ, Cr2/SZ, and Cr3/SZ catalysts. Reprinted with permission from Dr. Hauli, Ref. [72]. Copyright 2019 Trans Tech Publication.

## 6. Ni/ZrO$_2$-SO$_4$ Catalyst

### 6.1. FITR Characterization of Ni/ZrO$_2$-SO$_4$ Catalyst

Research on the ZrO$_2$-SO$_4$ catalyst was also carried out about its application in the hydrocracking of used cooking oil into liquid fuel. Modification of ZrO$_2$-SO$_4$ with Ni metal has been reported by Aziz et al. [73]. Figure 24 show the FTIR spectra of Ni-SZ 1, Ni-SZ 2, and Ni-SZ 3. Impregnated Cr metal to SZ caused the presence of a new peak in the area of 1103 and 1141 cm$^{-1}$, confirming the S-O-S stretching of from SO$_4$ ion from SZ coordinated with Ni metal [76,77]. However, the sulfate spectra at 1002–1219 cm$^{-1}$ disappear due to the Ni metal coverage on the SZ surface and the higher calcination temperature.

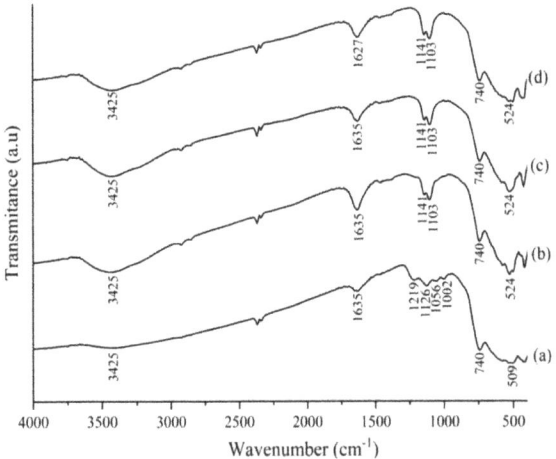

**Figure 24.** FTIR spectra of SZ (a), Ni-SZ 1 (b), Ni-SZ 2 (c), and Ni-SZ 3 (d) catalysts. Reprinted with permission from Aziz, Ref. [73]. Copyright 2020 Budapest University.

### 6.2. Acidity and SAA Analysis of Ni/ZrO$_2$-SO$_4$ Catalyst

The total acidity and pore characteristic of Ni-SZ catalysts is shown in Table 11. It can be seen that the sulfation of ZrO$_2$ (SZ catalyst) and Ni metal impregnated to SZ catalysts successfully increase the total acidity of the catalyst. Ni metal has a vacant p orbital that will accepting electron pair and acts as a Lewis acid site [78]. The Ni-SZ 3 catalyst was the catalyst with the highest total acidity of 4.235 mmol/g.

**Table 11.** The acidity and pore characteristics of the Ni-SZ catalysts [73].

| Sample | Acidity | Surface Area (m$^2$/g) | Pore Diameter (nm) | Pore Volume (cm$^3$/g) |
| --- | --- | --- | --- | --- |
| ZrO$_2$ | 0.355 | 5.63 | 30.72 | 0.04 |
| SZ | 2.004 | 4.04 | 35.84 | 0.04 |
| Ni-SZ 1 | 3.905 | 7.09 | 3.82 | 0.06 |
| Ni-SZ 2 | 4.061 | 8.75 | 3.82 | 0.05 |
| Ni-SZ 3 | 4.235 | 11.68 | 3.82 | 0.05 |

The pore characteristic of the catalyst shows the increase in surface area after the impregnation of Cr metal onto the SZ catalyst due to the highly dispersion of Ni metal on the pore and surface of the SZ catalyst [79]. However, the presence of Ni causing the form of pore-blocking that significantly decreases the pore diameter of Ni-SZ catalysts [72].

The selectivity of the Ni-ZrO$_2$-SO$_4$ catalyst is shown in Table 12. ZrO$_2$ and SZ catalysts were observed to have lower activity and selectivity than Ni-SZ due to the lowest acidity and surface area that will produce considerable amounts of coke (block the active site) and decreases the amount of liquid product [80,81]. The addition of Ni metal (Ni-SZ 3 catalyst) increased the acidity (4.23 mmol/g) and surface area (11.68 m$^2$/g) of ZrO$_2$, thereby increasing its activity and selectivity in the hydrocracking process [82,83]. The largest amount of liquid product produced was gasoline with the highest selectivity was produced by the Ni-SZ 3 catalyst with the diesel fraction (C$_5$–C$_{12}$) reached 100%.

Table 12. Selectivity of ZrO$_2$, SZ, Ni-SZ 1, Ni-SZ 2, and Ni-SZ 3 catalysts in the hydrocracking reaction of used cooking oil into liquid fuel [73].

| Catalyst | Selectivity (wt%) | | | |
|---|---|---|---|---|
| | (C$_1$–C$_4$) | (C$_5$–C$_{12}$) | (C$_{13}$–C$_{20}$) | Non-Hydrocarbon |
| ZrO$_2$ | 1.86 | 77.31 | 19.26 | 1.57 |
| SZ | 1.93 | 67.83 | 28.15 | 2.54 |
| Ni-SZ 1 | 0.00 | 51.59 | 15.16 | 33.25 |
| Ni-SZ 2 | 0.00 | 87.01 | 12.99 | 0.00 |
| Ni-SZ 3 | 0.00 | 100.00 | 0.00 | 0.00 |

## 7. Conclusions

Zirconia and its modified heterogeneous catalyst forms hold great potential in hydrocracking reaction applications to convert LDPE waste into liquid fuels with their excellent activity, selectivity, and stability. The sulfation process on ZrO$_2$ with various concentrations of sulfuric acid and calcination temperatures succeeded in increasing the acidity of ZrO$_2$. The ZrO$_2$-SO$_4$ catalyst treated with Platinum (Pt) and Chrome (Cr) transition metals had significantly increased acidity. Characterization analyses confirmed that Pt and Cr metals had been successfully impregnated on the SZ surface. The percentage of liquid fraction obtained with the use of the Pt/ZrO$_2$-SO$_4$ catalyst proofed better activity, selectivity, and stability than Cr/ZrO$_2$-SO$_4$ and ZrO$_2$-SO$_4$. The optimum amount of liquid fraction produced from the hydrocracking reaction LDPE plastic waste with the Pt/ZrO$_2$-SO$_4$ catalyst was 67.515%, while that from the Cr/ZrO$_2$-SO$_4$ catalyst was 40.99%. The development of Ni on the ZrO$_2$-SO$_4$ catalyst also demonstrated an increase in selectivity in the hydrocracking reaction of used cooking oil into liquid fuel. The selectivity of the Ni-SZ catalyst was found to be better than that of SZ. The percentages of gasoline fractions produced were 100%.

## 8. Future Suggestion

Zirconia-based nanocatalysts have bright prospects for application in various industrial areas such as petroleum cracking, biofuel synthesis, pharmaceutical, and the synthesis of various organic materials. The potential application of sulfated nanozirconia catalysts is shown in the Figure 25. Due to characteristics such as being non-toxic and easy to regenerate, as well as having a large surface area and high thermal and structural resistance, nanozirconia catalysts also have the potential to be used in the pharmaceutical industry. Preliminary studies in our laboratory have shown that this catalyst has the potential to be used as a solid acid catalyst in the synthesis of nitrobenzene from benzene.

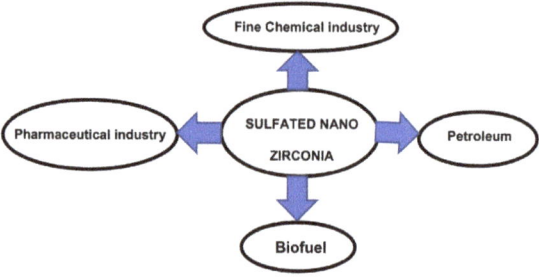

Figure 25. Potential application of sulfated nanozirconia catalysts.

**Author Contributions:** Conceptualization, R.A.P., S.J.S., M.U., L.H. and A.K.A.; validation, R.A.P., M.U., L.H. and A.K.A.; investigation, R.A.P.; resources, R.A.P.; data curation, R.A.P., M.U., L.H. and A.K.A.; writing—original draft preparation, R.A.P. and S.J.S.; writing—review and editing, R.A.P.; visualization, K.W.; supervision, R.A.P. and K.W.; project administration, K.W.; funding acquisition, K.W. All authors have read and agreed to the published version of the manuscript.

**Funding:** This research was funded by Postdoctoral Research Grant Universitas Gadjah Mada (Contract Number: 6144/UN1/DITLIT/DIT-LIT/PT/2021).

**Conflicts of Interest:** The authors declare no conflict of interest.

## References

1. Li, D.; Feng, W.; Chen, C.; Chen, S.; Fan, G.; Liao, S.; Wu, G.; Wang, Z. Transesterification of *Litsea cubeba* kernel oil to biodiesel over zinc supported on zirconia heterogeneous catalysts. *Renew. Energy* **2021**, *177*, 13–22. [CrossRef]
2. Hanafi, M.F.; Sapawe, N. Reusability study of zirconia catalyst toward photocatalytic degradation of Remazol brilliant blue dye. *Mater. Today Proc.* **2020**, *31*, 266–2688. [CrossRef]
3. Amin, A.K.; Trisunaryanti, W.; Wijaya, K. Effect of promoters and calcination temperature on surface and acidity of modified zirconia. *J. Nano Res.* **2018**, *57*, 31–39. [CrossRef]
4. Hauli, L.; Wijaya, K.; Armunanto, R. Preparation and characterization of sulfated zirconia from a commercial zirconia nanopowder. *Orient. J. Chem.* **2018**, *34*, 1559–1564. [CrossRef]
5. Yan, C.; Zheng, S.; Chen, N.; Yuan, S.; Chen, Y.; Li, B.; Zhang, Y. Sulfated zirconia catalysts supported on mesoporous Mg-SBA-15 with different morphologies for highly efficient conversion of fructose to 5-hydroxymethylfurfural. *Micropor. Mesopor. Mater.* **2021**, *328*, 111507. [CrossRef]
6. Utami, M.; Wijaya, K.; Trisunaryanti, W. Pt-promoted Sulfated Zirconia as Catalyst for Hydrocracking of LDPE Plastic Waste into Liquid Fuels. *Mater. Chem. Phys.* **2018**, *213*, 548–555. [CrossRef]
7. Yan, G.X.; Wang, A.; Wachs, I.E.; Baltrusaitis, J. Critical review on the active site structure of sulfated zirconia catalysts and prospects in fuel production. *Appl. Catal. A Gen.* **2019**, *572*, 210–225. [CrossRef]
8. Wijaya, K.; Kurniawan, M.A.; Saputri, W.D.; Trisunaryanti, W.; Mirzan, M.; Hariani, P.L.; Tikoalu, A.D. Synthesis of nickel catalyst supported on $ZrO_2/SO_4$ pillared bentonite and its application for conversion of coconut oil into gasoline via hydrocracking process. *J. Environ. Chem. Eng.* **2021**, *9*, 105399. [CrossRef]
9. Utami, M.; Trisunaryanti, W.; Shida, K.; Tsushida, M.; Kawakita, H.; Ohto, K.; Wijaya, K.; Tominaga, M. Hydrothermal preparation of a platinum-loaded sulphated nanozirconia catalyst for the effective conversion of waste low density polyethylene into gasoline-range hydrocarbons. *RSC Adv.* **2019**, *9*, 41392–41401. [CrossRef]
10. Hauli, L.; Wijaya, K.; Syoufian, A. Fuel production from LDPE-based plastic waste over chromium supported on sulfated zirconia. *Indones. J. Chem.* **2019**, *20*, 422–429. [CrossRef]
11. Amin, A.K.; Trisunaryanti, W.; Wijaya, K. The Catalytic Performance of $ZrO_2$-$SO_4$ and $Ni/ZrO_2$-$SO_4$ Prepared from Commercial $ZrO_2$ in Hydrocracking of LDPE Plastic Waste into Liquid Fuels. *Orient. J. Chem.* **2018**, *34*, 3070–3078. [CrossRef]
12. Treccani, L.; Klein, T.Y.; Meder, F.; Pardun, K.; Rezwan, K. Functionalized ceramics for biomedical, biotechnological and environmental applications. *Acta Biomater.* **2013**, *9*, 7115–7150. [CrossRef] [PubMed]
13. Skovgaard, M.; Gudik-Sorensen, M.; Almdal, K.; Ahniyas, A. Nanoporous zirconia microspheres prepared by salt-assisted spray drying. *SN Appl. Sci.* **2020**, *2*, 784. [CrossRef]
14. Utami, M. Platinum/Sulfated Nanozirconia: Preparation, Characterization, and Their Application in Conversion of LDPE Plastic Waste into Liquid Fuels. Ph.D. Thesis, Department of Chemistry, Universitas Gadjah Mada, Yogyakarta, Indonesia, 2019.
15. Le Ore, M.S.; Wijaya, K.; Trisunaryanti, W.; Saputri, W.D.; Heraldy, E.; Yuwana, M.W.; Hariani, P.L.; Budiman, A.; Sudiono, S. The Synthesis of $SO_4/ZrO_2$ and Zr/CaO catalysts via hydrothermal treatment and their application for conversion of low-grade coconut oil into biodiesel. *J. Environ. Chem. Eng.* **2020**, *8*, 104205. [CrossRef]
16. Burger, W.; Kiefer, G. Alumina, zirconia and their composite ceramics with properties tailored for medical applications. *J. Compos. Sci.* **2021**, *5*, 306. [CrossRef]
17. Siddiqui, M.R.H.; Al-Wassil, A.I.; Alotaibi, A.M.; Mahfouz, R. Effects of precursor on the morphology and size of $ZrO_2$ nanoparticles, synthesized by sol-gel method in non-aqueous medium. *Mater. Res.* **2012**, *15*, 986–989. [CrossRef]
18. Halter, W.; Eisele, R.; Rothenstein, D.; Bill, J.; Allgower, F. Moment dynamics of zirconia particle formation foe optimizing particle size distribution. *Nanomaterials* **2019**, *9*, 333. [CrossRef]
19. Maleki, F.; Pacchioni, G. Characterization of acid and basic sites on zirconia surfaces and nanoparticles by adsorbed probe molecules: A theoretical study. *Top. Catal.* **2020**, *63*, 1717–1730. [CrossRef]
20. Glorius, M.; Markovits, M.A.C.; Breitkopf, C. Design of specific acid-base-properties in $CeO_2$-$ZrO_2$-mixed oxides via templating and Au modification. *Catalysts* **2018**, *8*, 358. [CrossRef]
21. De Souza, E.F.; Appel, L.G. Oxygen vacancy formation and their role in the $CO_2$ activation on Ca doped $ZrO_2$ surface: An an-initio DFT study. *App. Surf. Sci.* **2021**, *553*, 149589. [CrossRef]
22. Song, L.; Cao, X.; Li, L. Engineering stable surface oxygen vacancies on $ZrO_2$ by hydrogen-etching technology: An efficient support of gold catalysts for water-gas shift reaction. *ACS Appl. Mater. Interfaces* **2018**, *109*, 31249–31259. [CrossRef] [PubMed]

23. Gautam, C.; Joyner, J.; Gautam, A.; Rao, J.; Vajtai, R. Zirconia Based Dental Ceramics: Structure, Mechanical Properties, Biocompatibility and Applications. *Dalton Trans.* **2016**, *45*, 19194–19215. [CrossRef] [PubMed]
24. Hauli, L.; Wijaya, K.; Syoufian, A. Hydrocracking of LDPE plastic waste into liquid fuel over sulfated zirconia from a commercial zirconia nanopowder. *Orient. J. Chem.* **2019**, *35*, 128–133. [CrossRef]
25. El-Desouki, D.S.; Ibrahim, A.H.; Abdelazim, S.M.; Aboul-Gheit, N.A.K.; Abdel-Hafizar, D.R. The optimum condition for methanol conversion to dimethyl ether over modified sulfated zirconia catalysts prepared by different methods. *J. Fuel Chem. Technol.* **2021**, *49*, 63–71. [CrossRef]
26. Yu, S.; Wu, S.; Li, L.; Ge, X. Upgrading bio-oil from waste cooking oil by esterification using $SO_4^{2-}/ZrO_2$ as catalyst. *Fuel* **2020**, *276*, 118019. [CrossRef]
27. Wang, P.; Yue, Y.; Wang, T.; Bao, X. Alkane isomerization over sulfated zirconia solid acid system. *Int. J. Energ. Res.* **2020**, *44*, 3270–3294. [CrossRef]
28. Rabee, A.I.M.; Mekhemer, G.A.H.; Osatiashtiani, A.; Isaacs, M.A.; Lee, A.F.; Wilson, K.; Zaki, M.I. Acidity-reactivity relationships in catalytic esterification over ammonium sulfate-derived sulfated zirconia. *Catalysts* **2017**, *7*, 204. [CrossRef]
29. Wang, P.; Zhang, J.; Han, C.; Yang, C.; Li, C. Effect of Modification Methods on the Surface Properties and *n*-Butane Isomerization Performance of La/Ni-Promoted $SO_4^{2-}/ZrO_2$-$Al_2O_3$. *Appl. Surf. Sci.* **2016**, *378*, 489–495. [CrossRef]
30. Saravanan, K.; Tyagi, B.; Bajaj, H.C. Esterifcation of caprylic acid with alcohol over nano-crystalline sulfated zirconia. *J. Sol-Gel Sci. Technol.* **2012**, *62*, 13–17. [CrossRef]
31. Utami, M.; Wijaya, K.; Trisunaryanti, W. Effect of Sulfuric Acid Treatment and Calcination on Commercial Zirconia Nanopowder. *Key Eng. Mater.* **2017**, *757*, 131–137. [CrossRef]
32. Yang, X.; Ma, X.; Yu, X.; Ge, M. Exploration of strong metal-support interaction in zirconia supported catalysts for toluene oxidation. *App. Catal. B Environ.* **2020**, *263*, 118355. [CrossRef]
33. Mirzan, M.; Syoufian, A.; Wijaya, K. Physico-chemical properties of nano $ZrO_2$-pillared bentonite with nickel as supporting metal. *Nano Hybrids Compos.* **2020**, *30*, 9–17. [CrossRef]
34. Loveless, B.T.; Gyanani, A.; Muggli, D.S. Discrepancy between TPD- and FTIR- based measurements of Brønsted and Lewis acidity for sulfated zirconia. *Appl. Catal. B Environ.* **2008**, *84*, 591–597. [CrossRef]
35. Li, L.; Yan, B.; Li, H.; Yu, S.; Liu, S.; Yu, H.; Ge, X. $SO_4^{2-}/ZrO_2$ as catalyst for upgrading of pyrolysis oil by esterification. *Fuel* **2018**, *226*, 190–194. [CrossRef]
36. Pratika, R.A.; Wijaya, K.; Trisunaryanti, W. Hydrothermal treatment of $SO_4/TiO_2$ and $TiO_2/CaO$ as heterogeneous catalysts for the conversion of Jatropha oil into biodiesel. *J. Environ. Chem. Eng.* **2021**, *9*, 106547. [CrossRef]
37. Utami, M.; Safitri, R.; Pradipta, M.F.; Wijaya, K.; Chang, S.W.; Ravindran, B.; Ovi, D.; Rajabathar, J.R.; Poudineh, N.; Gengan, R.M. Enhanced catalytic conversion of palm oil into biofuels by Cr-incorporated sulphated zirconia. *Mater. Lett.* **2022**, *309*, 131472. [CrossRef]
38. Hanifah, A.; Nadia, A.; Saputri, W.D.; Syoufian, A.; Wijaya, K. Performance of Ni-Mo sulfated nanozirconia catalyst for conversion of waste cooking oil into biofuel via hydrocracking process. *Mater. Sci. Forum* **2021**, *1045*, 79–89. [CrossRef]
39. Sari, E.P.; Wijaya, K.; Trisunaryanti, W.; Syoufian, A.; Hasanudin, H.; Saputri, W.D. The effective combination of zirconia superacid and zirconia-imregnated CaO in biodiesel manufacturing: Utilization of used coconut cooking oil (UCCO). *Int. J. Energ. Environ. Eng.* **2021**, 1–12. [CrossRef]
40. Wijaya, K.; Nadia, A.; Dinana, A.; Pratiwi, A.F.; Tikoalu, A.D.; Wibowo, A.C. Catalytic hydrocracking of fresh and waste frying oil over Ni- and Mo-based catalysts supported on sulfated silica for biogasoline production. *Catalysts* **2021**, *11*, 1150. [CrossRef]
41. Fernandez-Morales, J.M.; Castillejos, E.; Asedegbega-Nieto, E.; Dongil, A.B.; Rodrigues-Ramos, I.; Guerrero-Ruiz, A. Comparative study of different acidic surface structures in solid catalysts applied for the isobutene dimerization reaction. *Nanomaterial* **2020**, *10*, 1235. [CrossRef]
42. Purba, S.E.; Wijaya, K.; Trisunaryanti, W.; Pratika, R.A. Dealuminated and desilicated natural zeolite as a catalyst for hydrocracking of used cooking oil into biogasoline. *Mediterr. J. Chem.* **2020**, *11*, 74–83. [CrossRef]
43. Wijaya, K.; Malau, M.L.L.; Utami, M.; Mulijani, S.; Patah, A.; Wibowo, A.C.; Chandrasekaran, M.; Rajabathar, J.R.; Al-Lohedan, H. Synthesis, characterizations and catalysis of sulfated silica and nickel modified silica catalysts for diethyl ether (DEE) production from ethanol towards renewable energy applications. *Catalysts* **2021**, *11*, 1511. [CrossRef]
44. Serrano, D.P.; Garcia, R.A.; Linares, M.; Gil, B. Influence of the calcination treatment on the catalytic properties of hierarchical ZSM-5. *Catal. Today* **2012**, *179*, 91–101. [CrossRef]
45. Swamidoss, C.M.A.; Sheraz, M.; Anus, A.; Jeong, S.; Park, Y.-K.; Kim, Y.-M.; Kim, S. Effect of $Mg/Al_2O_3$ and calcination temperature on the catalytic decomposition of HFC-134a. *Catalysts* **2019**, *9*, 270. [CrossRef]
46. Abedin, M.A.; Kanitkar, S.; Bhattar, S.; Spivey, J.J. Methane dehydroaromatization using Mo supported on sulfated zirconia catalyst: Effect of promoters. *Catal. Today* **2021**, *365*, 71–79. [CrossRef]
47. Susi, E.P.; Wijaya, K.; Wangsa; Pratika, R.A.; Hariani, P.L. Effect of nickel concentration in natural zeolite as catalyst in hydrocracking process of used cooking oil. *Asian J. Chem.* **2020**, *32*, 2773–2777. [CrossRef]
48. Yin, P.; Hu, S.; Qian, K.; Wei, Z.; Zhang, L.-L.; Lin, Y.; Huang, W.; Xiong, H.; Li, W.X.; Liang, H.-W. Quantification of critical particle distance for mitigating catalyst sintering. *Nat. Commun.* **2021**, *12*, 4865. [CrossRef]
49. Prašnikar, A.; Pavlišič, A.; Ruiz-Zepeda, F.; Kovač, J.; Likozar, B. Mechanism of Copper-based catalyst deactivation during $CO_2$ reduction to methanol. *Ind. Eng. Chem. Res.* **2019**, *58*, 13021–13029. [CrossRef]

50. Rahmati, M.; Safdari, M.S.; Fletcher, T.H.; Argyle, M.D.; Bartholomew, C.H. Chemical and thermal sintering of supported metals with emphasis on cobalt catalysts during Fischer–Tropsch synthesis. *Chem. Rev.* **2020**, *120*, 4455–4533. [CrossRef]
51. Zhou, S.; Song, Y.; Zhao, J.; Zhou, X.; Chen, L. Study on the Mechanism of Water Poisoning Pt-Promoted Sulfated Zirconia Alumina in n-Hexane Isomerization. *Energy Fuels* **2021**, *35*, 14860–14867. [CrossRef]
52. Peng, S.; Li, M.; Yang, X.; Li, P.; Liu, H.; Xiong, W.; Peng, X. Atomic layer deposition of Pt nanoparticles on $ZrO_2$ based metal-organic frameworks for increased photocatalytic activity. *Ceram. Int.* **2019**, *45*, 18128–18134. [CrossRef]
53. Bikmetova, L.I.; Kazantsev, K.V.; Zatolokina, L.V.; Smolikov, M.D.; Belyi, A.S. A study on the synthesis steps of alumina-supported $Pt/SO_4/ZrO_2$ catalysts for isomerization of n-hexane. *AIP Conf. Proc.* **2020**, *2285*, 020003. [CrossRef]
54. Souza, I.C.A.; Manfro, R.L.; Souza, M.M.V.M. Hydrogen production from steam reforming of acetic acid over Pt-Ni bimetallic catalysts supported on $ZrO_2$. *Biomass Bioenergy* **2022**, *156*, 106317. [CrossRef]
55. Ullah, I.; Taha, T.A.; Alenad, A.M.; Uddin, I.; Hayat, A.; Hayat, A.; Sohail, M.; Irfan, A.; Khan, J.; Palamanit, A. Platinum-alumina modified $SO_4^{2-}$-$ZrO_2/Al_2O_3$ based bifunctional catalyst for significantly improved n-butane isomerization performance. *Surf. Interfaces* **2021**, *25*, 101227. [CrossRef]
56. Tamizhdurai, P.; Lavanya, M.; Meenakshisundaram, A.; Shanti, K.; Sivasanker, S. Isomerization of alaknes over Pt-sulphated zirconia supported on SBA-15. *Adv. Porous Mater.* **2017**, *5*, 169–174. [CrossRef]
57. Cho, H.J.; Kim, D.; Li, S.; Su, D.; Ma, D.; Xu, B. Molecular-level proximity of metal and acid sites in zeolite-encapsulated Pt nanoparticles for selective multistep tandem catalysis. *ACS Catal.* **2020**, *10*, 3340–3348. [CrossRef]
58. Gan, T.; Yang, J.; Morris, D.; Chu, X.; Zhang, P.; Zhang, W.; Zuo, Y.; Yan, W.; Wei, S.-H.; Liu, G. Electron donation of non-oxide supports boosts $O_2$ activation on nano-platinum catalysts. *Nat. Commun.* **2021**, *12*, 2741. [CrossRef] [PubMed]
59. Aboul-Gheit, A.K.; Gad, F.K.; Abdel-Aleem, G.M.; El-Desouki, D.S.; Abdel Hamid, S.M.; Ghoneim, S.S.; Ibrahim, A.H. Pt, Re and Pt-Re incorporation in sulfated zirconia as catalysts for n-pentane isomerization. *Egypt. J. Pet.* **2014**, *23*, 303–314. [CrossRef]
60. Zhao, P.-P.; Chen, J.; Yu, H.-B.; Cen, B.-H.; Wang, W.-Y.; Luo, M.-F.; Lu, J.-Q. Insight into propane combustion over $MoO_3$ promoted $Pt/ZrO_2$ catalysts: The generation of $Pt$-$MoO_3$ interface and its promotional role on catalytic activity. *J. Catal.* **2020**, *391*, 80–90. [CrossRef]
61. Kondratowicz, T.; Drozdek, M.; Michalik, M.; Gac, W.; Gajewska, M.; Kustroski, P. Catalytic activity of Pt species variously dispersed on hollow $ZrO_2$ spheres in combustion of volatile organic compounds. *App. Surf. Sci.* **2020**, *513*, 145788. [CrossRef]
62. Verga, L.G.; Rusell, A.E.; Skylaris, C.K. Ethanol, O, and CO adsorption on Pt nanoparticles: Effect of nanoparticle size and graphene support. *Phys. Chem. Chem. Phys.* **2018**, *20*, 25918–25930. [CrossRef] [PubMed]
63. Wang, S.; Pu, J.; Wu, J.; Liu, H.; Xu, H.; Li, X.; Wang, H. $SO_4^{2-}/ZrO_2$ as a solid acid for the esterification of palmitic acid with methanol: Effects of the calcination time and recycle method. *ACS Omega* **2020**, *5*, 30139–30147. [CrossRef] [PubMed]
64. Ren, K.; Kong, D.; Meng, X.; Wang, X.; Shi, L.; Liu, N. The effect of ammonium sulfate and sulfamic acid on the surface acidity of sulfated zirconia. *J. Saudi Chem. Soc.* **2019**, *23*, 198–204. [CrossRef]
65. Vance, B.C.; Kots, P.A.; Wang, C.; Hinton, Z.R.; Quinn, C.M.; Epps, T.H., III; Korley, L.T.J.; Vlachos, D.G. Single pot catalyst strategy to branched products via adhesive isomerization and hydrocracking of polyethylene over platinum tungstated zirconia. *App. Catal. B. Environ.* **2021**, *299*, 120483. [CrossRef]
66. Liu, S.; Lots, P.; Vance, B.C.; Danielson, A.; Vlachos, D.G. Plastic waste to fuels by hydrocracking at mild conditions. *Sci. Adv.* **2021**, *7*, eabf8283. [CrossRef] [PubMed]
67. Chen, J.; Zhang, Y.; Chen, X.; Dai, S.; Bao, Z.; Yang, Q.; Ren, Q.; Zhang, Z. Cooperative interplay of Bronsted acid and Lewis acid sites in MIL-101 (Cr) for cross-dehydrogenative coupling of C–H. Bonds. *ACS Appl. Mater. Interfaces* **2021**, *13*, 10845–10854. [CrossRef] [PubMed]
68. Hu, Z.-P.; Wang, Z.; Yuan, Z.-Y. $Cr/Al_2O_3$ catalysts with strong metal-support interactions for stable catalytic dehydrogenation of propane to propylene. *Mol. Catal.* **2020**, *493*, 111052. [CrossRef]
69. Annuar, N.H.R.; Triwahyono, S.; Jalil, A.A.; Basar, N.; Abdullah, T.A.T.; Ahmad, A. Effect of $Cr_2O_3$ loading on the properties and cracking activity of $Pt/Cr_2O_3$-$ZrO_2$. *App. Catal. A Gen.* **2017**, *541*, 77–86. [CrossRef]
70. Wang, H.; Lin, N.; Xu, R.; Yu, Y.; Zhao, X. First principles studies of electronic, mechanical and optical properties of Cr-doped cubic $ZrO_2$. *Chem. Phys.* **2020**, *539*, 110972. [CrossRef]
71. Hauli, L. Chromium/Sulfated Nanozirconia: Preparation, Characterization, and Their Application in Conversion of LDPE Plastic Waste into Liquid Fuels. Ph.D. Thesis, Department of Chemistry, Universitas Gadjah Mada, Yogyakarta, Indonesia, 2019.
72. Hauli, L.; Wijaya, K.; Armunanto, R. Preparation of Cr metal supported on sulfated zirconia catalyst. *Mater. Sci. Forum* **2019**, *948*, 221–227. [CrossRef]
73. Aziz, I.T.A.; Saputri, W.D.; Trisunaryanti, W.; Sudiono, S.; Syoufian, A.; Budiman, A.; Wijaya, K. Synthesis of Nickel-loaded sulfated zirconia catalyst and its application for converting used palm cooking oil to gasoline via hydrocracking process. *Period. Polytech. Chem. Eng.* **2021**, *66*, 101–113. [CrossRef]
74. Tao, Y.; Zhu, Y.; Liu, C.; Yue, H.; Ji, J.; Yuan, S.; Jiang, W.; Liang, B. A highly selective $Cr/ZrO_2$ catalyst for the reverse water-gas shift reaction prepared from simulated Cr-containing wastewater by a photocatalytic deposition process with $ZrO_2$. *J. Environ. Chem. Eng.* **2018**, *6*, 6761–6770. [CrossRef]
75. Serrano, D.; Escola, J.; Briones, L.; Arroyo, M. Hydroprocessing of the LDPE thermal cracking oil into transportation fuels over Pd supported on hierarchical ZSM-5 catalyst. *Fuel* **2017**, *206*, 190–198. [CrossRef]

76. Wong, S.L.; Ngadi, N.; Abdullah, T.A.T.; Inuwa, I.M. Conversion of low-density polyethylene (LDPE) over ZSM-5 zeolite to liquid fuel. *Fuel* **2017**, *192*, 71–82. [CrossRef]
77. Lyu, Y.; Xu, R.; Williams, O.; Wang, Z.; Sievers, C. Reaction paths of methane activation and oxidation of surface intermediates over NiO on Ceria-Zirconia catalysts studied by In-situ FTIR spectroscopy. *J. Catal.* **2021**, *404*, 334–347. [CrossRef]
78. Dahdah, E.; Estephane, J.; Gennequin, C.; Aboukais, A.; Abi-Aad, E.; Aouad, S. Zirconia supported nickel catalysts for glycerol steam reforming: Effect of zirconia structure on the catalytic performance. *Int. J. Hydrogen Energy* **2020**, *45*, 4457–4467. [CrossRef]
79. Kou, J.; Yi, L.; Li, G.; Cheng, K.; Wang, R.; Zhnag, D.; Jin, H.; Guo, L. Structural effect of $ZrO_2$ on supported Ni-based catalysts for supercritical water gasification of oil-containing water. *Int. J. Hydrogen Energ.* **2021**, *46*, 12874–12885. [CrossRef]
80. Li, X.; Shao, Y.; Zhang, S.; Wang, Y.; Xiang, J.; Hu, S.; Xu, L.; Hu, X. Pore diameters of $Ni/ZrO_2$ catalysts affect properties of the coke in steam reforming of acetic acid. *Int. J. Hydrogen Energy* **2021**, *46*, 23642–23657. [CrossRef]
81. Kou, J.; Feng, H.; Wei, W.; Wang, G.; Sun, J.; Jin, H.; Guo, L. Study on the detailed reaction pathways and catalytic mechanism of a $Ni/ZrO_2$ catalyst for supercritical water gasification of diesel oil. *Fuel* **2022**, *312*, 122849. [CrossRef]
82. Papageridis, K.N.; Charisiou, N.D.; Douvartzides, S.L.; Sebastian, V.; Hinder, S.J.; Baker, M.A.; Alkhoori, S.; Polychronopoulou, K.; Goula, M.A. Effect of operating parameters on the selective catalytic deoxygenation of palm oil to produce renewable diesel over Ni supported on $Al_2O_3$, $ZrO_2$, and $SiO_2$ catalysts. *Fuel Process. Technol.* **2020**, *209*, 106547. [CrossRef]
83. Munir, M.; Ahmad, M.; Saeed, M.; Waseem, A.; Nizami, A.-S.; Sultana, S.; Zafar, M.; Rehan, M.G.; Srinivasan, G.R.; Ali, A.M.; et al. Biodeisel production from novel non-edible caper (*Capparis spinosa* L.) seeds oil employing Cu–Ni doped $ZrO_2$ catalyst. *Renew. Sustain. Energy Rev.* **2021**, *138*, 110558. [CrossRef]

Article

# Modification of Silica Xerogels with Polydopamine for Lipase B from *Candida antarctica* Immobilization

Honghai Wang [1,2], Wenda Yue [1,2], Shuling Zhang [1,2], Yu Zhang [1,2], Chunli Li [1,2] and Weiyi Su [1,2,*]

[1] School of Chemical Engineering and Technology, Hebei University of Technology, Tianjin 300130, China; ctstwhh@hebut.edu.cn (H.W.); ctstyuewenda@163.com (W.Y.); zsl_hhsy@163.com (S.Z.); zyu1327@163.com (Y.Z.); lichunli_hebut@126.com (C.L.)
[2] The National and Local Joint Engineering Laboratory for Energy Conservation of Chemical Process Integration and Resources Utilization, Tianjin 300130, China
[*] Correspondence: suweiyi@hebut.edu.cn; Tel.: +86-022-185-0225-9209; Fax: +86-022-6020-2248

**Abstract:** Silica xerogels have been proposed as a potential support to immobilize enzymes. Improving xerogels' interactions with such enzymes and their mechanical strengths is critical to their practical applications. Herein, based on the mussel-inspired chemistry, we demonstrated a simple and highly effective strategy for stabilizing enzymes embedded inside silica xerogels by a polydopamine (PDA) coating through in-situ polymerization. The modified silica xerogels were characterized by scanning and transmission electron microscopy, Fourier tranform infrared spectroscopy, X-ray diffraction, X-ray photoelectron spectroscopy and pore structure analyses. When the PDA-modified silica xerogels were used to immobilize enzymes of *Candida antarctica* lipase B (CALB), they exhibited a high loading ability of 45.6 mg/g$_{support}$, which was higher than that of immobilized CALB in silica xerogels (28.5 mg/g$_{support}$). The immobilized CALB of the PDA-modified silica xerogels retained 71.4% of their initial activities after 90 days of storage, whereas the free CALB retained only 30.2%. Moreover, compared with the immobilization of enzymes in silica xerogels, the mechanical properties, thermal stability and reusability of enzymes immobilized in PDA-modified silica xerogels were also improved significantly. These advantages indicate that the new hybrid material can be used as a low-cost and effective immobilized-enzyme support.

**Keywords:** *Candida antarctica* lipase B; silica xerogel; enzyme immobilization; polydopamine; modification

## 1. Introduction

Biocatalysts play a vital role in various scientific fields due to their unique advantages, such as high substrate specificity, outstanding catalytic ability and mild reaction conditions. Biocatalysis, applied in ester synthesis, is useful and its synthetic products can be identical to natural products. Recently, a transesterification reaction catalyzed by lipase (triacylglycerol ester hydrolase, EC 3.1.1.3) has been performed to produce esters [1]. Lipase-catalyzed reactions have been applied to the synthesis of chiral drugs [2], wax esters [3], structural lipids [4] and biodiesel [5]. However, the main bottlenecks of enzyme application are its low thermal stability, poor operational stability and the difficulty of reusing enzymes. Therefore, significant effort has been devoted to exploiting immobilization strategies to stabilize enzymes and endow them with greater stability and reusability [6,7].

In immobilizing enzymes, it is necessary to select an appropriate support material, which can improve the properties of enzymes. Lipases are widely recognized to have a hydrophobic domain [8]. The hydrophobic immobilization of a lipase can act upon its domains, to increase its activity and stability, by interfacial activation [9,10]. Thus, materials comprised with ordered mesoporous organosilica, in which organic hydrophobic groups are homogeneously distributed within their frameworks, may be ideal supports for lipase immobilization. Silica xerogel, thanks to a high specific surface area, good mechanical strength, inertness and stability at high temperature, has attracted much attention in enzyme immobilization [11,12]. The xerogel synthesis of entrapped enzymes based on silicon-containing compounds has been widely used [13]. The formation principle of a silicon sol−gel matrix for enzyme immobilization consists in the transition of a silicon alkoxide sol into a gel as a consequence of hydrolysis and polycondensation reactions, with the subsequent transformation into a monolithic xerogel, powder or film coating [14]. This method retains the inherent structures biocomposites, showing enzymatic activity and an expanded range of conditions for catalysis [15]. Enzymes' inclusion in xerogel structures allows increasing the resistances thereof to different physical and chemical factors, such as temperature, pH, radiation and aggressive compounds. In general, xerogel-encapsulated enzyme technology is a method for preparing bioactive nanocomposites [16].

However, the limitation of xerogels' entrapping of enzymes is the shrinkage of their structures, which is not conducive to entrapping enzyme due to the large capillary force caused by continuous internal shrinkage [17]. In our previous work, we proposed a strategy of producing an immobilized enzyme-containing xerogel coating on metal packings for reactive distillation, but there was still a relatively weak interaction between the enzyme and the surface of the support, which often lead to the leakage of the enzyme from the support [18]. In addition, mesoporous silica shows a certain brittleness [19]. When a xerogel is subjected to an external force, the partial or complete rupture of its skeletal structure will also lead to the loss and leakage of an entrapped enzyme. Many studies have been conducted to improve the interactions between enzymes and supports, or the mechanical properties of the support. To date, polyacrylamide [20] and glutaraldehyde [21] have been used as crosslinking agents to increase enzyme loading, while glycerol has been used to prevent xerogel cracking [22]. In fact, a simple and effective method of the two problems latter is to coat an active protective layer on the surfaces of pre-xerogel polymers. Inspired by mussel adhesion proteins, polydopamine (PDA) technology has attracted extensive research [23,24]. Dopamine molecules have been shown to self-polymerize under alkaline conditions, leading to a facile deposition of PDA coating on a material's surface [25]. More importantly, the residual quinone on the surface of polydopamine or an intermediate displays a nucleophilic amino reaction that can be covalently connected with nucleophilic biological molecules, producing a polydopamine coating that is robust and durable [26]. This provides a new way of improving the interactions between enzymes and supports, and of enhancing the mechanical properties of xerogels. However, to the best of our knowledge, there are few reports on dopamine self-polymerization deposition on the surfaces of xerogel supports aimed at improving the mechanical properties of and interactions between enzymes and supports.

Surface modification with polydopamine has already become an efficient and feasible method of endowing inorganic materials with biological functionality since Messersmith et al. pioneered the single-step formation of polymer film-based dopamine on various substrates [27]. Meanwhile, this method is not involved in complex linkers and is free of organic solvents, making it suitable for biomaterial applications. Furthermore, the abundant functional groups (i.e., catechol and amine) existing on such modified surfaces could enhance enzymes' binding abilities [28]. Therefore, in the present work and based on this idea, we design a new hybrid support by modifying, with polydopamine, the surfaces of silica xerogels. Specifically, CALB was chosen as a model enzyme. Firstly, CALB was encapsulated in silica xerogels by the sol–gel method, denoted as $SiO_2$–$CH_3$–CALB. Second, in order to prevent enzyme leakage and improve enzyme stability, the polymer networks in xerogels were coated with polydopamine (denoted as $SiO_2$–$CH_3$–CALB@PDA). Finally, the xerogels' mechanical properties and the enzyme-immobilizing ability of $SiO_2$–$CH_3$–CALB and $SiO_2$–$CH_3$–CALB@PDA were investigated in detail. The results show that $SiO_2$–$CH_3$–CALB@PDA had a significant CALB-embedding ability. Compared with the $SiO_2$–$CH_3$–CALB, the results showed that PDA-modified $SiO_2$–$CH_3$–CALB had better mechanical properties, thermal stability, storage stability and reusability. This indicated that the new hybrid silica xerogel could be used as a low-cost and relatively effective immobilized-enzyme support.

## 2. Results and Discussion

### 2.1. Characterization

Figure 1a shows a FTIR spectral comparison of $SiO_2$–$CH_3$–CALB and $SiO_2$–$CH_3$–CALB@PDA. The adsorption peaks at 777 $cm^{-1}$ and 445 $cm^{-1}$ corresponded to the Si−O−Si group [29]. The band at 1277 $cm^{-1}$ was assigned to the characteristic peak of Si–$CH_3$, which proved that MTMS successfully deposited a methyl polymer layer on the silica surface. Other bands, at 3388 $cm^{-1}$ and 1643 $cm^{-1}$, belonged to the stretching and bending vibrations of −OH [30]. After modification by polydopamine, the vibration absorption peak at 3388 $cm^{-1}$ was significantly enhanced and widened, which was related to the catechol composition of polydopamine [31]. In addition, $SiO_2$–$CH_3$–CALB@PDA showed a peak at 1508 $cm^{-1}$, which was ascribed to the bending vibrations of indole-quinone groups [32]. The XRD patterns of $SiO_2$–$CH_3$–CALB and $SiO_2$–$CH_3$–CALB@PDA are illustrated in Figure 1b. There was a relatively wide peak at $2\theta = 22°$, which is characteristic of amorphous silica [33]. The peak at $2\theta < 10°$ was due to the siloxane network and the xerogel's structure, composed of ordered organic layers [34]. After modification by polydopamine, the intensity of the characteristic peak at 10° became weak, implying the microstructure of $SiO_2$–$CH_3$–CALB had changed due to the uniformly distributed deposition of polydopamine within the structure of the xerogel [35]. These results indicate that a polydopamine coating formed on the silica xerogel through self-polymerization.

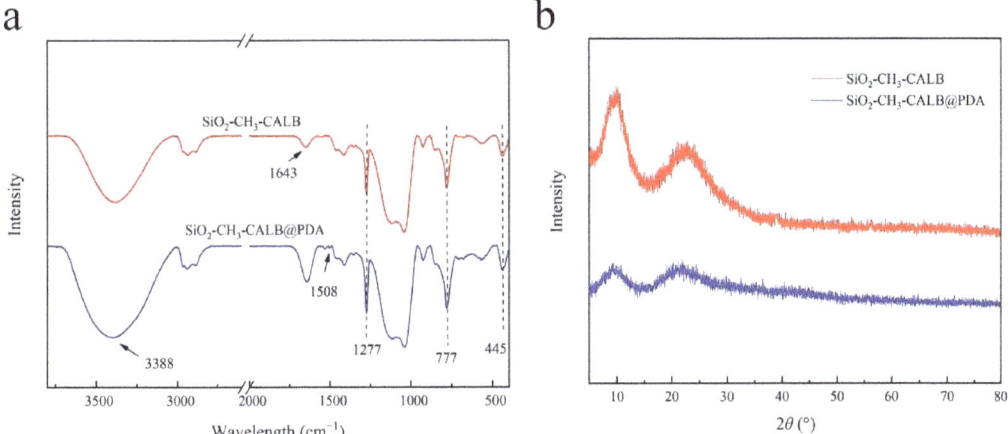

**Figure 1.** Physical properties of $SiO_2$–$CH_3$–CALB and $SiO_2$–$CH_3$–CALB@PDA. (**a**) FTIR spectra and (**b**) XRD $2\theta$ scans.

Microstructural images of $SiO_2$–$CH_3$–CALB@PDA are shown in Figure 2. Figure 2a,b shows the SEM images of the $SiO_2$–$CH_3$–CALB and $SiO_2$–$CH_3$–CALB@PDA prepared in this work, respectively by panel. They were constituted by the agglomeration of many silica clusters in uniform shape. Compared with $SiO_2$–$CH_3$–CALB, the surface of $SiO_2$–$CH_3$–CALB@PDA was rougher and looser between clusters, indicating that a polydopamine layer had formed on the Si−O−Si surface. In this structure, the formation of a protective enzyme barrier can absorb and disperse most of the energy from external forces, preventing the xerogel from breaking [36]. TEM images of monodispersed $SiO_2$–$CH_3$–CALB@PDA showed that its particles have a relatively uniform, nano-scale size (Figure 2c); a more intuitive expression is shown in Figure 2d. The shape of the $SiO_2$–$CH_3$–CALB@PDA particles irregularly spherical. Additionally, the $SiO_2$–$CH_3$–CALB@PDA surface had open-framework channels (Figure 2e) that facilitated the diffusion of the substrate and product molecules [37]. Elemental mapping analysis (Figure 2g−j) demonstrated that PDA was uniformly distributed within the xerogel (as these contained nitrogen), and oxygen, carbon and silicon were also found in the SiO$_2$–CH$_3$–CALB@PDA surface. Notably, the oxygen content was high.

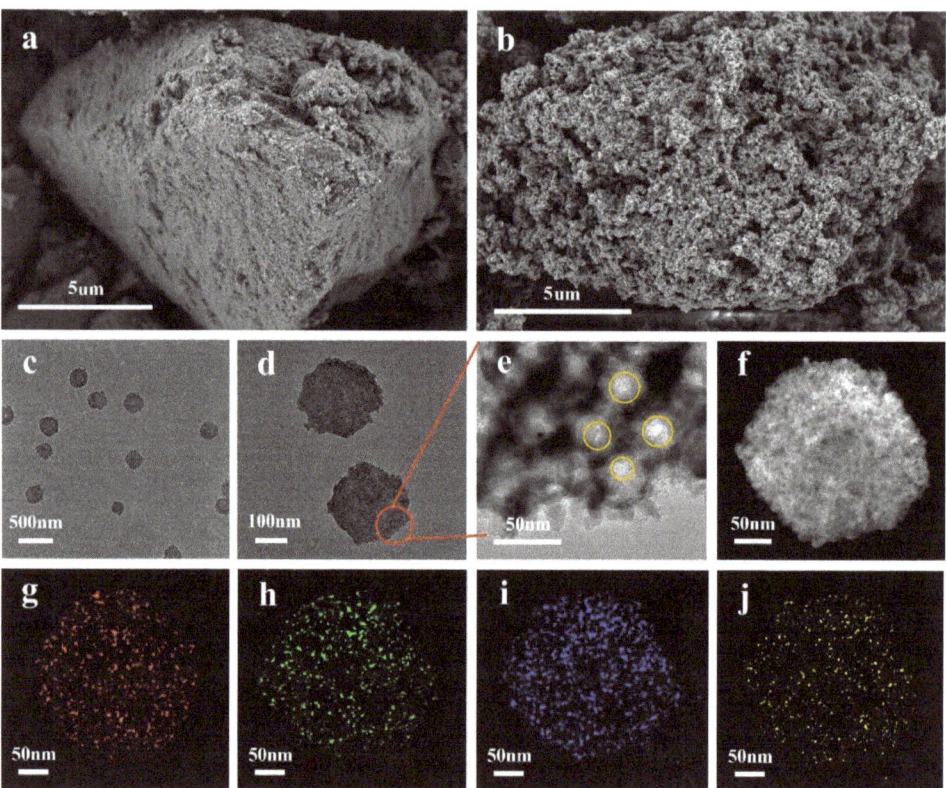

**Figure 2.** SEM images of (**a**) SiO$_2$–CH$_3$–CALB, (**b**) SiO$_2$–CH$_3$–CALB@PDA, (**c–e**) TEM images of SiO$_2$–CH$_3$–CALB@PDA and (**f–j**) elemental mapping analysis of SiO$_2$–CH$_3$–CALB@PDA (TEM).

XPS measurements in Figure 3 confirmed the existence of polydopamine coating on SiO$_2$–CH$_3$–CALB@PDA. XPS spectra of SiO$_2$–CH$_3$–CALB@PDA show the presence of C, N, O and Si (Figure 3a). The different chemical states of C, O and N in the regional spectra reveal the complex properties of polydopamine on SiO$_2$–CH$_3$–CALB@PDA (Figure 3b–d). The main peaks of C 1s spectra (Figure 3b) at 283.9, 284.5, 285.7, 286.4 and 287.2 eV, respectively, corresponding to C–C, C–N, C–O, C=O bands, O–C=O and feature for aromatic carbon species in the polydopamine. In the O1s peak (Figure 3c), two peaks were observed at 531.9 eV and 532.7 eV, respectively, which belonged to O atoms of polydopamine in the form of quinone and catechol [38]. The high-resolution spectra of N 1's peak are shown in Figure 3d. The main peak, at 399.4 eV, indicated the existence of R$_2$NH and RNH$_2$, while the peak at 401.2 eV was attributed to R$_3$N [39]. This result indicates that an adhesive polydopamine coating formed on the surface of the Si−O−Si network structure of the xerogel by self-polymerization.

Figure 4a−c and Figure 4d−f show the sol–gel process and finished products, SiO$_2$–CH$_3$–CALB and SiO$_2$–CH$_3$–CALB@PDA, respectively. They were no significant differences in the solution phase (sol), showing a slightly yellow liquid (Figure 4a,d). Then, they entered the gel phase, the SiO$_2$–CH$_3$–CALB showed a milky white gel block, while the SiO$_2$–CH$_3$–CALB@PDA showed a black transparent gel block (Figure 4b,e). This may be explained by dopamine's having begun to self-polymerize into polydopamine on the gel network. After the final drying stage, The final samples of $SiO_2$–$CH_3$–CALB and $SiO_2$–$CH_3$–CALB@PDA were obtained by grinding (Figure 4c,f).

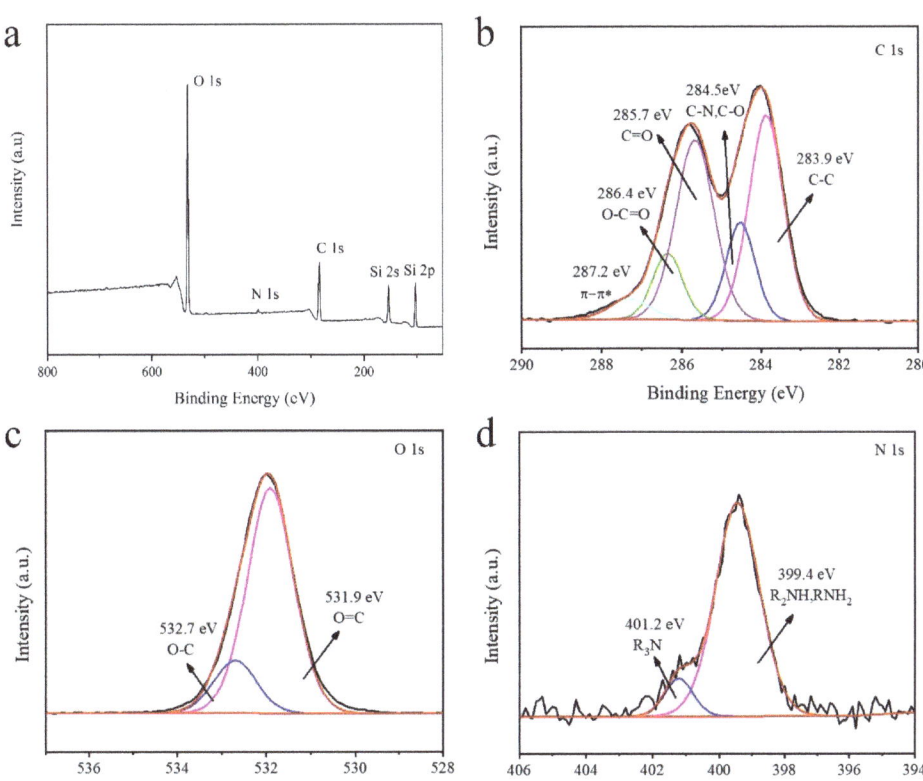

**Figure 3.** XPS analyses of $SiO_2$–$CH_3$–CALB@PDA. (**a**) Survey scan, (**b**) C 1s, (**c**) O 1s and (**d**) N 1s.

With the aim of further prove the PDA can delay shrinkage of xerogel, analysis of BET of $SiO_2$–$CH_3$–CALB and $SiO_2$–$CH_3$–CALB@PDA were taken into account. It can be seen from their adsorption–desorption curves, in Figure 4g, that they had strong interaction with $N_2$ at low pressure and presented typical Langmuir type IV curves. The H2 hysteresis loops were also observed, indicating the mesoporous structure and the characteristics of 'ink bottle' pores [40]. From the pore size distribution curve in Figure 4h, it can be seen that the pore size (15.02 nm) and BET surface area (165.84 $m^2 \cdot g^{-1}$) of $SiO_2$–$CH_3$–CALB@PDA were larger than those of $SiO_2$–$CH_3$–CALB (13.52 nm and 121.67 $m^2 \cdot g^{-1}$), which we believe to be due to the PDA coatings and deposits on the surface of the Si−O−Si network structure during the sol–gel process, delaying the gel shrinkage [41].

The immobilization capacity of $SiO_2$–$CH_3$–CALB@PDA was evaluated by investigating the CALB loading. As shown in Figure 4i, the amount of CALB immobilized on $SiO_2$–$CH_3$–CALB@PDA increased with increasing CALB concentration. When the CALB concentration was 14.5 mg/mL, the CALB loading increased to 45.6 mg/g. However, when the CALB concentration was more than 14 mg/mL, a decline in the activity recovery of the immobilized CALB was observed. The loading reached a maximum at a high enzyme concentration (~16 mg/mL), and there is a slightly continuous decrease in the enzyme activity when the enzyme concentration exceeds 14.5 mg/mL. This can be explained by the fact that excess CALB loading will easily lead to the congestion of the enzyme molecules. Therefore, the resulting spatial constraint can increase the mass transfer resistance of the substrate and product, which is expressed as reducing activity [42]. Therefore, the optimum CALB concentration was chosen as 14.5 mg/mL. In this case, the CALB loading is efficient (activity recovery higher than 93%) without sacrificing excess enzyme to unnecessary use. Meanwhile, compared with the enzyme loading of 28.5 mg/g on pristine $SiO_2$–$CH_3$–CALB at an initial CALB concentration of 14.5 mg/mL, the enzyme loading on $SiO_2$–$CH_3$–CALB@PDA reached as high as 45.6 mg/g, nearly twice as high as that on $SiO_2$–$CH_3$–CALB. As mentioned above, the modification of PDA provided a barrier for the enzyme, and covalent linking enhanced the interaction between the enzyme and the support, effectively preventing enzyme leakage.

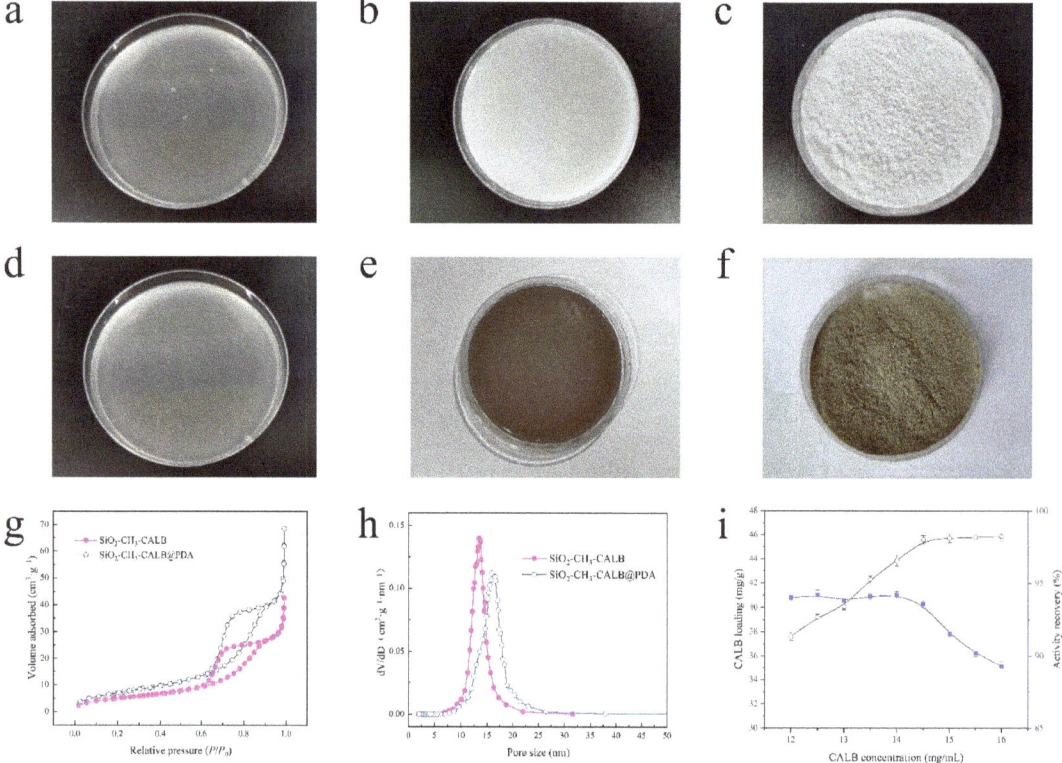

**Figure 4.** Sol–gel process of (**a**–**c**) $SiO_2$–$CH_3$–CALB, (**d**–**f**) $SiO_2$–$CH_3$–CALB@PDA, (**g**,**h**) pore structure and (**i**) enzyme loading.

## 2.2. Strategy for Immobilizing CALB and Possible Mechanism

Figure 5 shows the internal microstructure of the $SiO_2$–$CH_3$–CALB- and $SiO_2$–$CH_3$–CALB@PDA-immobilized enzyme and the mechanism of the polydopamine-modified immobilized enzyme. The Si–O–Si polymer network skeleton was obtained by hydrolysis and a condensation reaction with TMOS and MTMS as co-precursors, and the enzyme molecules were embedded in the Si–O–Si network. In hydrolysis reaction, the methyl group of MTMS was not involved in hydrolysis, replacing and cross-linking with hydroxyl groups on Si–O–Si network, which provided a necessary condition for the development of hydrophobic properties [43]. The polydopamine-modified immobilized enzyme was based on the synergistic sol–gel mechanism [44]. In short, dopamine nanoparticles were uniformly mixed into the sol. In this system, dopamine hydrochloride was self-polymerized into PDA under alkaline conditions and deposited on the surface of Si–O–Si network. Moreover, the residual quinone functional groups presented in the polydopamine coating were reactive toward nucleophilic groups, and CALB could couple covalently with polydopamine through Michael-type addition or Shiff-based formation [45,46]. We expected that the resulting SiO$_2$–CH$_3$–CALB@PDA xerogels would have excellent mechanical strengths and enzyme activity stabilities.

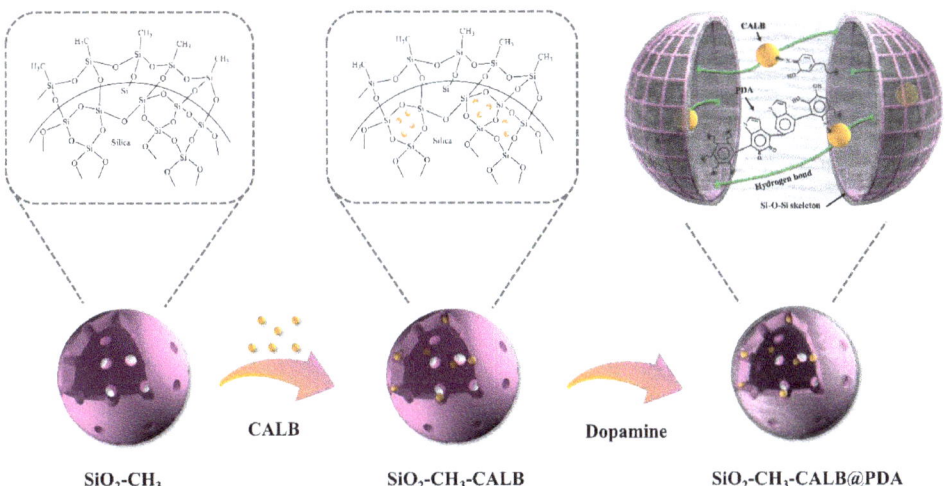

**Figure 5.** The internal microstructures of SiO$_2$–CH$_3$–CALB and SiO$_2$–CH$_3$–CALB@PDA and the possible mechanism of the polydopamine-modified immobilized enzyme.

### 2.3. Mechanical Properties

In practical applications, xerogel is prone to deformation under external force, resulting in enzyme leakage or inactivation. Therefore, strength is crucial for the application of xerogel in organic catalysis. We experimentally compared the strengths of SiO$_2$–CH$_3$–CALB and SiO$_2$–CH$_3$–CALB@PDA. The compressive stress–strain curves for SiO$_2$–CH$_3$–CALB and SiO$_2$–CH$_3$–CALB@PDA are presented in Figure 6. A macroscopic compression experiment showed that the SiO$_2$–CH$_3$–CALB@PDA xerogel model could withstand higher pressures (12.55 Mpa) than that of SiO$_2$–CH$_3$–CALB (9.00 Mpa), and the strain of SiO$_2$–CH$_3$–CALB@PDA (9.64%) was greater than that of SiO$_2$–CH$_3$–CALB (9.07%). In addition, the fracture modes of the two materials were also significantly different. The fracture mode of SiO$_2$–CH$_3$–CALB was similar to that of brittle materials, while the fracture mode of SiO$_2$–CH$_3$–CALB@PDA was similar to that of viscoelastic materials [47,48]. This can be ascribed to two factors. On the one hand, PDA was deposited on the surface of the Si−O−Si network, which reduced the capillary force generated by the shrinkage of the xerogel during drying [29]. On the other hand, the surface of the PDA contained a large number of functional groups that could interact with Si−O−Si chains, serving as crosslinking sites to increase the mechanical strength of the SiO$_2$–CH$_3$–CALB@PDA xerogel, preventing it from breaking under pressure [32]. Overall, the internal network structure of the xerogel and polydopamine coatings played a key role in the whole compression process, confirming the formation of a stable xerogel. After modification by PDA nanoparticles, the mechanical properties of the xerogels were improved. This occurred because polydopamine can interact with the xerogel matrix, increasing xerogel hardness and improving brittleness.

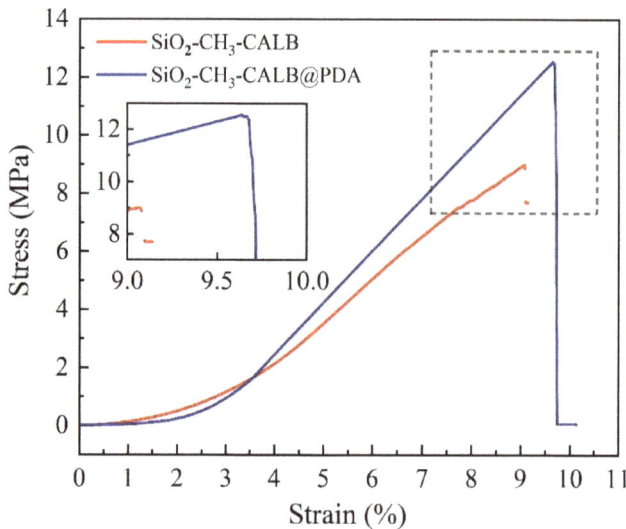

**Figure 6.** Stress–strain curves of SiO$_2$–CH$_3$–CALB and SiO$_2$–CH$_3$–CALB@PDA.

*2.4. Stability of Immobilized CALB*

The free and immobilized CALB was incubated at 60 °C for a certain time to investigate their thermal stability. The influence of temperature towards the stability of CALB is illustrated in Figure 7a. With the increasing of incubation time, the hydrolytic activity of free CALB and SiO$_2$–CH$_3$–CALB decreased, and free CALB was entirely deactivation after 3 h treatment. However, the SiO$_2$–CH$_3$–CALB@PDA exhibited better stability, which maintained 36.5% of its activity after 6 h of incubation. These results revealed that the better thermal stability of SiO$_2$–CH$_3$–CALB@PDA among free CALB and SiO$_2$–CH$_3$–CALB was attributed to the strong covalent bonds that formed through the reaction between the amine in the enzyme and the electrophilic groups in the PDA [49]. In addition, the PDA layer provides a stiffer external backbone to protect the CALB molecule from high temperatures [50]. Improvements in thermal stability will expand the range of applications for immobilized enzymes.

In order to investigate the storage stability of SiO$_2$–CH$_3$–CALB and SiO$_2$–CH$_3$–CALB@PDA, the examination was carried out at room temperature for 90 days. As shown in Figure 7b, SiO$_2$–CH$_3$–CALB exhibited 66.3% of its initial activity after 90 days, while SiO$_2$–CH$_3$–CALB@PDA exhibited approximately 71.4% under the same conditions. The high storage stability exhibited by CALB can be explained by the protective effect of the Si–O–Si network in the silica structure, which protects the enzyme activity inside, and further enhances its structural stability. The interactions of different geometries of the enzyme and support may have a significant influence on the enzyme activity. Generally, it is accepted that the highly curved surface reduces the possibility of enzyme denaturation and inhibits lateral interactions between adjacent enzymes, further leading to the structural stability and persistent activity of the adsorbed enzyme [51,52]. Additionally, multiple points of binding were observed between the PDA support and CALB in SiO$_2$–CH$_3$–CALB@PDA, which formed the PDA coating on the surface of the polymer network inside the xerogels, acting in a protective role [53,54]; this could explain their greater activity in external environments.

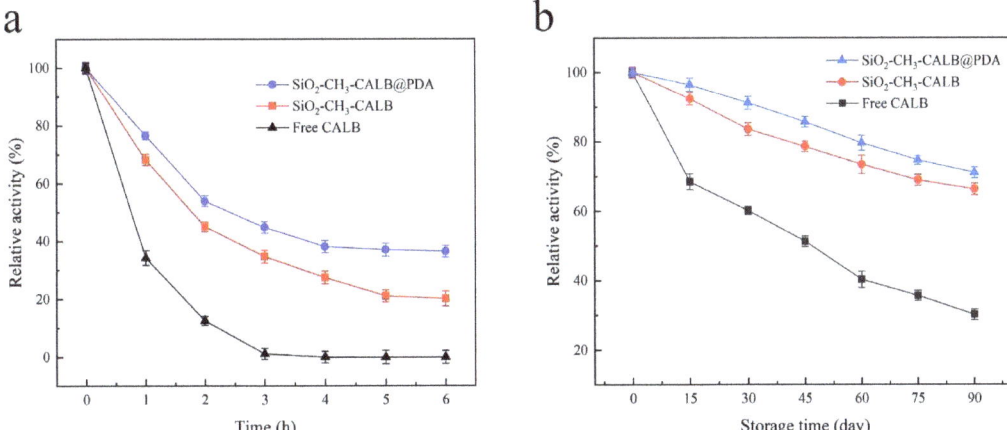

**Figure 7.** Thermal stability at 60 °C (**a**) and storage stability at 25 °C (**b**) of free CALB, SiO$_2$–CH$_3$–CALB and SiO$_2$–CH$_3$–CALB@PDA.

*2.5. Transesterification and Reusability*

Some enzymes have been used as biocatalysts to synthetize ester compounds, among which CALB can form high value-added ester products by transesterification reactions. As one of the major biocatalysts for ester synthesis, CALB can catalyze the transesterification of *n*-butanol with ethyl acetate to produce butyl acetate, which is an excellent organic solvent. Figure 8 shows the CALB-catalyzed synthesis of butyl acetate by transesterification of *n*-butanol with ethyl acetate. The reaction is a solvent-free system, and was carried out in a batch reactor at 70 °C. In a solvent-free system, the enzyme directly acts on the substrate, increases the substrate concentration, improving the reaction rate and selectivity and reducing the damage of organic solvents to the enzyme [55]. Therefore, we chose ethyl acetate as a reactant, as it also acts as a solvent in the CALB-catalyzed synthesis of butyl acetate.

*n*-butanol      ethyl acetate      butyl acetate      ethanol

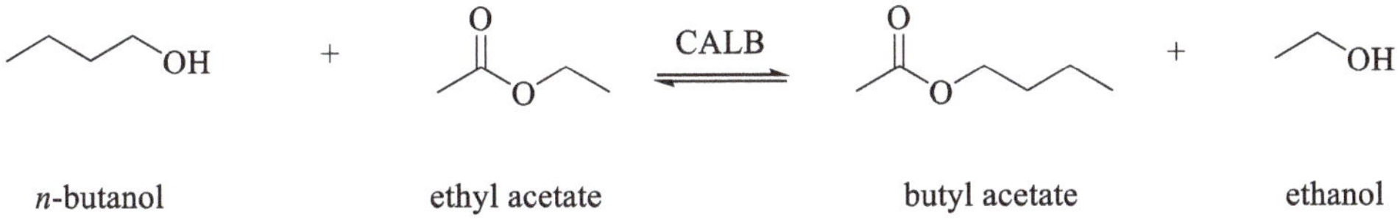

**Figure 8.** The reaction diagram of butyl acetate synthesis catalyzed by CALB.

The transesterification of *n*-butanol with ethyl acetate was selected as a target reaction to evaluate the conversion efficiency and reusability of immobilized CALB in the present work. The conversion of *n*-butanol and the reusability of SiO$_2$–CH$_3$–CALB and SiO$_2$–CH$_3$–CALB@PDA were compared under optimal active conditions. As shown in Figure 9, in the first cycle, the conversion of *n*-butanol of SiO$_2$–CH$_3$–CALB and SiO$_2$–CH$_3$–CALB@PDA retained 52.42% and 57.67%, respectively. For SiO$_2$–CH$_3$–CALB, CALB molecules were encapsulated in the xerogel polymer network by physical adsorption, Virgen-Ortíz et al. have reported some substrates/product may produce the enzyme's release from physically absorbed enzymes, so the leakage of CALB was prone to denaturation or inactivation during the reaction [56]. The decrease in conversion was observed in the first five cycles. After the fifth cycle, the activity began a slow deceleration state, lasting for the next three cycles. After eight cycles, SiO$_2$–CH$_3$–CALB@PDA retained more than a 30.84% conversion of *n*-butanol. SiO$_2$–CH$_3$–CALB retained a 25.04% conversion of *n*-butanol. The conversion of *n*-butanol loss could be due to enzyme leakage during washing and enzyme deactivation during repeated uses [57]. As the reaction produces a by-product of ethanol in the batch reactor system, resulting in enzyme inhibition, inhibition will reduce lipase activity. High concentrations of *n*-butanol inhibit the synthesis of butyl acetate catalyzed by immobilized CALB. This inhibitory effect has been found in the reaction among butyric acid and lauric acid with ethanol [58,59]. Therefore, operational stability of the enzyme is not too high. Considering $SiO_2$–$CH_3$–CALB@PDA had better reusability, storage stability and mechanical strength, $SiO_2$–$CH_3$–CALB@PDA is more applicable for practical applications.

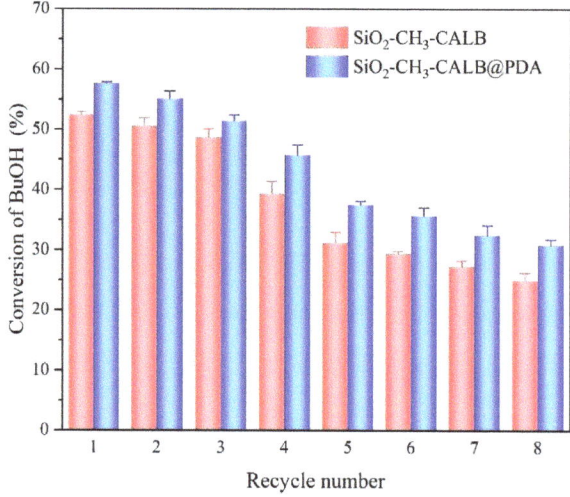

**Figure 9.** Conversion of *n*-butanol and repeatability of $SiO_2$–$CH_3$–CALB and $SiO_2$–$CH_3$–CALB@PDA.

*2.6. Comparison of Butyl Acetate Production Using Previous Lipase Biocatalysts*

The prepared catalyst of $SiO_2$–$CH_3$–CALB@PDA possessed the advantages of biocompatibility, environmental friendliness, operating convenience and safety. Compared with previous lipase catalysts, such as $SiO_2$–$CH_3$–CALB and Novozyme 435, the catalytic efficiency of $SiO_2$–$CH_3$–CALB@PDA (57.67%) in the transesterification reaction system was slightly higher than those of $SiO_2$–$CH_3$–CALB (52.42%) and Novozyme 435 (55.30%) [60]. Although the improvement in operational stability and catalytic performance is not obvious, the polydopamine modification strategy is worth adopting; it can improve the immobilized enzyme loading and balance the mechanical properties of the supports, which expands the application range of immobilized enzymes in some special cases.

### 3. Experimental Section

*3.1. Materials*

Tetramethoxysilane (TMOS), methyltrimethoxysilane (MTMS), methanol (MeOH), sodium fluoride (NaF), polyethylene glycol (PEG, MW 400), dopamine hydrochloride, 4-nitrophenyl palmitate (*P*-NPP), Coomassie Brilliant Blue G250, bovine serum albumin ($\geq$96%), *n*-butanol, ethyl acetate, *n*-propanol and sodium-phosphate buffer (PBS, 0.1 M, pH 7.5) were purchased from Aladdin (Shanghai, China). *Candida antarctica* lipase B (CALB) was provided from Novozymes (Copenhagen, Denmark) with a free enzyme activity of 510 $U \cdot g^{-1}$. All reagents except bovine serum albumin were analytically pure without further purification.

*3.2. Preparation of $SiO_2$–$CH_3$–CALB*

$SiO_2$–$CH_3$–CALB was prepared by sol–gel method. Firstly, TOMS (0.54 g), MTMS (1.934 g), methanol (3.39 g), PEG (0.14 g), NaF solution (0.49 g, 1 M), water (1.26 g) and

CALB enzyme solution (3.39 g) were mixed and stirred at 0 °C, and the mixture was transferred to a clean petri dish. Then, the petri dish was sealed and placed at room temperature for 2 h to form a gel network. Finally, the petri dish was opened to evaporate the water and solvent in the gel completely, and then dried at room temperature for 48 h.

*3.3. Preparation of $SiO_2$–$CH_3$–CALB@PDA*

Dopamine hydrochloride (0.02 g) was dispersed in methanol (3.39 g), then NaF solution (0.98 g, 1 M) was added and mixed for 10 min. The use of NaF solution rather than Tris buffer was due to the fact that primary amine group in Tris can covalently interact with PDA, which could affect the deposition of PDA and the continuous coupling of CALB with PDA.

To the obtained mixture we added TMOS (0.54 g) and 1.934 g of MTMS (1.934 g), PEG (0.14 g), water (1.26 g) and CALB enzyme solution (3.39 g), which was then mixed and stirred at 0 °C, and the mixture was transferred to a clean petri dish. Then, the petri dish was sealed and placed at room temperature for 4 h to form a gel network. Finally, the petri dish was opened to evaporate the water and solvent in the gel completely and then dried at room temperature for 48 h.

*3.4. Characterization*

The microstructures of the samples were observed using a transmission electron microscope (TEM, Talos F200S, Hillsboro, FL, USA) and scanning electron microscopy (SEM, Nova Nano SEM 450, Hillsboro, FL, USA). Fourier transform infrared (FT–IR) spectra of the samples were collected from 4000 to 400 cm$^{-1}$ on a Bruker Tensor 27 analyzer (Bremen, Germany) using KBr pellets method. X-ray diffraction (XRD) patterns were measured by a Bruker D8 Discover (Bremen, Germany) with scanning rate of 6° min$^{-1}$ under Cu K$\alpha$ radiation ($\lambda$ = 0.154056 nm). Samples were mounted on a low background silicon substrate and diffraction scans covered a $2\theta$ range of 5° to 80°. X-ray photoelectron spectra (XPS, Al-K$\alpha$) were recorded on an X-ray photoelectron spectrometer (ESCALAB 250Xi, Hillsboro, FL, USA), and the C 1 s of 284.8 eV was referred to for calibrating the binding energy. The $N_2$ adsorption–desorption isotherms were measured by a pore size-specific surface area analyzer (SSA–6000, Beijing, China) at 77 K. The pore size distribution and surface area were determined through calculating $N_2$ adsorption–desorption according to the Brunauer–Emmett–Teller (BET) method. A spectrophotometer (UV-2600, Shimadzu, Kyoto, Japan) was used to analyze the concentration and activity of the enzyme.

*3.5. Determination of Enzyme Loading*

The Bradford method was used to determine enzyme embedding in the silica xerogels by measuring of the protein concentrations in the initial enzyme solutions and immobilized enzyme phosphate detergents. A calibration curve was plotted, using Coomassie Brilliant Blue G-250 solutions as standards. The enzyme concentration in the solution was able to be determined with UV-vis spectrophotometry, by measuring the absorbance at 595 nm. The amount of enzyme embedded in silica xerogels was calculated by the following equation:

$$enzyme\,loading = \frac{C_0 - C_1}{C_0} \times 100\% \tag{1}$$

where $C_0$ is the initial enzyme concentration (mg/g), $C_1$ is the enzyme concentration in phosphate detergent (mg/g).

*3.6. Properties of Free CALB and the Immobilized CALB*

3.6.1. Assay of the CALB Activity

The free CALB and samples of immobilized CALB activities were determined by using $p$-NPP (5 mg/mL in ethanol) as the substrate. Typically, 200 µL of $p$-NPP solution was added to the solution consisting of the samples (2 mg) and PBS (0.1 M, pH 7.5, 3 mL). After reaction for 3 min, the filtrate of the reaction that contained 4-nitrophenol ($p$-NP), and the concentration of *p*-NP was quantified via absorbance at 410 nm on a spectrophotometer. One unit (U) of lipase hydrolytic activity was regard as the lipase mass that liberates 1 nmol of *p*-NP under these test conditions per minute. The relative enzymatic activity was related to a percentage of this highest activity (100% means the highest enzymatic activity). The activity recovery was calculated from the value of the activity of the initial CALB solution divided by the activity value of immobilized CALB obtained immediately after the immobilization procedure.

3.6.2. Thermal and Storage Stability of the Free CALB and Immobilized CALB

Free CALB, $SiO_2$–$CH_3$–CALB and $SiO_2$–$CH_3$–CALB@PDA were incubated in PBS (50 mM, pH 7.5) at 70 °C for 6 h to examine their thermal stabilities. The *p*-NPP assay was employed to measure residual activity as described in Section 3.6.1. To evaluate the storage stability, the residual activity of $SiO_2$–$CH_3$–CALB and $SiO_2$–$CH_3$–CALB@PDA was tested after a given treatment duration at 25 °C, respectively. The residual activity of each sample under treatment was measured at given time intervals and used for comparison with the original activity.

*3.7. Mechanical Performance Tests*

The mechanical performances of $SiO_2$–$CH_3$–CALB and $SiO_2$–$CH_3$–CALB@PDA were tested using a microcomputer control electron universal testing machines (CMT6104, Shenzhen, China) with a 5000-N load cell. To facilitate testing, samples were made into rectangular specimens. Compression strain tests of the samples (lengths, 23 mm; widths, 13.28 mm; thicknesses, 6 mm) were performed at a compression rate of 2 mm/min.

*3.8. Transesterification and Reusability*

The reaction for the transesterification of *n*-butanol with ethyl acetate was performed in a glass three-necked reactor with a volume of 250 mL at 343 K and 101.3 kPa. The electric stirring was controlled up to 3000 rpm to achieve uniform mixing of the reactive mixture. In the experiment, the mixture of reactants ethyl acetate and *n*-butanol (molar ratio of ethyl acetate to *n*-butanol was 1:1) were heated to 343 K in a water bath, then the catalysts (the catalyst dosage was 10% of the mass of *n*-butanol, and the catalysts were $SiO_2$–$CH_3$–CALB and $SiO_2$–$CH_3$–CALB@PDA) were set in the reactor to start the reaction. Samples were withdrawn from the reactor every 30 min with a syringe during the reaction for composition analysis until the 5 h. Finally, the catalysts were washed with PBS (0.1 M, pH 7.5) buffer and dried for 12 h before next cycle.

The composition of the product was analyzed by gas chromatography (GC-2010 Pro, Shimadzu, Kyoto, Japan) equipped with a flame ionization detector (FID) and an InertCap FFAP capillary column (30 m × 0.25 mm × 0.25 mm). Typically, *n*-propanol was used as the internal standard substance. $N_2$ with purity of 99.99 wt% was used as carrier gas at 1 mL/min. The temperature of the injection port and the detector were controlled at 473 K and 493 K, respectively. 0.4 µL sample was injected each time.

## 4. Conclusions

In this work, the immobilization of CALB in PDA-modified silica xerogels was successfully prepared by the self-polymerization of dopamine on the Si−O−Si network surfaces of silica xerogels. The modified silica xerogels showed an excellent embedding ability for CALB compared with conventional silica xerogels. They exhibited a high capacity of 45.6 mg/g$_{support}$ for CALB encapsulation. The mechanical strength and thermal and storage stability of the immobilized CALB were greatly elevated. Moreover, the immobilization of an enzyme in PDA-modified silica xerogels was utilized in the transesterification between *n*-butanol with ethyl acetate, which retained 30.84% conversion of *n*-butanol after eight cycles. In short, the $SiO_2$–$CH_3$–CALB@PDA catalyst was prepared by a simple and practical method, which is expected to overcome the related problems of shrinkage and weak binding force in conventional silica xerogels, and it has great application potential in the field of industrial catalysis.

**Author Contributions:** Conceptualization, H.W. and W.S.; methodology, H.W. and W.S.; validation, H.W., W.Y. and S.Z.; formal analysis, H.W.; investigation, H.W., W.Y.; resources, H.W.; data curation, Y.Z.; writing—original draft preparation, W.Y.; writing—review and editing, H.W., W.Y., Y.Z., C.L. and W.S.; visualization, H.W. and W.Y.; supervision, W.S.; project administration, W.S.; funding acquisition, H.W., W.S. All authors have read and agreed to the published version of the manuscript.

**Funding:** This research was funded by financial support of National Natural Science Foundation of China (No. 21878066 and No. 21878068), National Natural Science Foundation of Hebei Province (No. B2020202015) and Special Correspondent Project of Tianjin (No. 18JCTPJC56500).

**Conflicts of Interest:** The authors declare no conflict of interest.

## References

1. Mathpati, A.C.; Kalghatgi, S.G.; Mathpati, C.S.; Bhanage, B.M. Immobilized lipase catalyzed synthesis of n-amyl acetate: Parameter optimization, heterogeneous kinetics, continuous flow operation and reactor modeling. *J. Chem. Technol. Biotechnol.* **2018**, *93*, 2906–2916. [CrossRef]
2. Sanfilippo, C.; Paternò, A.A.; Patti, A. Resolution of racemic amines via lipase-catalyzed benzoylation: Chemoenzymatic synthesis of the pharmacologically active isomers of labetalol. *Mol. Catal.* **2018**, *449*, 79–84. [CrossRef]
3. Kuo, C.H.; Chen, H.H.; Chen, J.H.; Liu, Y.C.; Shieh, C.J. High yield of wax ester synthesized from cetyl alcohol and octanoic acid by lipozyme RMIM and Novozym 435. *Int. J. Mol. Sci.* **2012**, *13*, 11694–11704. [CrossRef] [PubMed]
4. Chojnacka, A.; Gładkowski, W. Production of structured phosphatidylcholine with high content of myristic acid by lipase-catalyzed acidolysis and interesterification. *Catalysts* **2018**, *8*, 281. [CrossRef]
5. Zhang, H.; Liu, T.; Zhu, Y.; Hong, L.; Li, T.; Wang, X.; Fu, Y. Lipases immobilized on the modified polyporous magnetic cellulose support as an efficient and recyclable catalyst for biodiesel production from Yellow horn seed oil. *Renew. Energy* **2020**, *145*, 1246–1254. [CrossRef]
6. Gao, J.; Kong, W.; Zhou, L.; He, Y.; Ma, L.; Wang, Y.; Yin, L.; Jiang, Y. Monodisperse core-shell magnetic organosilica nanoflowers with radial wrinkle for lipase immobilization. *Chem. Eng. J.* **2017**, *309*, 70–79. [CrossRef]
7. Fernandez-Lopez, L.; Pedrero, S.G.; Lopez-Carrobles, N.; Gorines, B.C.; Virgen-Ortiz, J.J.; Fernandez-Lafuente, R. Effect of protein load on stability of immobilized enzymes. *Enzym. Microb. Technol.* **2017**, *98*, 18–25. [CrossRef]
8. Gascon, V.; Diaz, I.; Blanco, R.M.; Marquez-Alvarez, C. Hybrid periodic mesoporous organosilica designed to improve the properties of immobilized enzymes. *RSC Adv.* **2014**, *4*, 34356–34368. [CrossRef]
9. Arana-Peña, S.; Rios, N.S.; Carballares, D.; Gonçalves, L.R.B.; Fernandez-Lafuente, R. Immobilization of lipases via interfacial activation on hydrophobic supports: Production of biocatalysts libraries by altering the immobilization conditions. *Catal. Today.* **2021**, *362*, 130–140. [CrossRef]
10. Manoel, E.A.; Dos Santos, J.C.S.; Freire, D.M.G.; Rueda, N.; Fernandez-Lafuente, R. Immobilization of lipases on hydrophobic supports involves the open form of the enzyme. *Enzym. Microb. Technol.* **2015**, *71*, 53–57. [CrossRef]
11. Xue, C.; Wang, J.; Tu, B.; Zhao, D. Hierarchically Porous Silica with Ordered Mesostructure from Confinement Self-Assembly in Skeleton Scaffolds. *Chem. Mater.* **2009**, *22*, 494–503. [CrossRef]
12. Wu, D.; Xu, F.; Sun, B.; Fu, R.; He, H.; Matyjaszewski, K. Design and preparation of porous polymers. *Chem. Rev.* **2012**, *112*, 3959–4015. [CrossRef]
13. Hwang, E.T.; Gu, M.B. Enzyme stabilization by nano/microsized hybrid materials. *Eng. Life Sci.* **2013**, *13*, 49–61. [CrossRef]
14. Znaidi, L. Sol–gel-deposited ZnO thin films: A review. *Mater. Sci. Eng. B* **2010**, *174*, 18–30. [CrossRef]
15. Vinogradov, V.V.; Avnir, D. Exceptional thermal stability of therapeutical enzymes entrapped in alumina sol-gel matrices. *J. Mater. Chem. B* **2014**, *2*, 2868–2873. [CrossRef] [PubMed]
16. Pierre, A.C. The sol-gel encapsulation of enzymes. *Biocatal. Biotransform.* **2009**, *22*, 145–170. [CrossRef]
17. Rivas Murillo, J.S.; Bachlechner, M.E.; Campo, F.A.; Barbero, E.J. Structure and mechanical properties of silica aerogels and xerogels modeled by molecular dynamics simulation. *J. Non-Cryst. Solids.* **2010**, *356*, 1325–1331. [CrossRef]
18. Wang, H.; Liu, W.; Gao, L.; Lu, Y.; Chen, E.; Xu, Y.; Liu, H. Synthesis of n-butyl acetate via reactive distillation column using Candida Antarctica lipase as catalyst. *Bioproc. Biosyst. Eng.* **2020**, *43*, 593–604. [CrossRef]
19. Latella, B.A.; Ignat, M.; Barbé, C.J.; Cassidy, D.J.; Li, H. Cracking and Decohesion of Sol-Gel Hybrid Coatings on Metallic Substrates. *J. Sol-Gel. Sci. Technol.* **2004**, *31*, 143–149. [CrossRef]
20. Yan, M.; Ge, J.; Liu, Z.; Ouyang, P. Encapsulation of single enzyme in nanogel with enhanced biocatalytic activity and stability. *J. Am. Chem. Soc.* **2006**, *128*, 11008–11009. [CrossRef]
21. Lee, H.R.; Chung, M.; Kim, M.I.; Ha, S.H. Preparation of glutaraldehyde-treated lipase-inorganic hybrid nanoflowers and their catalytic performance as immobilized enzymes. *Enzym. Microb. Technol.* **2017**, *105*, 24–29. [CrossRef]
22. Rodrigues, E.G.; Pereira, M.F.R.; Órfão, J.J.M. Glycerol oxidation with gold supported on carbon xerogels: Tuning selectivities by varying mesopore sizes. *Appl. Catal. B* **2012**, *115*, 1–6. [CrossRef]

23. Huang, Q.; Chen, J.; Liu, M.; Huang, H.; Zhang, X.; Wei, Y. Polydopamine-based functional materials and their applications in energy, environmental, and catalytic fields: State-of-the-art review. *Chem. Eng. J.* **2020**, *387*, 124019. [CrossRef]
24. Cheng, W.; Zeng, X.; Chen, H.; Li, Z.; Zeng, W.; Mei, L.; Zhao, Y. Versatile Polydopamine Platforms: Synthesis and Promising Applications for Surface Modification and Advanced Nanomedicine. *ACS Nano.* **2019**, *13*, 8537–8565. [CrossRef] [PubMed]
25. Zhang, C.; Gong, L.; Mao, Q.; Han, P.; Lu, X.; Qu, J. Laccase immobilization and surface modification of activated carbon fibers by bio-inspired poly-dopamine. *RSC Adv.* **2018**, *8*, 14414–14421. [CrossRef]
26. Samyn, P. A platform for functionalization of cellulose, chitin/chitosan, alginate with polydopamine: A review on fundamentals and technical applications. *Int. J. Biol. Macromol.* **2021**, *178*, 71–93. [CrossRef] [PubMed]
27. Kang, S.M.; Hwang, N.S.; Yeom, J.; Park, S.Y.; Messersmith, P.B.; Choi, I.S.; Langer, R.; Anderson, D.G.; Lee, H. One-Step Multipurpose Surface Functionalization by Adhesive Catecholamine. *Adv. Funct. Mater.* **2012**, *22*, 2949–2955. [CrossRef]
28. Lee, H.; Scherer, N.F.; Messersmith, P.B. Single-molecule mechanics of mussel adhesion. *Proc. Natl. Acad. Sci. USA* **2006**, *103*, 12999–13003. [CrossRef] [PubMed]
29. Fidalgo, A.; Ilharco, L.M. Correlation between physical properties and structure of silica xerogels. *J. Non-Cryst. Solids* **2004**, *347*, 128–137. [CrossRef]
30. Jiang, Y.; Wang, Y.; Wang, H.; Zhou, L.; Gao, J.; Zhang, Y.; Zhang, X.; Wang, X.; Li, J. Facile immobilization of enzyme on three dimensionally ordered macroporous silica via a biomimetic coating. *N. J. Chem.* **2015**, *39*, 978–984. [CrossRef]
31. Qu, K.; Zheng, Y.; Dai, S.; Qiao, S.Z. Graphene oxide-polydopamine derived N,S-codoped carbon nanosheets as superior bifunctional electrocatalysts for oxygen reduction and evolution. *Nano Energy* **2016**, *19*, 373–381. [CrossRef]
32. Zeng, Q.; Qian, Y.; Huang, Y.; Ding, F.; Qi, X.; Shen, J. Polydopamine nanoparticle-dotted food gum hydrogel with excellent antibacterial activity and rapid shape adaptability for accelerated bacteria-infected wound healing. *Bioact. Mater.* **2021**, *6*, 2647–2657. [CrossRef]
33. Guzel Kaya, G.; Yilmaz, E.; Deveci, H. A novel silica xerogel synthesized from volcanic tuff as an adsorbent for high-efficient removal of methylene blue: Parameter optimization using Taguchi experimental design. *J. Chem. Technol. Biotechnol.* **2019**, *94*, 2729–2737. [CrossRef]
34. Moriones, P.; Echeverria, J.C.; Parra, J.B.; Garrido, J.J. Phenyl siloxane hybrid xerogels: Structure and porous texture. *Adsorption* **2019**, *26*, 177–188. [CrossRef]
35. Zhang, H.; Hu, Q.; Zheng, X.; Yin, Y.; Wu, H.; Jiang, Z. Incorporating phosphoric acid-functionalized polydopamine into Nafion polymer by in situ sol-gel method for enhanced proton conductivity. *J. Membr. Sci.* **2019**, *570*, 236–244. [CrossRef]
36. Tian, Y.; Cao, Y.; Wang, Y.; Yang, W.; Feng, J. Realizing ultrahigh modulus and high strength of macroscopic graphene oxide papers through crosslinking of mussel-inspired polymers. *Adv. Mater.* **2013**, *25*, 2980–2983. [CrossRef] [PubMed]
37. Ganonyan, N.; Benmelech, N.; Bar, G.; Gvishi, R.; Avnir, D. Entrapment of enzymes in silica aerogels. *Mater. Today* **2020**, *33*, 24–35. [CrossRef]
38. Chao, C.; Liu, J.; Wang, J.; Zhang, Y.; Zhang, B.; Zhang, Y.; Xiang, X.; Chen, R. Surface modification of halloysite nanotubes with dopamine for enzyme immobilization. *ACS Appl. Mater. Interfaces* **2013**, *5*, 10559–10564. [CrossRef]
39. Zangmeister, R.A.; Morris, T.A.; Tarlov, M.J. Characterization of polydopamine thin films deposited at short times by autoxidation of dopamine. *Langmuir* **2013**, *29*, 8619–8628. [CrossRef]
40. Sing, K.S.W. Reporting physisorption data for gas/solid systems with special reference to the determination of surface area and porosity (Recommendations 1984). *Pure Appl. Chem.* **1985**, *57*, 603–619. [CrossRef]
41. Lai, Y.; Xia, W.; Li, J.; Pan, J.; Jiang, C.; Cai, Z.; Wu, C.; Huang, X.; Wang, T.; He, J. A confinement strategy for stabilizing two-dimensional carbon/CoP hybrids with enhanced hydrogen evolution. *Electrochim. Acta* **2021**, *375*, 137966. [CrossRef]
42. Liu, Y.; Zeng, Z.; Zeng, G.; Tang, L.; Pang, Y.; Li, Z.; Liu, C.; Lei, X.; Wu, M.; Ren, P.; et al. Immobilization of laccase on magnetic bimodal mesoporous carbon and the application in the removal of phenolic compounds. *Bioresour. Technol.* **2011**, *102*, 3653–3661. [CrossRef]
43. Nadargi, D.; Gurav, J.; Marioni, M.A.; Romer, S.; Matam, S.; Koebel, M.M. Methyltrimethoxysilane (MTMS)-based silica-iron oxide superhydrophobic nanocomposites. *J. Colloid Interface Sci.* **2015**, *459*, 123–126. [CrossRef] [PubMed]
44. Owens, G.J.; Singh, R.K.; Foroutan, F.; Alqaysi, M.; Han, C.-M.; Mahapatra, C.; Kim, H.-W.; Knowles, J.C. Sol–gel based materials for biomedical applications. *Prog. Mater. Sci.* **2016**, *77*, 1–79. [CrossRef]
45. Chao, C.; Zhang, B.; Zhai, R.; Xiang, X.; Liu, J.; Chen, R. Natural Nanotube-Based Biomimetic Porous Microspheres for Significantly Enhanced Biomolecule Immobilization. *ACS Sustain. Chem. Eng.* **2013**, *2*, 396–403. [CrossRef]
46. Ni, K.; Lu, H.; Wang, C.; Black, K.C.; Wei, D.; Ren, Y.; Messersmith, P.B. A novel technique for in situ aggregation of Gluconobacter oxydans using bio-adhesive magnetic nanoparticles. *Biotechnol. Bioeng.* **2012**, *109*, 2970–2977. [CrossRef]
47. Niu, Z.; He, X.; Huang, T.; Tang, B.; Cheng, X.; Zhang, Y.; Shao, Z. A facile preparation of transparent methyltriethoxysilane based silica xerogel monoliths at ambient pressure drying. *Microporous Mesoporous Mater.* **2019**, *286*, 98–104. [CrossRef]
48. Léonard, A.; Blacher, S.; Crine, M.; Jomaa, W. Evolution of mechanical properties and final textural properties of resorcinol–formaldehyde xerogels during ambient air drying. *J. Non-Cryst. Solids* **2008**, *354*, 831–838. [CrossRef]
49. Sedo, J.; Saiz-Poseu, J.; Busque, F.; Ruiz-Molina, D. Catechol-based biomimetic functional materials. *Adv. Mater.* **2013**, *25*, 653–701. [CrossRef]
50. Zhao, Z.; Liu, J.; Hahn, M.; Qiao, S.; Middelberg, A.P.J.; He, L. Encapsulation of lipase in mesoporous silica yolk–shell spheres with enhanced enzyme stability. *RSC Adv.* **2013**, *3*, 22008–22013. [CrossRef]

51. Mu, Q.; Liu, W.; Xing, Y.; Zhou, H.; Li, Z.; Zhang, Y.; Ji, L.; Wang, F.; Si, Z.; Zhang, B.; et al. Protein Binding by Functionalized Multiwalled Carbon Nanotubes Is Governed by the Surface Chemistry of Both Parties and the Nanotube Diameter. *J. Phys. Chem. C* **2008**, *112*, 3300–3307. [CrossRef]
52. Zhang, C.; Luo, S.; Chen, W. Activity of catalase adsorbed to carbon nanotubes: Effects of carbon nanotube surface properties. *Talanta* **2013**, *113*, 142–147. [CrossRef]
53. Sathishkumar, P.; Chae, J.C.; Unnithan, A.R.; Palvannan, T.; Kim, H.Y.; Lee, K.J. Laccase-poly(lactic-co-glycolic acid) (PLGA) nanofiber: Highly stable, reusable, and efficacious for the transformation of diclofenac. *Enzym. Microb. Technol.* **2012**, *51*, 113–118. [CrossRef]
54. Bao, C.; Xu, X.; Chen, J.; Zhang, Q. Synthesis of biodegradable protein–poly(ε-caprolactone) conjugates via enzymatic ring opening polymerization. *Polym. Chem.* **2020**, *11*, 682–686. [CrossRef]
55. Huang, S.M.; Huang, H.Y.; Chen, Y.M.; Kuo, C.H.; Shieh, C.J. Continuous Production of 2-Phenylethyl Acetate in a Solvent-Free System Using a Packed-Bed Reactor with Novozym®435. *Catalysts* **2020**, *10*, 714. [CrossRef]
56. Virgen-Ortíz, J.J.; Tacias-Pascacio, V.G.; Hirata, D.B.; Torrestiana-Sanchez, B.; Rosales-Quintero, A.; Fernandez-Lafuente, R. Relevance of substrates and products on the desorption of lipases physically adsorbed on hydrophobic supports. *Enzym. Microb. Technol.* **2017**, *96*, 31–35. [CrossRef]
57. Jiang, Y.; Zhai, J.; Zhou, L.; He, Y.; Ma, L.; Gao, J. Enzyme@silica hybrid nanoflowers shielding in polydopamine layer for the improvement of enzyme stability. *Biochem. Eng. J.* **2018**, *132*, 196–205. [CrossRef]
58. Pires-Cabral, P.; Da Fonseca, M.; Ferreira-Dias, S. Synthesis of ethyl butyrate in organic media catalyzed by Candida rugosa lipase immobilized in polyurethane foams: A kinetic study. *Biochem. Eng. J.* **2009**, *43*, 327–332. [CrossRef]
59. Gawas, S.D.; Jadhav, S.V.; Rathod, V.K. Solvent free lipase catalysed synthesis of ethyl laurate: Optimization and kinetic studies. *Appl. Biochem. Biotechnol.* **2016**, *180*, 1428–1445. [CrossRef]
60. Wang, H.; Duan, B.; Li, H.; Li, S.; Lu, Y.; Liu, Z.; Su, W. PEGylation and macroporous carrier adsorption enabled long-term enzymatic transesterification. *N. J. Chem.* **2020**, *44*, 3463–3470. [CrossRef]

*Article*

# Cold-Active Lipase-Based Biocatalysts for Silymarin Valorization through Biocatalytic Acylation of Silybin

Giulia Roxana Gheorghita [1], Victoria Ioana Paun [2], Simona Neagu [2], Gabriel-Mihai Maria [2], Madalin Enache [2], Cristina Purcarea [2], Vasile I. Parvulescu [1] and Madalina Tudorache [1,*]

[1] Department of Organic Chemistry, Biochemistry and Catalysis, Faculty of Chemistry, University of Bucharest, 4-12 Regina Elisabeta Blvd., 030016 Bucharest, Romania; giulia.gheorghita@s.unibuc.ro (G.R.G.); vasile.parvulescu@chimie.unibuc.ro (V.I.P.)

[2] Department of Microbiology, Institute of Biology Bucharest, Romanian Academy, 296 Splaiul Independentei, 060031 Bucharest, Romania; ioana.paun@ibiol.ro (V.I.P.); simona.neagu@ibiol.ro (S.N.); gabriel.maria@ibiol.ro (G.-M.M.); madalin.enache@ibiol.ro (M.E.); cristina.purcarea@ibiol.ro (C.P.)

* Correspondence: madalina.sandulescu@g.unibuc.ro

**Citation:** Gheorghita, G.R.; Paun, V.I.; Neagu, S.; Maria, G.-M.; Enache, M.; Purcarea, C.; Parvulescu, V.I.; Tudorache, M. Cold-Active Lipase-Based Biocatalysts for Silymarin Valorization through Biocatalytic Acylation of Silybin. *Catalysts* **2021**, *11*, 1390. https://doi.org/10.3390/catal11111390

Academic Editor: Gloria Fernandez-Lorente

Received: 24 October 2021
Accepted: 15 November 2021
Published: 17 November 2021

**Publisher's Note:** MDPI stays neutral with regard to jurisdictional claims in published maps and institutional affiliations.

**Copyright:** © 2021 by the authors. Licensee MDPI, Basel, Switzerland. This article is an open access article distributed under the terms and conditions of the Creative Commons Attribution (CC BY) license (https://creativecommons.org/licenses/by/4.0/).

**Abstract:** Extremophilic biocatalysts represent an enhanced solution in various industrial applications. Integrating enzymes with high catalytic potential at low temperatures into production schemes such as cold-pressed silymarin processing not only brings value to the silymarin recovery from biomass residues, but also improves its solubility properties for biocatalytic modification. Therefore, a cold-active lipase-mediated biocatalytic system has been developed for silybin acylation with methyl fatty acid esters based on the extracellular protein fractions produced by the psychrophilic bacterial strain *Psychrobacter SC65A.3* isolated from Scarisoara Ice Cave (Romania). The extracellular production of the lipase fraction was enhanced by 1% olive-oil-enriched culture media. Through multiple immobilization approaches of the cold-active putative lipases (using carbodiimide, aldehyde-hydrazine, or glutaraldehyde coupling), bio-composites (S1–5) with similar or even higher catalytic activity under cold-active conditions (25 °C) have been synthesized by covalent attachment to nano-/micro-sized magnetic or polymeric resin beads. Characterization methods (e.g., FTIR DRIFT, SEM, enzyme activity) strengthen the biocatalysts' settlement and potential. Thus, the developed immobilized biocatalysts exhibited between 80 and 128% recovery of the catalytic activity for protein loading in the range 90–99% and this led to an immobilization yield up to 89%. The biocatalytic acylation performance reached a maximum of 67% silybin conversion with methyl decanoate acylating agent and nano-support immobilized lipase biocatalyst.

**Keywords:** cold-active lipolytic activity; lipase immobilization; biocatalysis; silymarin; silybin

## 1. Introduction

*Silybum marianum* is a very effective natural remedy known in the popular consciousness as milk thistle and originates from Southern and Southeastern Europe. Being associated with daisies and artichokes [1], *S. marianum* belongs to the aster family, *Asteraceae* or *Compositae*. From a biochemical point of view, *S. marianum* comprises a group of flavonolignans collectively known as the silymarin mixture. Milk thistle extract contains 30% to 65% silymarin: 20–45% silydianin, 40–65% silybin A and B, and 10–20% isosilybin A and B [2]. The flavonolignans from the extract are natural polyphenols, biogenetically related to lignans due to their similar synthetic pathways [3]. Two phenylpropanoid units linked to another complex structural part ensure the binding of the $C_6C_3$ ring with the flavonoid nucleus in different positions. These compounds show multiple chirality, which leads to the existence of several stereoisomers in nature [4,5].

The main component and the most biologically active of silymarin is silybin [6]. Structurally, the chromone ring of silybin is responsible for the weak acidic properties and the antioxidant response. It is given by the phenolic hydroxyls at 3,4- and 4,5- positions, which form complexes with various metal ions [7]. Moreover, silybin exists in two stereoisomeric forms, A (2R, 3R, 10R, 11R) and B (2R, 3R, 10S, 11S), and has reduced solubility either in water or lipid media [8].

The cold-press silymarin extraction method proposed by Duran et al. is a more economical and energy efficient technique than either hot water extraction [9] or solvent-solvent partition [3,10]. The popularity of the cold-press method has increased since neither heat nor chemical treatment is used, and all beneficial nutritional properties of the raw material are transmitted to the oils. However, a problem arises in the light of the valorization of biomass waste, as silymarin abundantly passed into the residue [11]. Silymarin with anti-oxidative [12], anti-fibrotic, anti-inflammatory, membrane stabilizing, immunomodulatory, and liver regenerating properties plays an important role in experimental liver diseases [13]. There are also significant responses of flavonolignans to mushroom (*Amanita* sp.) poisoning, hepatitis, cirrhosis, and liver fibrosis [3], and the low water solubility of silymarin (430 mg/L) restricts therapeutic efficacy even if it is clinically safe at high doses (>1500 mg/day for humans) [14].

A new study on the bioavailability improvement was achieved through selective acylation of silybin in the 3–OH position with a 60% product yield in the presence of $CeCl_3 \cdot 7H_2O$ Lewis acid catalyst [8]. In order to obtain phosphorylated conjugates, Zarelli et al. initially acetylated silybin in a complex three-step organic synthetic process with a 75% yield [15]. Moreover, there are encapsulated silymarin variants in matrices less toxic and much better accepted by the human organism [10,16]. Silymarin could be included in natural β-cyclodextrins forming complexes, often used because of their solubilization potential within the body barriers without metabolic degradation [14]. Liposomal systems are known to find immediate access to reticulo-endothelial system (RES) rich sites like the liver and spleen, and this self-targeted nature of liposomal carriers can be exploited for silymarin distribution to hepatic sites [14,17].

Besides chemical modification, or chemo-enzymatic or biocatalytic approaches, Xanthakis et al. showed that a lipase-based acylation of silybin with vinyl butyrate via Novozyme 435 biocatalyst strongly depends on the organic medium, where the conversion degree increased from 78 to 100% [18]. In favor of a faster synthesis, Vavrikova et al. used divinyl esters to modify silybin through *Candida antartica* B lipase mediation and achieved silybin dimers with a 24–44% yield range [19].

Microorganisms inhabiting low-temperature environments including bacteria, yeasts, fungi, and algae [20] have adapted to produce cold-active enzymes that ensure the minimum rate of chemical and metabolic reactions [21,22]. The optimum temperatures for cold-adapted enzyme activities are usually in the range 20–30 °C, which could be considered close to that of the neutro-/thermophiles [21]. The ability of these enzymes to adjust their activity to low temperatures is based on their structural flexibility [23]. There are two strengths related to the active enzymes' operation: the manipulation of thermo-labile and sensitive substrates at low temperature and the inactivation of the enzyme when the temperature increases [20]. Few genomes of psychrophiles have been deposited in public databases [13].

Scarisoara Ice Cave is located in the Carpathian Mountains (Bihor, Romania) and has been accumulating perennial ice deposits for over 10,000 years. The constant negative temperatures throughout the year lead to a continuous deposit in glacial alternating organic-inorganic sediment layers. Thus, Scarisoara Ice Cave is known to host the world's oldest and largest underground perennial ice block [24]. During subsequent ice sampling from the Great Hall area of the cave, Paun et al. successfully determined the prokaryotic diversity [25] and isolated bacterial strains from a 13,000-year-old ice core [26]. The plethora of potentially active bacterial communities comprises 25 phyla, 58 classes, and 325 genera. Thus, a new *Psychrobacter SC65A.3* was taxonomically assigned and characterized by C. Purcarea's research group (unpublished data).

For drug and cosmetic industries, a cold-active lipase-mediated system [27,28] for the acylation of silybin A/B with proper methyl fatty acid esters [29,30] has been designed to improve the initial liposolubility [31,32] of silybin and to enhance silymarin bioavailability (Scheme 1).

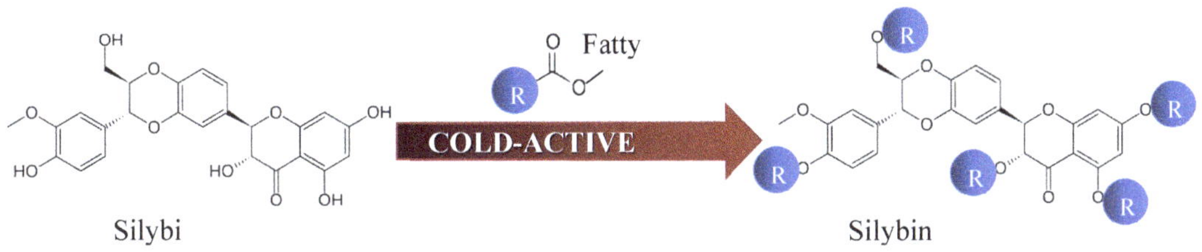

**Scheme 1.** Acylation reaction of silybin with methyl fatty acid esters based on lipase biocatalysis.

In this context, the extracellular enzymatic fraction of the *Psychrobacter SC65A.3* strain recently isolated from the ice deposits of Scarisoara Ice Cave (Romania) was exploited for proper identification and characterization of the cold-active lipolytic activity. Therefore, new lipase-based biocatalysts were designed and used in a silybin conversion system at low temperature for silymarin valorization.

## 2. Results and Discussion

### 2.1. Evaluation of Psychrobacter SC65A.3 Lipolytic Activity on Different Substrates

Formation of white calcium oleate microcrystals around the inoculum spot is a direct consequence of extracellular lipolytic activity, hence for the ability of the strain to hydrolyze the ester bond from a long hydrocarbon chain (C18, Tween 80). The diameter of the hydrolysis zone is directly proportional to the amount of enzymatic activity (Figure 1), so the larger the zone diameter, the more lipolytic activity. After seven days of 15 °C, the hydrolysis area expended near the reaction spot.

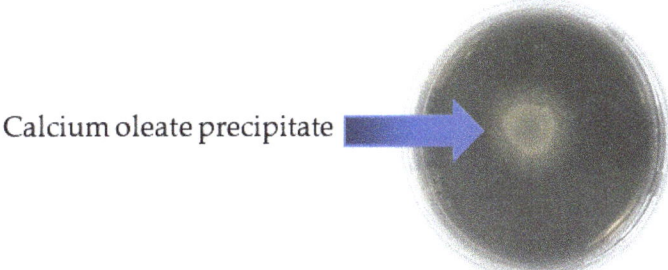

**Figure 1.** Lipolytic activity on Tween 80 substrate by formation of calcium oleate precipitate.

Screening on vegetable oils was performed to estimate the optimal substrate concentration that increases extracellular lipase production of *Psychrobacter SC65A.3*. The hydrolysate of oil substrate reacted with 0.001% Rhodamine B to create a red-orange, fluorescent halo under 350 nm UV light (Figure 2). However, the highest lipase activity for *Psychrobacter SC65A.3* was observed to be a concentration of 1% sunflower and 1% olive oil.

The *Psychrobacter SC65A.3* was cultivated at 15 °C for 3 days for fully grown cell colonies to produce and extracellularly release proteins into the culture medium supplemented with 1% olive oil, which is the preferred carbon source for lipolytic hydrolysis [33].

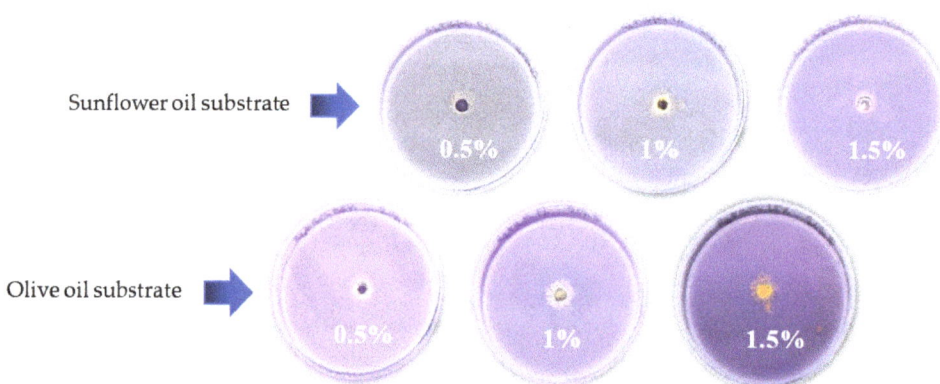

**Figure 2.** Lipolytic activity on vegetable oil substrates by formation of fluorescent complex.

*2.2. Characterization of the Immobilized Cold-Active Lipase*

Catalytic activity of cold-active lipase from the protein extracellular fraction was determined for both free and immobilized forms. A total of five bio-composites (S1–5) were prepared by covalent attachment of the protein extract on either magnetic particles or resin polymer supports using different immobilization approaches (see Section 3,). Protein loading of the bio-composites was calculated and the corresponding values are displayed in Table 1. Recovery of the lipase activity and also the immobilization yield were determined regarding the catalytic activity of the free lipase fraction (4.4 U × mg protein$^{-1}$). As a general remark, more than 95% protein loading was achieved for all the bio-composites with the immobilization yield in the range 74–89%. Additionally, the catalytic activity was preserved after the immobilization (around 80% recovery of the catalytic activity). A slight increase in the catalytic activity after immobilization (105 ± 5.1% recovery of the catalytic activity) can be observed for the GA approach (S5), characterized by relatively good protein loading (90 ± 4.3%) and high immobilization yield (89 ± 4.0%). The lipase fraction underwent an "apparent" activation phenomenon, which is a common aspect of lipase behavior already reported in the literature [34]. In addition, the GA linker can exhibit a positive effect on the catalytic activity by placing the protein molecules away from the support surface to avoid the steric inhibition from protein–support interactions. Moreover, an increase in lipase activity was registered for the S2 bio-composite and can also be explained based on lipase "apparent" activation (see Table 1). In this case, the immobilization was performed based on the EDC approach, which led to high protein loading and significant immobilization yield. The greater availability of the support surface nano-design area for S2 allowed larger immobilization of the protein amount and avoided the steric inhibition by protein–protein interaction. Similarly, the periodate method exhibited good protein loading for low immobilization yield.

**Table 1.** Evaluation of the immobilization efficiency of the prepared bio-composites. (The experimental data were determined in triplicates).

|    | Recovery of Lipase Activity (%) | Immobilization Yield (%) | Protein Loading (%) |
|----|---------------------------------|--------------------------|---------------------|
| S1 | 82 ± 5.1                        | 80 ± 4.5                 | 99 ± 5.3            |
| S2 | 128 ± 6.3                       | 84 ± 4.7                 | 99 ± 4.8            |
| S3 | 80 ± 5.4                        | 76 ± 4.3                 | 93 ± 4.6            |
| S4 | 82 ± 5.5                        | 74 ± 4.8                 | 96 + 4.9            |
| S5 | 105 ± 5.1                       | 89 ± 4.0                 | 90 ± 4.3            |

FTIR DRIFT analysis of free and immobilized lipase were performed. Interferograms of the supports were considered as the reference (Supplementary Material, Figure S1).

A simple comparison of the spectra confirmed the enzyme attachment to the support surface. For the immobilized specimens, the bands around 1550–1700 cm$^{-1}$ specific for peptide bonds (Amide I and II region) [35] are recovered from the free biocatalyst. For S1–4, the bands in the range 570–650 cm$^{-1}$ stand for Fe-O specific vibrations in the magnetic particles [36], while methacrylate support (S5) gives a particular band at 1730 cm$^{-1}$ for the C=O stretching carboxyl, as well as two bands between 3300 and 3500 cm$^{-1}$ for primary amino groups projecting out from the polymer outer face. Additionally, FTIR spectra allowed us to analyze the changes in protein secondary structure as a direct effect of the immobilization.

SEM characterization of the immobilized specimens is presented in Figure 3. As a method of control, SEM micrographs were obtained for purified lipase from Aspergillus niger (Supplementary Material, Figure S2). The protein aggregates dispersed freely in the liquid phase had similar profiles to the ones recovered from each support surface. Therefore, the protein content on the functionalized particle layer was highlighted. The aspect of magnetic particles is defined and characteristically crystallized to magnetite and maghemite (Figure 3). In the case of the free support based on methacrylate polymer, the beads have a clear, smooth surface. For immobilized lipase, changes in the morphology of the bio-composites occurred. A brighter contrast displays protein deposits that adhere covalently to the support surface [37,38]. In this way, the immobilization of the lipase on the solid support was confirmed one more time according to the results of FTIR analysis.

Kinetics measurements were performed for the immobilized lipase fraction and compared to a free lipase biocatalyst. The kinetic constants $K_m$ and $V_{max}$, and particularly $k_{cat}$, were calculated for 25 and 37 °C to confirm the cold-active behavior of the lipase. The corresponding values are presented in Table 2. Lipolytic activity of free and immobilized enzymes was detected for both tested temperatures (25 and 37 °C). According to the $K_m$ values at 25 °C, lipase biocatalysts exhibited cold-active behavior, without substantial changes to the substrate's apparent affinity for a higher temperature (37 °C). In the case of the catalytic constant ($k_{cat}$), similar values were recorded for both temperatures. For catalytic efficiency ($k_{cat}/K_m$), the close values for both temperatures support the structural and functional stability of the proteinaceous material.

**Table 2.** Kinetic parameters of cold-active lipase for both free and immobilized forms at 25 °C and 37 °C, using 1/4 $v/v$ lipase biocatalysts (see Section 3). (The experimental data were determined in triplicates.)

| Biocatalyst | $K_m$ (mM) | | $k_{cat}$ (min$^{-1}$) | | $k_{cat}/K_m$ (mM$^{-1}$ min$^{-1}$) | |
|---|---|---|---|---|---|---|
| | 25 °C | 37 °C | 25 °C | 37 °C | 25 °C | 37 °C |
| Free | 2.18 ± 0.15 | 1.85 ± 0.09 | 1.63 ± 0.08 | 1.95 ± 0.09 | 0.75 ± 0.05 | 1.05 ± 0.09 |
| S1 | 1.61 ± 0.08 | 1.17 ± 0.09 | 0.99 ± 0.05 | 0.97 ± 0.06 | 0.62 ± 0.04 | 0.83 ± 0.04 |
| S2 | 0.56 ± 0.04 | 0.65 ± 0.03 | 0.93 ± 0.05 | 0.99 ± 0.06 | 1.67 ± 0.09 | 1.54 ± 0.07 |
| S3 | 4.18 ± 0.21 | 1.41 ± 0.07 | 2.29 ± 0.09 | 1.25 ± 0.07 | 0.55 ± 0.03 | 0.88 ± 0.04 |
| S4 | 0.63 ± 0.03 | 0.78 ± 0.05 | 0.65 ± 0.03 | 0.92 ± 0.05 | 1.02 ± 0.09 | 1.17 ± 0.09 |
| S5 | 0.94 ± 0.05 | 0.64 ± 0.03 | 0.85 ± 0.08 | 1.2 ± 0.09 | 0.90 ± 0.05 | 1.88 ± 0.09 |

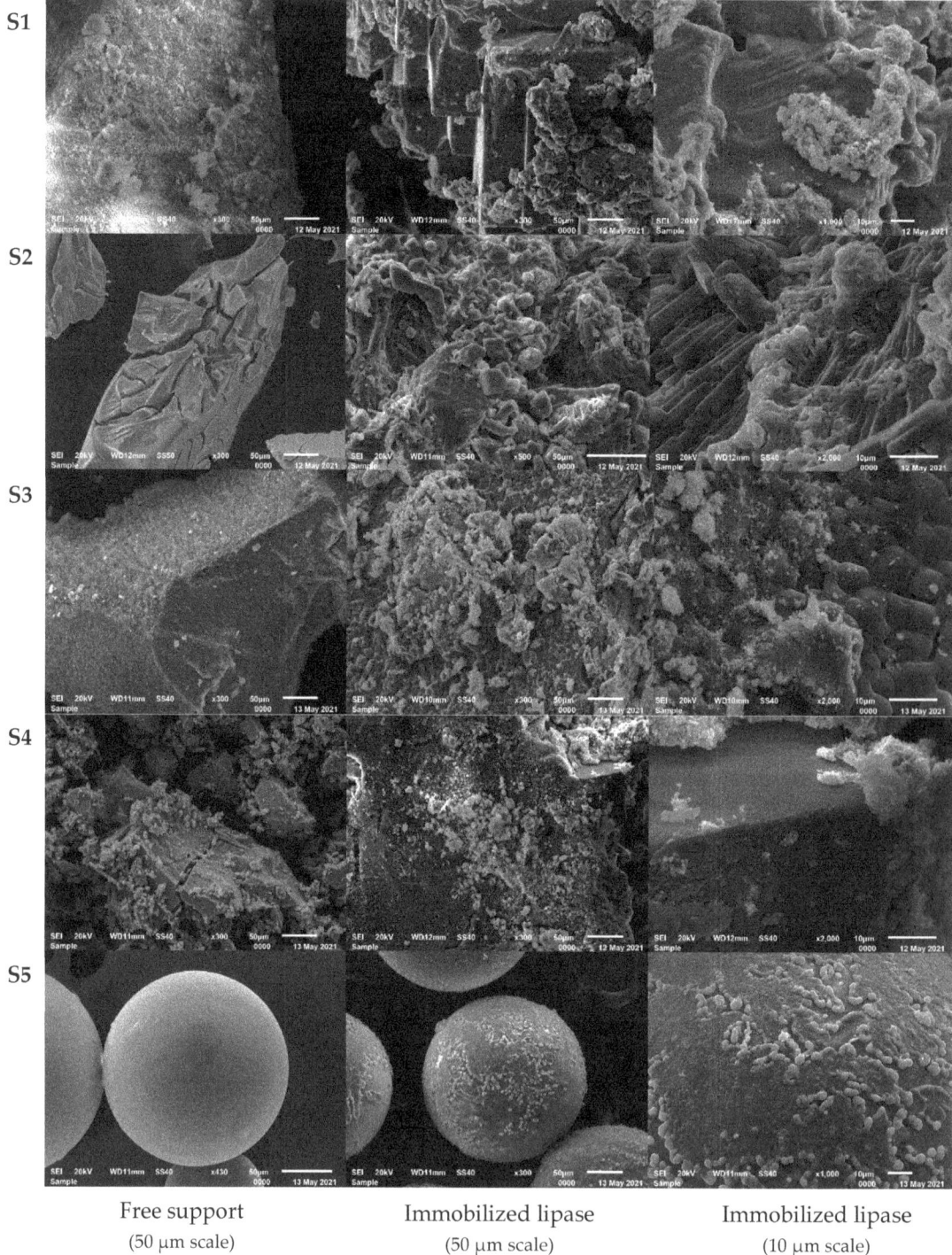

**Figure 3.** SEM micrographs of the bio-composites and corresponding supports.

## 2.3. Cold-Active Lipase Catalyzing Silybin Acylation

Free and immobilized cold-active lipase fractions were tested for the acylation of silybin substrate. Methyl fatty acid esters as acylating donors were used for the transesterification process that leads to silybin ester products (Scheme 1). The experiments were performed at 25 °C in line with the cold-active behavior of the expressed lipase. Additionally, the same enzymatic system was tested at 40 °C for comparison. For both temperatures, silybin was acylated with different acylation reagents (e.g., methyl decanoate, methyl laurate, methyl myristate, and methyl palmitate) and the reaction system was assisted by free lipase and S5 biocatalyst. THF was chosen as the organic solvent of the system (the activity of the cold-active lipase fraction was not affected by the THF—data are not included). The experimental results are presented in Figure 4.

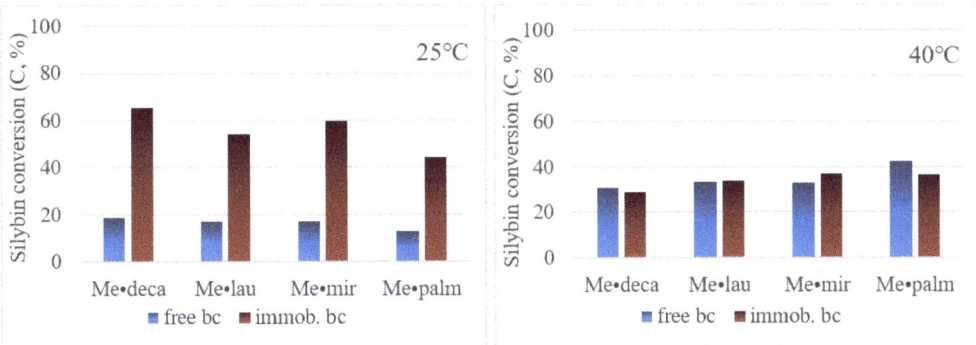

**Figure 4.** System performance in terms of silybin conversion by using different temperatures: 25 °C and 40 °C, with free lipase (■) or S5 biocatalyst (□). Experimental conditions: 2 mM Silybin A/B, 40 mM acylation reagent, 4.72 mg/mL free lipase/4.25 mg/mL S5 biocatalyst in 1 mL THF, at 25 °C or 40 °C and 1000 rpm for 24 h.

From the beginning, it is important to specify that the lipase fraction exhibited catalytic activity at 25 °C sustaining the cold-active character in the developed biocatalysts. For free lipase biocatalyst, the system performance was improved when the temperature increased from 25 to 40 °C. As an example, the silybin conversion was 19% for 25 °C and 31% for 40 °C using methyl decanoate for the acylation process. Therefore, increasing the temperature led to "an apparent" activation of the free lipase fraction.

Catalytic performance of the system was strongly improved when the immobilized lipase was used, especially at 25 °C. For the methyl decanoate agent, silybin conversion reached a maximum value of 65%, which is more than three times higher compared to the free lipase biocatalyst. Similar behavior was observed for all the methyl fatty acid esters tested. Therefore, the developed lipase fraction exhibited preponderantly cold-active catalytic activity in the immobilized forms.

The experiments were extended for all the prepared bio-composites (S1–5). Acylation of silybin was performed separately with different methyl fatty acid esters at 25 °C. The case of a free lipase biocatalyst was also chosen as a reference. The experimental results are presented in Figure 5.

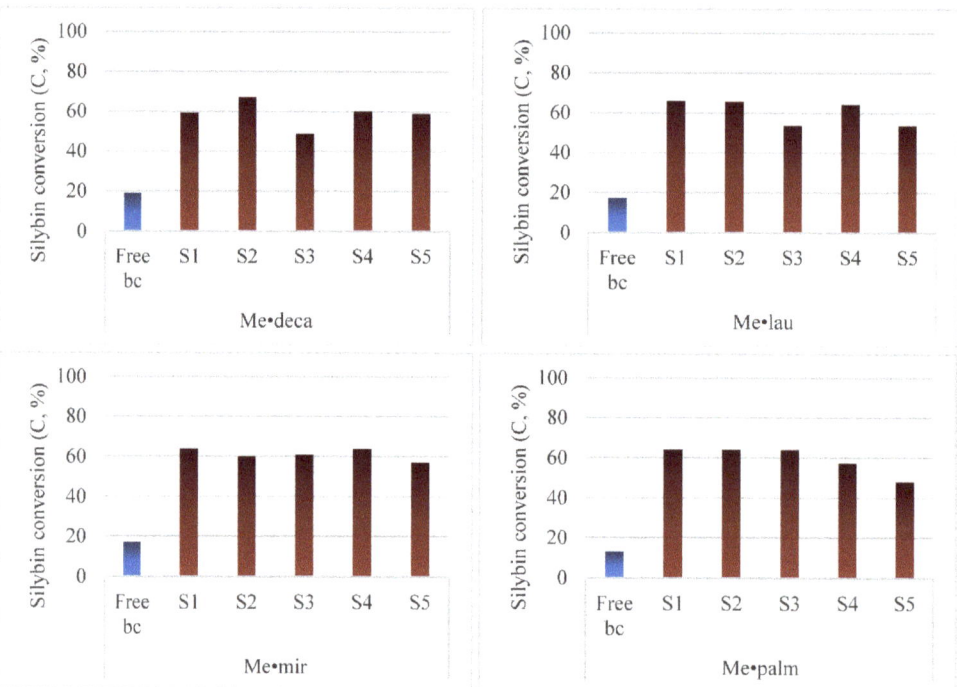

**Figure 5.** System performance in terms of silybin conversion for every biocatalyst specimen. Experimental conditions: 2 mM Silybin A/B, 45 mM Me-decanoate (Me•deca), 40 mM Me-laurate (Me•lau), 35 mM Me-miristate Me•mir), 30 mM Me-plamitate (Me•palm), 1:10 $v/v$ biocatalyst, free (■) or immobilized (□), in 1 mL THF, at 25 °C and 1000 rpm for 24 h.

In the case of the free lipase specimen, the performance range in terms of substrate conversion varied between 13 and 19% (Figure 5). A slight variation in the silybin conversion with the hydrocarbon chain of the acylating agent was achieved, e.g., 19% for the methyl decanoate system and 13% for the methyl palmitate system. It is true that when the total protein content is embedded generically in the free biocatalyst, steric hindrances are a risky factor in achieving the synthesis, the lipase catalytic lid being restricted by the neighboring interactions.

For the immobilized lipase fraction, silybin conversion varied between 41 and 66% (Figure 5). The first important observation is based on catalytic activity improved up to 53% while comparing to the free enzyme. Thus, lipase is apparently activated by immobilization even though covalent attachment was performed. It has already been reported in the literature that lipases have such unusual behavior, but this is very rewarding when they are found at the interface between two different environments. Most studies are based on the hyper-activation phenomenon of lipases between organic and aqueous media [39,40]. However, the heterogeneity of the system given by the immobilized form and the liquid reaction medium was also reported as an alternative for lipase activation [41,42]. Now, our experimental results confirm again the lipase activation after immobilization on the solid support.

A second observation that correlated with the immobilized lipase biocatalysts (S1 and S3) is given by the increased biocatalytic potential in a constant trend with the increase in the hydrocarbon chain provided by the acylation agent. This behavior is obvious for the S3 bio-composite (49, 53, 61, and 64% silybin conversion for methyl decanoate, methyl laurate, methyl myristate, and methyl palmitate, respectively). Basically, the protein conformation sustained by immobilization allowed the facial reception of the substrate at the catalytic site [43].

Moreover, the highest catalytic performance was recorded for the S2 biocatalyst and the lowest for S5, independently of the acylation agent. Maximum conversion of silybin (67%) was achieved for the S2 and methyl decanoate couple, while the minimum value (48%) was recorded for S5 and methyl palmitate. Therefore, the nano-scale biocatalyst (S2) offered a good catalytic performance in the developed system. This conclusion is also supported by the data from Tables 1 and 2.

## 3. Materials and Methods

### 3.1. Chemicals and Solutions

MES buffer (2-(N-morpholino) ethane sulfonic acid, 0.1 M stock solution, pH = 5), PBS buffer (phosphate buffered saline, 0.1 M stock solution, pH = 7.4) and TRIS-HCl buffer (2-Amino-2-(hydroxymethyl)-1,3-propanediol hydrochloride, 0.1 M stock solution, pH = 8) were standardly prepared. All the reagents used for the buffer preparation were purchased from Sigma-Aldrich (Taufkirchen, Germany).

For microbiological manipulations, R2B culture media were purchased from Melford Biolaboratories Ltd. (Ipswich, UK), while 1% Tween 80, 0.001% Rhodamine B, and 0.01% $CaCl_2 \cdot 2H_2O$ were from Sigma Aldrich-Merck, Darmstadt, Germany. The vegetal oils were achieved from commercial market and filtered before using.

For the biochemical methods were used acetone, ethanol, bovine serum albumin, *p*-nitrophenol and *p*-nitrophenyl butyrate, sodium bicarbonate, all purchased from Sigma Aldrich-Merk, Darmstadt, Germany. To build up the immobilized biocatalysts, 1-ethyl-3-(3-dimethylaminopropyl) carbodiimide, sodium periodate, glutaraldehyde, potassium bromide and sodium azide were purchased from Sigma-Aldrich Merck, Darmstadt, Germany, while the magnetic particles and amino-$C_2$-methacrylate type resin were purchased from Chemicell Company (Rostock, Germany) and Purolite Life Sciences Company (Wales, UK), respectively.

Acetone, acetonitrile, tetrahydrofuran of HPLC purity (99.99%), and reagents for the biocatalytic system comprising silybin A/B (silybin), methyl decanoate, methyl laurate, methyl myristate, and methyl palmitate, were obtained from Sigma-Aldrich Merck, Darmstadt, Germany.

### 3.2. Plate Screening Assays

*Psychrobacter SC65A.3* strain was isolated from the perennial ice accumulated in Scarisoara Ice Cave (Romania) [26] and identified by 16S rRNA gene sequencing and genome sequencing (Macrogen, Seoul, Korea) (unpublished data). The strain was cultivated in R2B media at 15 °C for 3 days. For plate screening assays, 150 µL of culture in logarithmic phase growth [33] was used to inoculate R2A growth medium containing 1% Tween 80, and 0.01% $CaCl_2 \cdot 2H_2O$ or 1% olive/sunflower oils and 0.001% Rhodamine B. The plates were incubated at 15 °C for 3 days, and the presence of a lipolytic activity was evaluated by the specific products formed after hydrolysis: calcium oleate (as white precipitate) or fluorescent Rhodamine B-hydrolysate complex.

### 3.3. Preparation of the Protein Extract

*Psychrobacter SC65A.3* was cultivated in R2B medium (500 mL) at 15 °C for 72 h, and the bacterial biomass was separated from the supernatant containing extracellular proteins by centrifugation at 4 °C for 20 min at 4800 rpm. The bacterial secreted proteins were precipitated with 80% acetone (2:1, $v/v$ to culture medium) and recovered after centrifugation at 10,000 rpm for 10 min in 100 mM TRIS-HCl pH 8 [44]. The concentration of the resulted extracellular protein extract was measured spectrophotometrically at 280 nm by the BSA method [45].

*3.4. Enzyme Assay*

Lipase activity was measured by the colorimetric method [46], using *p*-nitrophenyl butyrate (*p*-NPB) as the substrate. The reaction mixture containing 2.5 mM *p*-NPB dissolved in ethanol, 1:4 $v/v$ protein extract, and 32.5 mM Tris-HCl (pH 7.5) was incubated for 30 min at 25 °C or 37 °C, and the reaction was terminated by addition of 20 mM $Na_2CO_3$ blocking solution (10 min incubation). The activity was measured spectrophotometrically at 347 nm using 0.1–0.5 mM *p*-nitrophenol in ethanol as the calibration curve. The activity was expressed as μM/min/mg protein (units per mg of protein extract, where 1 U corresponds to μM product concentration obtained in 1 min). Kinetic parameters $k_{cat}$ and $K_m$ were calculated from substrate saturation curves carried out in the presence of 0.25–2.5 mM *p*-nitro phenylbutyrate substrate.

*3.5. Lipase Immobilization*

3.5.1. Preparation of the Immobilized Lipase Biocatalyst

The enzyme from the protein extract was attached to the solid support to improve the efficiency of the enzyme biocatalyst. For the immobilization, different approaches were applied based on covalent attachment of the enzyme via $-NH_2$ or –COOH residues. Magnetic particles and resin beads were considered as the solid support.

Covalent immobilization of the enzyme on the functionalized surface of magnetite particles was performed using the carbodiimide approach. The carbodiimide method involved the use of EDC (1-ethyl-3-(3-dimethylaminopropyl) carbodiimide, 0.25 M), PBS (0.1 M, pH = 7.4)/MES (0.1 M, pH = 5) washing buffers and blocking, and storage solution prepared in PBS with 0.1% bovine serum albumin and 0.05% sodium azide. Immobilization protocols were reproduced after the manufacturer's suggestions [47]. An amount of 1 mL of commercial suspension of Si-MAG-amine, Si-MAG-carboxyl, and fluid-MAG-ARA magnetic particles was gathered with a magnetic separator and foremost mixed with 0.25 M EDC in 0.1 M MES buffer (pH = 5). Then, 200 μL of protein extract was added and the resulting mixture was gently shaken for 2 h at room temperature.

For the $NaIO_4$ method, 500 μL protein extract was incubated in the dark with 5 mg $NaIO_4$ at room temperature for 30 min in the presence of Si-MAG-Hydrazine particles. The magnetic beads were placed to react with 0.25 mL of newly oxidized protein extract for 6 h at room temperature. PBS buffer (pH = 7) containing 0.1% BSA and 0.05% $NaN_3$ was used to block the ongoing reaction.

Covalent immobilization via glutaraldehyde (GA) linker was performed on amino-$C_2$-methacrylate type resin in MES buffer (10 mM, pH 5). The support, spherical beads of 150–300 μm diameter, was originally functionalized with -$NH_2$ groups using methacrylate crosslinker polymer. An amount of 0.1 g amino–$C_2$–methacrylate support was dispersed in the mixture containing 0.1% glutaraldehyde and 100 μL protein extract. After 2 h under gentle stirring, the immobilized enzyme was separated by centrifugation at 2000 rpm for 10 min. A washing step and resuspension were performed with/in PBS pH 7.4 buffer 3 times [48].

Finally, five different enzyme composites were obtained as indicated in Table 3:

**Table 3.** Immobilized lipase biocatalysts.

|  | Support | Immobilization Method |
|---|---|---|
| S1 | sMP-COOH | EDC |
| S2 | fMP-COOH | EDC |
| S3 | sMP-$NH_2$ | EDC |
| S4 | sMP-NH-$NH_2$ | NaIO4 |
| S5 | MTC-$NH_2$ | GA |

### 3.5.2. Characterization of the Immobilized Lipase Biocatalyst

FTIR spectra of bio-composites and simple/functionalized supports were recorded using a Vertex 70 spectrophotometer (Bruker, Ettlingen, Germany) equipped with the Diffuse Reflectance Infrared Fourier Transform cell. A minimum amount of sample was diluted with potassium bromide matrix. Sixty-four scans were collected with a resolution of 4 cm$^{-1}$ in the wavenumber range 4000–400 cm$^{-1}$.

SEM micrographs were obtained through scanning electron microscopy, performed on a Jeol instrument (JSM-6610LV). Before SEM imaging, a thin layer of gold was dispersed over the sample in a high-vacuum chamber.

The activity of the immobilized lipase fraction was evaluated using the same protocol as for the free lipase fraction (see Section 3.4. Enzyme assay). The results of the colorimetric analysis (p-NPB protocol) allowed us to characterize the efficiency of the immobilization methods based on protein loading, recovery of lipase activity, and immobilization yield [49].

$$protein\ loading\ (\%) = \frac{m_i - m_s}{m_i} \times 100$$

$$recovery\ of\ the\ lipase\ activity\ (\%) = \frac{LA_{immob}}{LA_i} \times 100$$

$$immobilization\ yield\ (\%) = \frac{LA_i - LA_s}{LA_i} \times 100$$

$m_i$—initial protein amount (after concentration of the extracellular extract);
$m_s$—protein amount in the supernatant after immobilization step;
$LA_{immob}$—lipase activity of the immobilized protein fraction;
$LA_i$—lipase activity of the solution used for immobilization;
$LA_s$—lipase activity of the supernatant after immobilization.

### 3.6. Biocatalytic System for Silybin Acylation

Biocatalytic acylation of silybin was performed in a reaction phase with 2 mM silybin dissolved in THF and 45 mM Me-decanoate (Me•deca), 40 mM Me-laurate (Me•lau), 35 mM Me-miristate (Me•mir), or 30 mM Me-plamitate (Me•palm). The incubated mix of silybin with fatty esters was assisted by the lipase biocatalyst (10%, $v/v$) at 25 °C and 1000 rpm for 24 h. After reaction, the mixture was centrifuged for 15 min at 1500 rpm, allowing the separation of (free or immobilized) protein by collecting the supernatant separately. The supernatant was filtrated through 0.22 μm Millipore filters into HPLC vials and left in the oven at 70 °C for complete solvent evaporation. These dried samples were dissolved in the mobile phase of the HPLC-DAD system to finally reveal the sample content. For this, a 1260 Infinity HPLC modular system from Agilent Technologies was equipped with a Poroshell 120 EC-C18 column and a diode array-type detector (DAD). Analysis parameters implied the 1 mL/min flow rate mobile phase (41:59 = acetonitrile: acetone) and 25 μL injection volume. The signal was recorded at 210 nm for 30 min.

Using standards, identification of the sample components led to the following retention times: silybin at 2.4 min, methyl decanoate at 3.1 min, methyl laurate at 3.3 min, methyl myristate at 3.7 min, and methyl palmitate at 4.1 min. Silybin conversion was calculated based on the chromatographic data, by using the well-known formula:

$$Conversion,\ C\ (\%) = \frac{mass\ of\ converted\ silybin}{initial\ mass\ of\ silybin} \times 100$$

## 4. Conclusions

A valuable biocatalytic process based on the enzymatic activity of a cold-active lipase fraction was successfully elaborated for silymarin for improved bioavailability. A cold-active lipase fraction was obtained from a novel *Psychrobacter SC65A.3* strain isolated from Scarisoara Ice Cave (Romania). Enzyme immobilization allowed us to generate bio-composites (S1–5) with preserved or even higher catalytic activity. Additionally, the kinetic parameters of the bio-composites sustained the cold-active character of the immobilized lipase biocatalysts (see S2 bio-composite) when measuring at 25 °C and 37 °C.

The developed biocatalytic process based on cold-active lipase activity involved the acylation of silybin as the main component of silymarin with methyl fatty acid esters (e.g., methyl decanoate, methyl laurate, methyl myristate, and methyl palmitate) for both free and immobilized (S1–5) biocatalysts. A maximum of 67% silybin conversion was reached for the S2 biocatalyst and methyl decanoate as the acylating agent.

This biocatalytic system could offer several advantages for silymarin valorization such as (i) energy saving by the cold-active behavior of the biocatalyst, (ii) derivatization of the silybin structure with fatty acid residue leading to a more hydrophobic product, (iii) curative features of silybin esters when omega acid residues are used as the acylation agent, and (iv) improving the efficiency of the cold-pressed milk thistle oil technology by considering the corresponding residues as a silybin and fatty acid source.

**Supplementary Materials:** The following are available online at https://www.mdpi.com/article/10.3390/catal11111390/s1, Figure S1: FTIR DRIFT spectra, Figure S2: SEM micrograph of Aspergillus niger.

**Author Contributions:** Conceptualization, G.R.G., C.P. and M.T.; methodology, G.R.G., V.I.P. (Vasile I. Parvulescu), S.N., G.-M.M. and M.E.; formal analysis, G.R.G., V.I.P. (Victoria Ioana Paun). and S.N.; investigation, G.R.G., V.I.P. (Victoria Ioana Paun) and G.-M.M.; data curation, G.R.G. and V.I.P. (Vasile I. Parvulescu); writing—original draft preparation, G.R.G. and M.T.; writing—review and editing, G.R.G. and M.T.; visualization, C.P.; supervision, M.T. and C.P.; project administration, M.T. and C.P.; funding acquisition, M.T. and C.P. All authors have read and agreed to the published version of the manuscript.

**Funding:** This research was funded by the UEFISCDI (Romania), grant number PN-III-P2-2.1-PED-2019-2461, 356PED/2020.

**Data Availability Statement:** The data presented in this study are available in the article and Supplementary Material.

**Conflicts of Interest:** The authors declare no conflict of interest.

## References

1. Flora, K.; Hahn, M.; Rosen, H.; Benner, K. Milk Thistle (*Silybum marianum*) for the Therapy of Liver Disease. *Am. J. Gastroenterol.* **1998**, *93*, 139–143. [CrossRef]
2. Kuki, Á.; Nagy, L.; Deák, G.; Nagy, M.; Zsuga, M.; Kéki, S. Identification of Silymarin Constituents: An Improved HPLC–MS Method. *Chromatographia* **2011**, *75*, 175–180. [CrossRef]
3. Begum, S.A.; Sahai, M.; Ray, A.B. Non-conventional Lignans: Coumarinolignans, Flavonolignans, and Stilbenolignans. *Fortschr. Chem. Org. Nat.* **2010**, *93*, 1–70.
4. Csupor, D.; Csorba, A.; Hohmann, J. Recent advances in the analysis of flavonolignans of Silybum marianum. *J. Pharm. Biomed. Anal.* **2016**, *130*, 301–317. [CrossRef]
5. Vostálová, J.; Tinková, E.; Biedermann, D.; Kosina, P.; Ulrichová, J.; Svobodová, A.R. Skin Protective Activity of Silymarin and its Flavonolignans. *Molecules* **2019**, *24*, 1022. [CrossRef]
6. Bijak, M. Silybin, a Major Bioactive Component of Milk Thistle (*Silybum marianum* L. Gaernt.)—Chemistry, Bioavailability, and Metabolism. *Molecules* **2017**, *22*, 1942. [CrossRef] [PubMed]
7. Biedermann, D.; Vavříková, E.; Cvak, L.; Křen, V. Chemistry of silybin. *Nat. Prod. Rep.* **2014**, *31*, 1138–1157. [CrossRef]
8. Drouet, S.; Doussot, J.; Garros, L.; Mathiron, D.; Bassard, S.; Favre-Réguillon, A.; Molinié, R.; Lainé, É.; Hano, C. Selective Synthesis of 3-O-Palmitoyl-Silybin, a New-to-Nature Flavonolignan with Increased Protective Action against Oxidative Damages in Lipophilic Media. *Molecules* **2018**, *23*, 2594. [CrossRef]
9. Duan, L.; Carrier, D.J.; Clausen, E.C. Silymarin Extraction from Milk Thistle Using Hot Water. *Appl. Biochem. Biotechnol.* **2004**, *114*, 559–568. [CrossRef]
10. Theodosiou, E.; Purchartová, K.; Stamatis, H.; Kolisis, F.; Křen, V. Bioavailability of silymarin flavonolignans: Drug formulations and biotransformation. *Phytochem. Rev.* **2013**, *13*, 1–18. [CrossRef]
11. Duran, D.; Ötleş, S.; Karasulu, E. Determination amount of silymarin and pharmaceutical products from milk thistle waste obtained from cold press. *Acta Pharm. Sci.* **2019**, *57*, 85. [CrossRef]

12. Fidrus, E.; Ujhelyi, Z.; Fehér, P.; Hegedűs, C.; Janka, E.A.; Paragh, G.; Vasas, G.; Bácskay, I.; Remenyik, É. Silymarin: Friend or Foe of UV Exposed Keratinocytes? *Molecules* **2019**, *24*, 1652. [CrossRef] [PubMed]
13. Hackett, E.S.; Twedt, D.C.; Gustafson, D.L. Milk Thistle and Its Derivative Compounds: A Review of Opportunities for Treatment of Liver Disease. *J. Vet. Intern. Med.* **2013**, *27*, 10–16. [CrossRef] [PubMed]
14. Di Costanzo, A.; Angelico, R. Formulation Strategies for Enhancing the Bioavailability of Silymarin: The State of the Art. *Molecules* **2019**, *24*, 2155. [CrossRef]
15. Di Fabio, G.; Zarrelli, A.; Romanucci, V.; Della Greca, M.; De Napoli, L.; Previtera, L. New Silybin Scaffold for Chemical Diversification: Synthesis of Novel 23-Phosphodiester Silybin Conjugates. *Synlett* **2012**, *24*, 45–48. [CrossRef]
16. Chambers, C.S.; Biedermann, D.; Valentová, K.; Petrásková, L.; Viktorová, J.; Kuzma, M.; Křen, V. Preparation of Retinoyl-Flavonolignan Hybrids and Their Antioxidant Properties. *Antioxidants* **2019**, *8*, 236. [CrossRef] [PubMed]
17. Kesharwani, S.S.; Jain, V.; Dey, S.; Sharma, S.; Mallya, P.; Kumar, V.A. An Overview of Advanced Formulation and Nanotechnology-based Approaches for Solubility and Bioavailability Enhancement of Silymarin. *J. Drug Deliv. Sci. Technol.* **2020**, *60*, 102021. [CrossRef]
18. Xanthakis, E.; Theodosiou, E.; Magkouta, S.; Stamatis, H.; Loutrari, H.; Roussos, C.; Kolisis, F. Enzymatic transformation of flavonoids and terpenoids: Structural and functional diversity of the novel derivatives. *Pure Appl. Chem.* **2010**, *82*, 1–16. [CrossRef]
19. Vavříková, E.; Gavezzotti, P.; Purchartová, K.; Fuksová, K.; Biedermann, D.; Riva, S.; Křen, V. Regioselective Alcoholysis of Silychristin Acetates Catalyzed by Lipases. *Int. J. Mol. Sci.* **2015**, *16*, 11983–11995. [CrossRef]
20. Brenchley, J. Psychrophilic microorganisms and their cold-active enzymes. *J. Ind. Microbiol. Biotechnol.* **1996**, *17*, 432–437. [CrossRef]
21. Mangiagalli, M.; Brocca, S.; Orlando, M.; Lotti, M. The "cold revolution". Present and future applications of cold-active enzymes and ice-binding proteins. *New Biotechnol.* **2020**, *55*, 5–11. [CrossRef] [PubMed]
22. Santiago, M.; Ramírez, C.; Zamora, R.; Parra, L.P. Discovery, Molecular Mechanisms, and Industrial Applications of Cold-Active Enzymes. *Front. Microbiol.* **2016**, *7*, 1408. [CrossRef] [PubMed]
23. Kavitha, M. Cold active lipases—An update. *Front. Life Sci.* **2016**, *9*, 226–238. [CrossRef]
24. Persoiu, A.; Lauritzen, S.E. Scarisoara Ice Cave. In *Ice Caves*, 1st ed.; Elsevier: Amsterdam, The Netherlands, 2018; Chapter 25.3.2.3, pp. 520–527.
25. Paun, V.I.; Lavin, P.; Chifiriuc, M.C.; Purcarea, C. First report on antibiotic resistance and antimicrobial activity of bacterial isolates from 13,000-year old cave ice core. *Sci. Rep.* **2021**, *11*, 54. [CrossRef] [PubMed]
26. Paun, V.I.; Icaza, G.; Lavin, P.; Marin, C.; Tudorache, A.; Perşoiu, A.; Dorador, C.; Purcarea, C. Total and Potentially Active Bacterial Communities Entrapped in a Late Glacial Through Holocene Ice Core From Scarisoara Ice Cave, Romania. *Front. Microbiol.* **2019**, *10*, 1193. [CrossRef]
27. de María, P.D.; Carboni-Oerlemans, C.; Tuin, B.; Bargeman, G.; van der Meer, A.; van Gemert, R. Biotechnological applications of Candida antarctica lipase A: State-of-the-art. *J. Mol. Catal. B Enzym.* **2005**, *37*, 36–46. [CrossRef]
28. Stergiou, P.-Y.; Foukis, A.; Filippou, M.; Koukouritaki, M.; Parapouli, M.; Theodorou, L.G.; Hatziloukas, E.; Afendra, A.; Pandey, A.; Papamichael, E.M. Advances in lipase-catalyzed esterification reactions. *Biotechnol. Adv.* **2013**, *31*, 1846–1859. [CrossRef]
29. Das, U.N. Essential fatty acids: Biochemistry, physiology and pathology. *Biotechnol. J.* **2006**, *1*, 420–439. [CrossRef]
30. Lawerence, G.D. *The Fats of Life. Essential Fatty Acids in Health and Disease*; Rutgers University Press: New Brunswick, NJ, USA, 2010; Chapter 2; pp. 15–20.
31. Gandhi, N.N.; Patil, N.S.; Sawant, S.B.; Joshi, J.B.; Wangikar, P.P.; Mukesh, D. Lipase-Catalyzed Esterification. *Catal. Rev.* **2000**, *42*, 439–480. [CrossRef]
32. Illanes, A. Chimioselective Esterification of Wood Sterols with Lipases. In *Enzyme Biocatalysis*; Springer: New York, NY, USA, 2008; Chapter 6.3; pp. 292–308.
33. Ramnath, L.; Sithole, B.; Govinden, R. Identification of lipolytic enzymes isolated from bacteria indigenous to Eucalyptus wood species for application in the pulping industry. *Biotechnol. Rep.* **2017**, *15*, 114–124. [CrossRef]
34. Peña, S.A.; Rios, N.S.; Carballares, D.; Sánchez, C.M.; Lokha, Y.; Gonçalves, L.R.B.; Fernandez-Lafuente, R. Effects of Enzyme Loading and Immobilization Conditions on the Catalytic Features of Lipase From Pseudomonas fluorescens Immobilized on Octyl-Agarose Beads. *Front. Bioeng. Biotechnol.* **2020**, *8*, 36. [CrossRef]
35. Barth, A. Infrared spectroscopy of proteins. *Biochim. Biophys. Acta (BBA) Bioenerg.* **2007**, *1767*, 1073–1101. [CrossRef]
36. Stoia, M.; Istratie, R.; Păcurariu, C. Investigation of magnetite nanoparticles stability in air by thermal analysis and FTIR spectroscopy. *J. Therm. Anal. Calorim.* **2016**, *125*, 1185–1198. [CrossRef]
37. Liu, X.; Guan, Y.; Shen, R.; Liu, H. Immobilization of lipase onto micron-size magnetic beads. *J. Chromatogr. B* **2005**, *822*, 91–97. [CrossRef]
38. Xu, Y.-Q.; Zhou, G.-W.; Wu, C.-C.; Li, T.-D.; Song, H.-B. Improving adsorption and activation of the lipase immobilized in amino-functionalized ordered mesoporous SBA-15. *Solid State Sci.* **2011**, *13*, 867–874. [CrossRef]
39. Rehm, S.; Trodler, P.; Pleiss, J. Solvent-induced lid opening in lipases: A molecular dynamics study. *Protein Sci.* **2010**, *19*, 2122–2130. [CrossRef] [PubMed]
40. Adlercreutz, P. Immobilisation and application of lipases in organic media. *Chem. Soc. Rev.* **2013**, *42*, 6406–6436. [CrossRef] [PubMed]

41. Sheldon, R.A.; van Pelt, S. Enzyme immobilisation in biocatalysis: Why, what and how. *Chem. Soc. Rev.* **2013**, *42*, 6223–6235. [CrossRef] [PubMed]
42. Bassi, J.J.; Todero, L.M.; Lage, F.A.; Khedy, G.I.; Ducas, J.D.; Custódio, A.P.; Pinto, M.A.; Mendes, A.A. Interfacial activation of lipases on hydrophobic support and application in the synthesis of a lubricant ester. *Int. J. Biol. Macromol.* **2016**, *92*, 900–909. [CrossRef] [PubMed]
43. Peña, S.A.; Rios, N.S.; Carballares, D.; Gonçalves, L.R.; Fernandez-Lafuente, R. Immobilization of lipases via interfacial activation on hydrophobic supports: Production of biocatalysts libraries by altering the immobilization conditions. *Catal. Today* **2021**, *362*, 130–140. [CrossRef]
44. Neagu, S.; Preda, S.; Anastasescu, C.; Zaharescu, M.; Enache, M.; Cojoc, R. The functionalization of silica and titanate nanostructures with halotolerant proteases. *Rev. Roum. Chim.* **2014**, *59*, 97–103.
45. Mach, H.; Middaugh, C.; Lewis, R.V. Statistical determination of the average values of the extinction coefficients of tryptophan and tyrosine in native proteins. *Anal. Biochem.* **1992**, *200*, 74–80. [CrossRef]
46. Moreno, M.D.L.; Garcia, M.T.; Ventosa, A.; Mellado, E. Characterization of Salicola sp. â€ƒIC10, a lipase- and protease-producing extreme halophile. *FEMS Microbiol. Ecol.* **2009**, *68*, 59–71. [CrossRef] [PubMed]
47. Available online: http://www.chemicell.com/home/index.html (accessed on 14 November 2021).
48. Lite, C.; Ion, S.; Tudorache, M.; Zgura, I.; Galca, A.C.; Enache, M.; Maria, G.-M.; Parvulescu, V.I. Alternative lignopolymer-based composites useful as enhanced functionalized support for enzymes immobilization. *Catal. Today* **2020**, *379*, 222–229. [CrossRef]
49. Hill, A.; Karboune, S.; Mateo, C. Investigating and optimizing the immobilization of levansucrase for increased transfructosylation activity and thermal stability. *Process. Biochem.* **2017**, *61*, 63–72. [CrossRef]

Article

# Lipase Catalyzed Synthesis of Enantiopure Precursors and Derivatives for β-Blockers Practolol, Pindolol and Carteolol

Morten Andre Gundersen, Guro Buaas Austli, Sigrid Sløgedal Løvland, Mari Bergan Hansen, Mari Rødseth and Elisabeth Egholm Jacobsen *

Department of Chemistry, Norwegian University of Science and Technology, Høgskoleringen 5, 7491 Trondheim, Norway; morten_gundersen@hotmail.no (M.A.G.); guro.austli@gmail.com (G.B.A.); Sigrid.lovland@gmail.com (S.S.L.); mari.bergan@live.no (M.B.H.); marrods@stud.ntnu.no (M.R.)
* Correspondence: elisabeth.e.jacobsen@ntnu.no; Tel.: +47-73596256

**Citation:** Gundersen, M.A.; Austli, G.B.; Løvland, S.S.; Hansen, M.B.; Rødseth, M.; Jacobsen, E.E. Lipase Catalyzed Synthesis of Enantiopure Precursors and Derivatives for β-Blockers Practolol, Pindolol and Carteolol. *Catalysts* **2021**, *11*, 503. https://doi.org/10.3390/catal11040503

Academic Editor: Francisco Plou

Received: 1 March 2021
Accepted: 12 April 2021
Published: 16 April 2021

**Publisher's Note:** MDPI stays neutral with regard to jurisdictional claims in published maps and institutional affiliations.

**Copyright:** © 2021 by the authors. Licensee MDPI, Basel, Switzerland. This article is an open access article distributed under the terms and conditions of the Creative Commons Attribution (CC BY) license (https://creativecommons.org/licenses/by/4.0/).

**Abstract:** Sustainable methods for producing enantiopure drugs have been developed. Chlorohydrins as building blocks for several β-blockers have been synthesized in high enantiomeric purity by chemo-enzymatic methods. The yield of the chlorohydrins increased by the use of catalytic amount of base. The reason for this was found to be the reduced formation of the dimeric by-products compared to the use of higher concentration of the base. An overall reduction of reagents and reaction time was also obtained compared to our previously reported data of similar compounds. The enantiomers of the chlorohydrin building blocks were obtained by kinetic resolution of the racemate in transesterification reactions catalyzed by *Candida antarctica* Lipase B (CALB). Optical rotations confirmed the absolute configuration of the enantiopure drugs. The β-blocker (*S*)-practolol ((*S*)-*N*-(4-(2-hydroxy-3-(isopropylamino)propoxy)phenyl)acetamide) was synthesized with 96% enantiomeric excess (*ee*) from the chlorohydrin (*R*)-*N*-(4-(3-chloro-2 hydroxypropoxy)phenyl)acetamide, which was produced in 97% *ee* and with 27% yield. Racemic building block 1-((1*H*-indol-4-yl)oxy)-3-chloropropan-2-ol for the β-blocker pindolol was produced in 53% yield and (*R*)-1-((1*H*-indol-4-yl)oxy)-3-chloropropan-2-ol was produced in 92% *ee*. The chlorohydrin 7-(3-chloro-2-hydroxypropoxy)-3,4-dihydroquinolin-2(1*H*)-one, a building block for a derivative of carteolol was produced in 77% yield. (*R*)-7-(3-Chloro-2-hydroxypropoxy)-3,4-dihydroquinolin-2(1*H*)-one was obtained in 96% *ee*. The *S*-enantiomer of this carteolol derivative was produced in 97% *ee* in 87% yield. Racemic building block 5-(3-chloro-2-hydroxypropoxy)-3,4-dihydroquinolin-2(1*H*)-one, building block for the drug carteolol, was also produced in 53% yield, with 96% *ee* of the *R*-chlorohydrin (*R*)-5-(3-chloro-2-hydroxypropoxy)-3,4-dihydroquinolin-2(1*H*)-one. (*S*)-Carteolol was produced in 96% *ee* with low yield, which easily can be improved.

**Keywords:** (*S*)-practolol; paracetamol; (*S*)-pindolol; (*S*)-carteolol; *Candida antarctica* Lipase B; chiral chromatography; dimer formation; absolute configuration

## 1. Introduction

Chiral compounds with one or several stereogenic centers consist of pairs of enantiomers. Many drugs on the market today have one or several stereogenic centers, β-blockers normally have one stereogenic center, and then consist of two enantiomers. The enantiomers may have the same effect on the patient, or the enantiomers may have different effects, or worse, one enantiomer may have several unwanted side effects. FDA considers the "wrong" enantiomer as an impurity and demands for pure enantiomers as the active pharmaceutical ingredient (API) in the marketed drugs, not racemates. The demand for enantiomerically pure drugs has increased year-by-year since the 1990s when FDA demanded manufacturers to evaluate the pharmacokinetics of a single enantiomer or mixture of enantiomers in a chiral drug. Quantitative assays for individual enantiomers should be developed for studies in in vivo samples early in drug development. It is postulated that the lower the effective dose of a drug, the greater the difference in the pharmacological effect of the optical isomers when drug receptor interactions are considered [1]. The ratio of the more active enantiomer (eutomer) compared to the less active enantiomer (distomer) is known as the eudismic ratio. The higher the eudismic ratio, the higher the effectiveness of the drug [2] and the "right" enantiomer should be provided.

High blood pressure (hypertension) and heart failure are big global health problems. In Norway, approximately 30% of all deaths are caused by cardiovascular diseases [3]. Approximately 26 million people worldwide live with heart failure, and approximately seven million deaths are caused by hypertension annually [4]. Heart failure cannot be cured but may be treated by medications in order to increase quality of life for the patients. Both heritage (genetic) and lifestyle may lead to increased risk of cardiovascular diseases. Risk factors for cardiovascular diseases, such as reduced activity, eating more sugar and salt, increased stress, obesity and overweight, may be reasons for the increasing health problems worldwide [5].

A class of drugs that have been used in treatment of cardiovascular diseases for a long time, β-adrenergic receptor antagonists, are so-called β-blockers. The $β_1$-receptors in the human body are mainly located in the heart and regulate the contraction of the heart muscle. The $β_1$-blockers are antagonists that affect the $β_1$-receptors in the heart and are suitable for the treatment of hypertension and heart failure. When a β-blocker inhibits the binding of adrenaline and noradrenaline, the stress hormone level in the body will decrease, and thus the blood pressure and heart rate will decrease. $β_2$-receptors are mainly found in the lungs, and most $β_2$-agonists are mainly used to treat asthma by relaxing the smooth muscles in the lungs [6].

Some β-blockers are non-selective and will inhibit both $β_1$- and $β_2$-receptors, while others are selective to either $β_1$- or $β_2$-receptors. Propranolol is an example of a so-called non-selective β-blocker. A problem for the non-selective β-blockers is that they may cause negative effects to the lungs, especially for asthma patients. β-receptors are found in all parts of the nervous system, and β-blockers will therefore give side effects even if the right drug is chosen. Selective β-blockers lead to fewer side effects and are safer in use. Choosing the right β-blocker for every type of treatment is therefore important [6].

We have for some time worked with optimization of biocatalytic processes for several of these drugs. Lipase-catalyzed kinetic resolution of racemates has been successful for several of these [7,8]. In attempts to obtain enantiopure building blocks for the β-agonist (R)-clenbuterol, asymmetrizations of ketones have been performed with ketoreductases giving high enantiomeric excess [9].

In general, the S-enantiomers of the β-blockers are known to be more active than the R-enantiomers, and have the opposite stereo configuration for the β-agonists. Common for these compounds is a side chain on an aromatic group which consists of a secondary alcohol and an amine on the omega carbon. Traditionally, most of these drugs have been manufactured in racemic form, however, e.g., (S)-atenolol, (S)-propranolol and (S)-metoprolol are also on the market as single enantiomers. When referring to a drug, one should refer to the API [10].

Because enantiomers are mirror images of each other and not identical, each of them may act differently on the receptor. The pharmacological effect, clinical effect and toxicity of both enantiomers need to be investigated separately according to the U.S. Food and Drug Administration (FDA) [4]. Today, chiral drugs are promoted increasingly with enantiopure API. For enantiopure drugs there is a requirement for enantiomeric excess (*ee*) > 96%. Therefore, development of environmentally friendly synthetic methods for each of the enantiomers is important.

Practolol is a selective $β_1$-antagonist and was the first $β_1$-selective β-blocker used in the treatment of cardiovascular diseases at the beginning of the 1970s [11]. Practolol showed effective treatment of heart failure and arrhythmic heart rate [12]. The drug showed later some critical side effects, such as culomucocutaneous syndrome, in some patients, and was withdrawn from the market [13].

Mulik et al. reported in 2016 a four-step synthesis of (S)-practolol in 100% *ee* with the use of *Pseudomonas cepacia* sol-gel AK lipase as enantioselective catalyst. However, they report that the produced ester from the transesterification reaction is hydrolyzed and aminated to give the enantiopure building block with S-configuration [14]. According to previous reports, it is the slower reacting enantiomer, (R)-N-(4-(3-chloro-2-hydroxypropoxy)phenyl)acetamide, which is aminated to give (S)-practolol. They also claim that the configuration of the chlorohydrin is inverted in the amination step, which will not be the case since the amine is not attacking the stereocenter, but the primary carbon with the chloro atom. Ader and Schneider reported in 1992 enzymatic kinetic resolution of racemic N-(4-(3-chloro-2-hydroxypropoxy)phenyl)acetamide with *Pseudomonas* sp. to give (R)-N-(4-(3-chloro-2-hydroxypropoxy)phenyl)acetamide which was subsequently aminated to yield (S)-practolol in 30% yield and >99% *ee* [15]. However, no optical rotation values of the building blocks or the drug have been reported. In order to develop greener and more sustainable processes for drugs with secondary alcohol side chains, we wanted to include the synthesis of practolol despite the regulations of the drug.

Pindolol, 1-(1H-Indol-4-yloxy)-3-(isopropylamino)-2-propanol, was first released for clinical use in USA in 1982. It is a non-selective β-blocker, used in the treatment of high blood pressure, chest pain and irregular heartbeat. Pindolol has its substituents on the aromatic ring in the *ortho*- and *meta*-position, which is common for non-selective β-blockers. Selective β-blockers usually have a substituent in the *para*-position on the aromatic ring [4]. Pindolol is also a partial agonist and will therefore slow the resting heart rate less than other β-blockers like atenolol or metoprolol [16]. Pindolol is usually sold under the brand names Visken (Sandoz) or Barbloc (Alpha) and is often used to treat high blood pressure during pregnancy because it does not affect the fetal heart function or blood flow. Although pindolol is a non-selective β-blocker, other uses for the drug have been reported. It has been tested in the treatment of fibromyalgia and related fatigue diseases, as well as in the treatment of depression in combination with selective serotonin reuptake inhibitors [17,18]. Pindolol is a rapidly absorbed drug and after oral ingestion it can be detected in the blood after 30 min. In patients with normal renal function, pindolol has a plasma half-life of three to four hours. The drug is also lipophilic and enters the central nervous system rapidly. Reported side effects include unwanted lowering of heart function or changes in the respiratory system. These side effects are related to its β-adrenergic blocking activity; other side effects have also been reported, such as dizziness, vivid dreams, feeling of weakness or fatigue, muscle cramps, as well as nausea [16]. Precursors of β-blocker pindolol were synthesized by biocatalysis in 2017 by Lima et al. They performed hydrolysis of 2-acetoxy-1-(1H-indol-4-yloxy)-3-chloropropane using lipase from *Pseudomonas fluorescens* which yielded (2S)-1-(1H-indol-4-yloxy)-3-chloro-2-propanol in 96% *ee* and (2R)-2-acetoxy-1-(1H-indol-4-yloxy)-3-chloropropane in 97% *ee*, which was hydrolysed giving 97% *ee* of the R-chlorohydrin for the synthesis of (S)-pindolol with retention of *ee* [17]. However, we have some doubts about the stereochemistry in this report which will be discussed.

Carteolol is another β-adrenergic antagonist (β-blocker) manufactured mostly with racemic API and administered as eye-drops for reduction of aqueous production in the eye (glaucoma) [19]. In these patients, an elevated intraocular pressure (IOP) leads to damage to the optic nerve, reducing the visual field gradually until the patient is completely blind. It is the second most common cause of irreversible blindness after age-related macular degeneration in western Europe. In 2010, 2.1 million people worldwide went irreversibly blind because of glaucoma [20]. In 2019, the majority of Norwegian glaucoma patients (68%) were treated with β-blockers betaxolol or timolol, either as single drugs or in combination with other drugs such as prostaglandin analogues or carboanhydrase inhibitors [21].

With the aim of sustainable production of enantiopure β-blockers, we have performed several synthetic strategies with lower amounts of reactants and shorter reaction times than previously reported. The general mechanism of base catalyzed deprotonation of phenolic protons with subsequent nucleophilic attachment of epichlorohydrin has been studied with different bases and different concentrations of epichlorohydrin. Lipase B from *Candida antarctica* has been shown to catalyze reactions of similar compounds with high *ee* of both product and remaining starting material (hydrolysis and transesterification reactions) [22,23].

## 2. Results

Chlorohydrin building blocks (*R*)-**1a-4a** for the synthesis of enantiomers of the β-blockers practolol ((*S*)-**1c**), pindolol and derivatives of carteolol ((*S*)-**3c-4c**) have been synthesized in 92–97% *ee* by chemo-enzymatic methods (Scheme 1). The highest yields of the racemic chlorohydrins were obtained with 0.3–1 equivalents of base in the deprotonation step of the starting materials **1–4**, 2 equivalents of epichlorohydrin, 12–26 h reaction time and 30 °C reaction temperature. The intermediate epoxides **1e–4e** were protonated with acetic acid and then opened with lithium chloride. Recently, we reduced the amount of acetic acid from 10 to 5 equivalents giving the same yields of the chlorohydrins. Kinetic resolutions of the racemic halohydrins were performed in different solvents with lipase B from *Candida antarctica* and vinyl butanoate as the acyl donor. Amination of the *R*-chlorohydrins (*R*)-**1a** and (*R*)-**3a-4a** gave the *S*-β-blockers with preserved or increased *ee*. Due to the low *ee* (92%), the amination step of (*R*)-**2a** was not performed. Previously, we have published the synthesis of the building block for (*S*)-atenolol in >98% *ee* by a similar protocol [7].

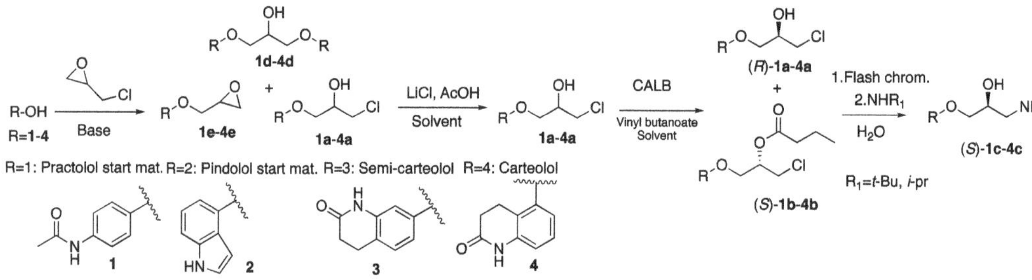

**Scheme 1.** Building blocks (*R*)-**1a-4a** synthesized in 92–97% *ee* for use in synthesis of the *S*-enantiomers of the β-blockers practolol, pindolol and derivatives of carteolol ((*S*)-**1c-4c**).

Analysis of the reaction mixtures from the syntheses of **1a–4a** on LC-MS showed that the most abundant by-products in these reactions were the dimers **1d–4d** (Scheme 1) of the deprotonated starting materials **1–4**. In order to ensure full conversion of the starting materials and to avoid the formation of the dimer by-products in the syntheses, concentrations of base and 2-(chloromethyl)oxirane (epichlorohydrin), reaction time and temperature have been varied. When high concentration of base was used, the intramolecular cyclization of the anions of **1a–4a** was observed to boost by increased reaction time; otherwise, the acid and lithium chloride were added immediately after full conversion of the starting materials. Investigations of reaction conditions in synthesis of the racemic practolol precursor **1a** are shown in Table 1. When 0.5–1.0 equivalents of sodium hydroxide dissolved in water was used in the reaction of **1** with epichlorohydrin with a reaction temperature of 80 °C for 24 h, only the dimer *N,N*-(((2-hydroxypropane-1,3-diyl)bis(oxy))-bis(4,1-phenylene))diacetamide (**1d**) was obtained in addition to a small fraction of the epoxide **1e** (Table 1). Dimer **1d** was characterized by NMR-, MS- and IR-analyses. The chemical shifts for **1d** were assigned using $^1$H-NMR-, $^{13}$C-NMR-, COSY-, HSQC- and HMBC-spectra. The analyses were performed on a 600 MHz NMR instrument from Bruker, Germany, with deuterated dimethyl sulfoxide as solvent.

Table 1. A reaction temperature of 80 °C and 0.5–1.0 eq of NaOH dissolved in water favored the formation of the dimer **1d** in the reaction of **1** with 2 eq of epichlorohydrin (Scheme 1).

| Base | Equivalent | Rx. Temp | Rx. Time | 1a (%) | 1e (%) | 1d (%) |
|---|---|---|---|---|---|---|
| NaOH | 0.5 | 80 °C | 24 | 0 | 3 | 95 |
| NaOH | 1.0 | 80 °C | 16 | 0.4 | 0.3 | 99 |
| NaOH | 5.0 | 80 °C | 16 | 1 | 11 | 6 |
| NaOH | 1.0 | 0 °C | 32 | – | – | – |
| $K_2CO_3$ | 1.0 | r.t. | 27 | – | 2 | – |
| $K_2CO_3$ | 1.0 | 60 °C | 27 | 2 | 12 | 72 |

The strategy for avoiding the formation of the dimers **1d–4d** and also to generate a high ratio of halohydrin to epoxide was to use only 0.3 equivalents of base. The chlorohydrins **1a–4a** were synthesized in 59–77% yield.

The starting material 5-(3-chloro-2-hydroxypropoxy)-3,4-dihydroquinolin-2(1H)-one (**4**) for the synthesis of carteolol is quite expensive, so we wanted to investigate similar reactions of 7-(3-chloro-2-hydroxypropoxy)-3,4-dihydroquinolin-2(1H)-one (**3**) as a model substrate.

Table 2 entry 3 shows that the highest relative rate of the formation of halohydrin **1a** over the formation of the epoxide **1e** is obtained with one eq of base.

Table 2. Synthesis of N-(4-(oxiran-2-ylmethoxy)phenyl)acetamide, **1a**, and N-(4-(3-chloro-2-hydroxypropoxy)phenyl)acetamide, **1e**, from paracetamol, **1**, with epichlorohydrin and NaOH at room temperature The table shows reaction conditions, equivalents and starting amounts of base (mmol), and the ratio of **1a** and **1e** in the product mixture, calculated from HPLC chromatograms on Eclipse XDB-C18-column and gradient program ($H_2O$:MeCN 90:10–$H_2O$:MeCN 75:25 over 20 min, 0.5 mL/min flow).

| Entry | Paracetamol (mmol) | Equiv. NaOH | NaOH (mmol) | $H_2O$ (mL) | Rx. Time (h) | Appearance of rx Mixture | 1a (%) ($t_R$ = 11.0 min) | 1e (%) ($t_R$ = 13.0 min) |
|---|---|---|---|---|---|---|---|---|
| 1 | 6.62 | 0.1 | 0.661 | 1.9 | 48 | Pink viscous liquid | 29 | 38 |
| 2 | 6.62 | 0.5 | 3.31 | 4.7 | 8 | Pink viscous liquid | 37 | 43 |
| 3 | 6.62 | 1 | 6.62 | 5.8 | 7.5 | White solid | 81 | 12 |
| 4 | 19.8 | 1 | 19.8 | 10.0 | 5 | White solid | 65 | 1 |
| 5 | 3.31 | 2 | 6.62 | 4.7 | 7 | White solid | 39 | 0 |
| 6 | 0.662 | 10 | 6.62 | 5.0 | 18 | Brown viscous liquid | 0 | 3 |

We saw the same trend in the synthesis of **2a/2e**. However, in the synthesis of **3a/3e** and **4a/4e**, 0.3 eq of base gave the highest yield. When catalytic amounts of base are used, the anions of the halohydrins formed will likely deprotonate a water molecule which regenerates the base for new deprotonations of the starting materials. A plausible mechanism for regeneration of the base in these reactions is shown for the reaction of 7-(3-chloro-2-hydroxypropoxy)-3,4-dihydroquinolin-2(1H)-one (**3**, Scheme 2). After deprotonation of phenol **3** by the base, **3a$_{Anion}$** can react with the least substituted epoxide-carbon in epichlorohydrin (path a) forming **3a$_{Anion}$**. 7-(Oxiran-2-ylmethoxy)-3,4-dihydroquinolin-2(1H)-one, **3e**, is formed by a Williamson ether synthesis reaction between **3** and epichlorohydrin, as shown in path b. An internal cyclization of **3a$_{Anion}$** also forms epoxide **3e** (path c), while protonation of **3a$_{Anion}$** yields chlorohydrin **3a** and regenerates the base allowing for the use of catalytic amounts of base (<1 equivalent).

Addition of lithium chloride and acetic acid before any work-up of all the reactions forming **1a–4a** gave higher yields than when the reactions were stopped after the nucleophilic attack of epichlorohydrin in the first step.

As the hydroxide ion may also attack epichlorohydrin directly both on carbon 1 in the oxirane and on the α-carbon, the by-products 2-(chloromethyl)oxirane (**6**), 3-chloropropane-1,2-diol (**7**) and propane-1,2,3-triol (**8**) may also be formed, see Scheme 3. All impurities were removed by flash chromatography.

Kinetic resolutions of **1a–4a** have been performed in different solvents catalyzed by CALB with moderate to high *E*-values (calculated by *E&K Calculator*, 2.1b0 PPC) [24], giving moderate to high *ee*-values of the chiral building blocks. (Figure 1). Optical rotation values of (*R*)-**1a**, (*S*)-**1a**, (*S*)-**1b**, (*R*)-**3a**, (*S*)-**3b**, (*S*)-**3c**, (*R*)-**4a** and (*S*)-**4b** have not been reported previously, and the determination of optical rotation and predictions of absolute configuration of these pure enantiomers should be of interest to both academia and industry. Compounds with >96% *ee* are regarded as "enantiopure" by pharmaceutical means. Due to the relatively low *ee* of (*R*)-**2a** in our hands (92% *ee*) we did not proceed with the synthesis of the pindolol enantiomer; however, we would have expected to retain the *ee* from the amination of (*R*)-**2a** giving (*S*)-pindolol of 92% *ee*.

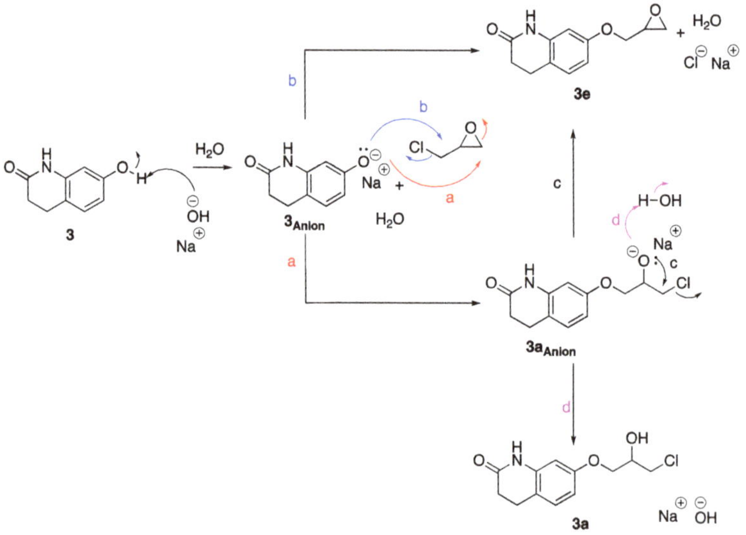

**Scheme 2.** Mechanism for base-catalyzed deprotonation of starting material **3** in the synthesis of halohydrin **3a** and the intermediate epoxide **3e** used in the synthesis of semi-carteolol, (*S*)-**3c**. The deprotonated **3**$_{Anion}$ reacts with the least substituted epoxide-carbon in epichlorohydrin (path a) forming **3a**$_{Anion}$. 7-(Oxiran-2-ylmethoxy)-3,4-dihydroquinolin-2(1H)-one, **3e**, is formed by a Williamson ether synthesis reaction between **3** and epichlorohydrin, as shown in path b. An internal cyclization of **3a**$_{Anion}$ also forms epoxide **3e** (path c), while protonation of **3a**$_{Anion}$ yields chlorohy-drin **3a** and regenerates the base allowing for the use of catalytic amounts of base (<1 equivalent).

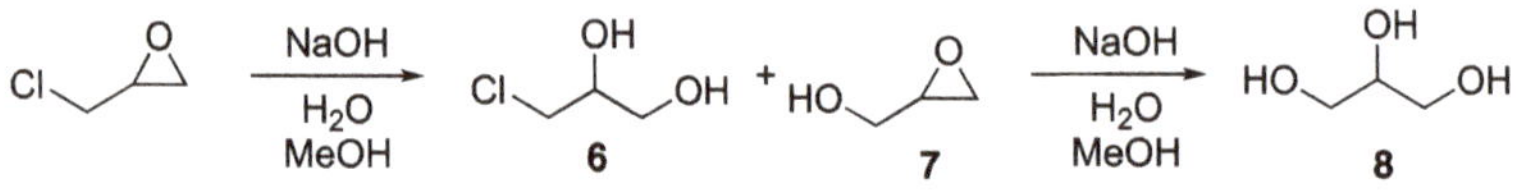

**Scheme 3.** 2-(Chloromethyl)oxirane (**6**), 3-chloropropane-1,2-diol (**7**) and propane-1,2,3-triol (**8**) may be formed in side reactions between epichlorohydrin and sodium hydroxide found in previous studies of these reactions in our group (LC-MS). Small amounts of 1,3-dichloro-2-propanol have been found from GC-MS analyses in the synthesis of **2a**.

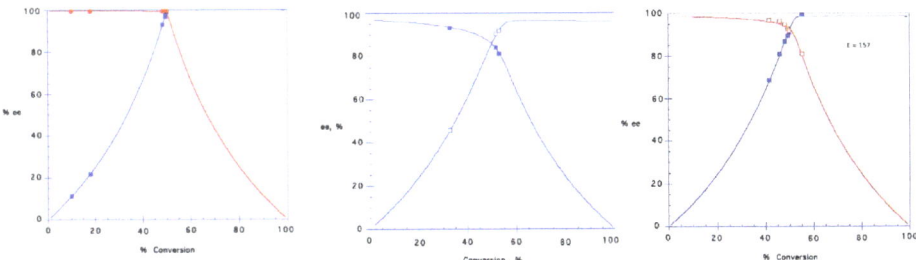

**Figure 1.** Graphical illustration of reaction progress of kinetic resolution at different degrees of conversion. **Left panel**: **1a** $ee_p$ (red filled squares) and $ee_s$ (blue filled squares); E-value >200. **Middle panel**: kinetic resolution of **2a** $ee_p$ (filled squares) and $ee_s$ (open squares) E = 66. **Right panel**: **3a** $ee_p$ (open red squares) and $ee_s$ (filled blue squares); E = 157. All three reactions used CALB from SyncoZymes as catalyst and vinyl butanoate as acyl donor in different solvents. All reactions were performed at 30 °C. E-values calculated from E&K Calculator 2.1b0 PPC [24].

In the amination reactions of (R)-**1a** and (R)-**3a**, the ee was retained and (S)-practolol ((S)-**1c**) and (S)-semi-carteolol ((S)-**3c**) were produced in 96 and 97% ee, respectively. (R)-**4a** was obtained in 96% ee, and (S)-**4c** was obtained in 96% ee from amination of (R)-**4a** (Table 3). Reaction times of the kinetic resolutions varied by the amounts of lipase added; however, 12 h should be a proper reaction time. Danilewicz and Kemp (1973) reported that the optical rotation of (R)-practolol was +4.3° at 25 °C [25]. We conclude that the negative rotation of (S)-practolol determined by us (−3.9°) accounts for the S-configuration.

**Table 3.** E-values, ee-values and yields of racemic building blocks **1-4** and of R-alcohols (R)-**1a-4a** and the drugs (S)-**1c-4c** from the reactions of **1-4** with vinyl butanoate as acyl donor to produce the butanoates (S)-**1b-4b** leaving the R-enantiomers unreacted. Different solvents were used, and the kinetic resolutions were catalyzed by lipase B from *Candida antarctica*. Optical rotations $[\alpha]_D^T$ were determined at 20–23 °C in different solvents with c = 1; for additional parameters, see the experimental section.

| Halohydrin | Yield | E-Value | Ester, ee, Yield, % | $[\alpha]_D^{20}$(Solvent) | Halohydrin, ee, Yield, % | $[\alpha]_D^T$(Solvent) | Drug, ee, Yield, % | $[\alpha]_D^{20}$(Solvent) |
|---|---|---|---|---|---|---|---|---|
| **1a** | 68 | 51 | (S)-**1b**, 84, 47 | +16.3 (c 1.1 MeCN, 23 °C) | (R)-**1a**, 97, 38 (S)-**1a**, 81, 26 | −1.0 (i-PrOH) +11.99 (i-PrOH) | (S)-**1c**, 96, 16 | −3.99 (EtOH) |
| **2a** | 59 | 66 | | | (R)-**2a**, 92 | | | |
| **3a** | 77 | 157 | (S)-**3b**, 86, 51 | +8.0 (MeOH) | (R)-**3a**, 96, 34 | −9.98 (MeOH) | (S)-**3c**, 97, 82 | −16.0 (MeOH). |
| **4a** | 61 | 30 | (S)-**4b**, 77, 43 | +9.99 (DMSO) | (R)-**4a**, 96, 38 | −9.99 (DMSO) | (S)-**4c**, 96, 70 | |

We noticed that other research groups have reported data for synthesis of enantiopure practolol and pindolol [14,17]. Reaction time for the enzymatic hydrolyses of the acetate of 1-(1H-indol-4-yloxy)-3-chloro-2-propanol reported by Lima et al. varied from 12 to 25 h and the authors reported an E-value of 30 in the hydrolytic kinetic resolution of the racemic acetylester of **2a** using Novozym® 435 [17]. By using lipase from *Pseudomonas fluorescens* in acetylation of racemic **2a**, an E-value of 11 was obtained, with an $ee_s$ of the chlorohydrin **2a** of 72% $ee_s$ and 69% $ee_p$ for 51% conversion, 24 h reaction time at 40 °C. However, there is a misunderstanding of stereochemistry in Lima´s report. The authors report that hydrolysis of the racemic acetate and transesterification reaction of the chlorohydrin **2a** give enantiomers with opposite stereochemistry. The product and the remaining alcohol will have the same configuration in hydrolysis of the acetyl ester and transesterification of the chlorohydrin **2a**. Hydrolysis of the ester enantiomer from the hydrolysis of racemic ester should not be the R-acetate of **2a**, but the S-ester. We claim that they have produced (S)-**2a** in 97% ee instead of (R)-**2a** in their hydrolysis of the acetic ester of 1-((1H-indol-4-yl)oxy)-3-chloropropan-2-ol. The authors must have used the S-halohydrin in the synthesis and would then have achieved (*R*)-pindolol. The optical rotation value is not reported. In our project we used CALB from SyncoZymes as catalyst and obtained an *E*-value of 66 in the esterification of **2a**. At 53% conversion (24 h), $ee_s$ and $ee_p$ values were 92% and 81%, respectively. The optimization of this synthesis is underway. We encourage researchers to report both optical rotation values and determination of absolute configuration of enantiomers.

## 3. Materials and Methods

All chemicals used in this project are commercially available, of analytical grade and were purchased from Sigma-Aldrich Norway (Oslo, Norway) or vwr Norway (Oslo, Norway). HPLC grade of solvents were used for the HPLC-analyses. Dry MeCN was acquired from a solvent purifier, MBraun MD-SPS800 (München, Germany), and stored in a flask containing molecular sieves (4Å).

### 3.1. Enzymes

*Candida antarctica* Lipase B (CALB) (activity $\geq$ 10,000 PLU/g, lot#20170315) immobilized at high hydrophobic macroporous resin, produced in fermentation with genetically modified *Pichia pastoris*. The enzyme is a gift from SyncoZymes Co, Ltd. (Shanghai, China).

### 3.2. Chromatographic Analyses

All analyses were performed on an Agilent HPLC 1100 (Santa Clara, CA, USA). Manual injector (Rheodyne 77245i/Agilent 10 μL loop) and a variable wavelength detector (VWD) set to 254 nm were used.

#### 3.2.1. Achiral HPLC

Separation was performed on a XDB C18-column (250 × 4.6 mm ID, 5 μm particle size, 80 Å, Phenomenex, Oslo, Norway).

Dimer **1d**: eluent gradient: $H_2O$:acetonitrile (90:10)-$H_2O$:acetonitrile (75:15) over 20 min: $t_R$ 15.71 min.

#### 3.2.2. Chiral HPLC

Separation of enantiomers was performed on a Chiralcel OD-H column (250 × 4.6 mm ID, 5 μm particle size, Daicel, Chiral Technologies Europe, Gonthier d'Andernach, Illkirch, France). Baseline separation was obtained; if not, $R_s$ is given. Chlorohydrin **1a**: $t_R$ (*S*)-**1a** = 31.1 min, $t_R$ (*R*)-**1a** = 38.8 min, butanoate ester **1b**: $t_R$ (*S*)-**1b** = 18.9 min and (*R*)-**1b** 23.7 min, eluent: hexane: *i*-PrOH, 83:17, 1 mL/min flow, $R_s$ = 12.05 and $t_R$ (*S*)-**1b** = 24.8 min and (*R*)-**1b** 31.6 min with eluent: hexane: *i*-PrOH, 85:15, 1 mL/min flow, $R_s$ = 8.91. Practolol **1c**, $t_R$ (*R*)-**1c** = 13.6 min, $t_R$ (*S*)-**1c** = 17.2 min, eluent: hexane: *i*-PrOH, 83:17, 1 mL/min flow, $R_s$ = 6.17. Chlorohydrin **2a**: $t_R$ (*S*)-**2a** = 50.7 min, $t_R$ (*R*)-**2a** = 22.5 min, butanoate ester **2b**: $t_R$ (*S*)-**2b** = 14.0 min, $t_R$ (*R*)-**2b** = 14.7 min, eluent: hexane: *i*-PrOH, 80:20, 1 mL/min flow. Chlorohydrin **3a**: $t_R$ (*S*)-**3a** = 20.10 min, $t_R$ (*R*)-**3a** = 25.83 min, eluent: hexane:*i*-PrOH:DEA, 60:40:0.4, 0.4 mL/min flow, butanoate ester **3b**: $t_R$ (*S*)-**3b** = 56.30 min, $t_R$ (*R*)-**3b** = 63.47 min, eluent: hexane: *i*-PrOH:DEA, 90:10:0.4, 0.4 mL/min flow. Chlorohydrin **4a**: $t_R$ (*S*)-**4a** = 63.43 min, $t_R$ (*R*)-**4a** = 68.68 min, eluent: hexane:EtOH:TFA, 90:10:0, 0.4 mL/min flow, $R_s$ = 1.70. Butanoate ester **4b**: $t_R$ (*S*)-**4b** = 110.33 min, $t_R$ (*R*)-**4b** = 118.44 min, $R_s$ = 1.65, eluent: hexane: *i*-PrOH:DEA, 95:5:0.4, 0.4 mL/min flow. Selected chromatograms can be found from Supplementary Materials.

### 3.3. TLC-Analyses and Column Chromatography

TLC-analyses were performed on Merck silica 60 $F_{254}$ (Sigma-Aldrich Norway, (Oslo, Norway) and detection with UV at $\lambda$ = 254 nm. Flash chromatography was performed using silica gel from Sigma-Aldrich Norway, (Oslo, Norway) (pore size 60Å, 230–400 mesh particle size, 40–63 μm particle size).

A New Brunswick G24 Environmental Incubator Shaker (New Brunswick co. Inc., Edison, NJ, USA) was used for enzymatic reactions.

*3.4. Analyses*

NMR-analyses were recorded on a Bruker 400 MHz Avance III HD instrument equipped with a 5 mm SmartProbe Z-gradient probe operating at 400 MHz for $^1$H and 100 MHz for $^{13}$C, respectively, or on a Bruker 600 MHz Avance III HD instrument equipped with a 5 mm cryogenic CP-TCI Z-gradient probe operating at 600 MHz for $^1$H and 150 MHz for $^{13}$C (Bruker, Rheinstetten, Germany). Chemical shifts are in ppm relative to TMS and coupling constants are in hertz (Hz). $^1$H and $^{13}$C NMR spectra can be found from Supplementary Materials. Infrared spectroscopy was performed at a Nexus FT-IR instrument (Madison, WI, USA). Exact masses were analyzed with a Synapt G2-S Q-TOF mass spectrometer from Waters™ (Waters Norway, Oslo, Norway). Ionization of samples was done with an ASAP probe (APCI), and calculation of exact masses and spectra processing were performed with Waters™ Software (Masslynxs V4.1 SCN871). IR and MS spectra details can be found from Supplementary Materials.

*3.5. Optical Rotations*

Optical rotations were measured on an Anton Paar (MCP 5100) polarimeter with a 2.5-centimeter-long cell (Dipl. Ing. Houm AS, Oslo, Norway). The analyses were performed at 20–23 °C and the samples were dissolved in different solvents, $c$ = 1.0g/100 mL, if not stated otherwise. Wavelength of the light was 589 nm (D).

*3.6. Absolute Configurations*

Absolute configuration of (*S*)-**1c** was determined by comparing the optical rotation with previously reported data. Optical rotation values of (*R*)-**1a**, (*S*)-**1a**, (*S*)-**1b**, (*R*)-**3a**, (*S*)-**3b**, (*S*)-**3c**, (*R*)-**4a** and (*S*)-**4b** have not been reported previously and the absolute configurations were determined by the enantioselectivity of CALB which we have reported previously [22,23].

*3.7. Synthesis Protocols*

N-(4-(3-chloro-2-hydroxypropoxy)phenyl)acetamide (**1a**)

*N*-(4-Hydroxyphenyl)acetamide (paracetamol, 151.163 g/mol, 1.998 g, 13.22 mmol) was dissolved in NaOH solution (0.3 eq), and 2-(chloromethyl)oxirane (epichlorohydrin) (26.44 mmol, 2 equiv.) was added dropwise with stirring. The reaction mixture was stirred at room temperature until TLC ($CH_2Cl_2$: MeOH, 10:1) showed full conversion of paracetamol. The reaction mixture was filtrated and washed with MeCN or distilled $H_2O$. The filtrate was extracted with EtOAc and the organic layer was dried over $MgSO_4$, filtrated and concentrated under reduced pressure. The product mixture was analyzed by HPLC with EclipseXDB C18-column and gradient $H_2O$:MeCN, 90:10-$H_2O$:MeCN, 75:25 over 20 min, 0.5 mL/min flow, showing both the chlorohydrin *N*-(4-(3-chloro-2-hydroxypropoxy)phenyl)acetamide, **1a**, and the epoxide *N*-(4-(oxiran-2-ylmethoxy)phenyl)acetamide, **1e**. The product mixture was reacted further without purification. The amount of reagents in step 2 are calculated from the assumption that the starting material contains only *N*-(4-(oxiran-2-ylmetoxy)phenyl)acetamide, **1e**, even if the starting material also contained *N*-(4-(3-chloro-2-hydroxypropoxy)phenyl)acetamide, **1a**. The mixture of **1a/1e** (0.681 g, 3.29 mmol) was dissolved in MeCN (10 mL), and LiCl (0.912 g, 21.5 mmol) and AcOH (3.0 mL, 53.1 mmol) were added. The reaction mixture was stirred at room temperature for 26 h and TLC ($CH_2Cl_2$:MeOH, 10:1) showed only the chlorohydrin **1a**, $R_f$ = 0.50. The reaction was stopped by adding $Na_2CO_3$ until reaching neutral pH. The precipitated salt was filtrated off. The reaction mixture was then extracted with EtOAc and washed with satd. NaCl solution. The organic layer was dried over $MgSO_4$, filtrated, and concentrated under reduced pressure. A yellow-brown viscous liquid was collected, which was purified by flash chromatography ($CH_2Cl_2$:MeOH, 10:1,

$v/v$). After purification, the product **1a** was collected as a white solid (0.543 g, 2.24 mmol, 68% yield, >99% purity). (HPLC, $t_R$ = 13.4 min). TLC (CH$_2$Cl$_2$:MeOH, 10:1) $R_f$ = 0.43 for N-(4-(3-chloro-2-hydroxypropoxy)phenyl) acetamide. $^1$H NMR (400 MHz, DMSO$_{d6}$): δ 9.78 (s, 1H, NH), 7.47 (d, 2H, AR-H, $J$ = 10.2 Hz), 6.88 (d, 2H, Ar-H, $J$ = 9.6 Hz), 5.52 (d, 1H, OH, $J$ = 5.4 Hz), 4.01 (sext, 1H, CH, $J$ = 5.3 Hz), 3.92 (d, 2H, CH$_2$, $J$ = 5.6 Hz), 3.74 (dd, 1H, CH$_2$, $J_1$ = 5.4 Hz, $J_2$ = 11.8 Hz), 3.66 (dd, 1H, CH$_2$, $J_1$ = 5.4 Hz, $J_2$ = 10.8 Hz), 2.00 (s, 3H, CH$_3$). $^{13}$C NMR (100 MHz, DMSO$_{d6}$): δ 168.1, 154.6, 133.2, 120.9, 114.9, 69.6, 69.1, 47.2, 24.3. HRMS (TOF-ASAP$^+$): [M+H]$^+$ = 244.0739 $m/z$ (calc. mass: 244.0740, C$_{11}$H$_{15}$NO$_3$Cl).

1-((1H-indol-4-yl)oxy)-3-chloropropan-2-ol (**2a**) [17]

1H-Indol-4-ol (0.51 g, 3.80 mmol) was dissolved in 1,4-dioxane (3 mL) and NaOH (0.16 g, 3.93 mmol, 1 eq) was dissolved in water (5 mL) and epichlorohydrin (2.98 mL, 38 mmol) was added. The mixture was stirred at rt for 5 h until TLC showed full conversion of starting material (CH$_2$Cl$_2$, R$_f$ = 0.21). The product was extracted using CH$_2$Cl$_2$ (50 mL) and washed with EtOAc (3 × 30 mL) and water (3 × 30 mL). The CH$_2$Cl$_2$-phase was dried over anhydrous MgSO$_4$ and evaporated under reduced pressure, yielding 0.48 g of a mixture of 1-((1H-indol-4-yl)oxy)-3-chloropropan-2-ol (**2a**) and 4-(oxiran-2-ylmethoxy)-1H-indole (**2e**) as a brown oil.

A mixture of **2a/2e** (0.48 g, 2.56 mmol) was dissolved in THF (8 mL). AcOH (1.46 mL, 25.6 mmol) and LiCl (0.22 g, 5.12 mmol) were added. The mixture was stirred at rt for 72 h. NaCO$_3$ was added until neutral pH was obtained. The product was extracted with CH$_2$Cl$_2$ (50 mL) and washed with satd. NaCl solution (3 × 30 mL). The CH$_2$Cl$_2$ phase was then dried over anhydrous MgSO$_4$ and evaporated. After purification by flash chromatography with CH$_2$Cl$_2$ as eluent, the product was obtained as a slightly yellow oil of 1-((1H-indol-4-yl)oxy)-3-chloropropan-2-ol (**2a**) (0.5701 g, 2.52 mmol, 98.5% yield). Spectroscopic data for **2a**: $^1$H NMR (400 MHz, DMSO$_{d6}$) δ 11.07 (s, 1H, NH), 7.22 (t, 1H, CH, $J$ = 2.6 Hz), 7.02 (m, 2H, Ar-H), 6.47 (t, 1H, CH, $J$ = 2.3 Hz), 6.49 (dd, 1H, Ar-H, $J_1$ = 1.2 Hz, $J_2$ = 7.1 Hz), 5.5 (s, 1H, OH), 4.09 (m, 3H, CH and CH$_2$), 3.84 (dd, 1H, CH, $J_1$ = 5.2 Hz, $J_2$ = 11.3 Hz), 3.75 (dd, 1H, CH, $J_1$ = 4.3 Hz, $J_2$ = 11.3 Hz). $^{13}$C NMR (100 MHz, DMSO$_{d6}$): δ 152.4, 137.9, 124.1, 122.5, 118.9, 105.6, 100.4, 98.7, 69.3, 69.2, 47.5. HRMS (TOF ASAP$^+$): [M+H]$^+$ = 226.0632 $m/z$.

1-((1H-indol-4-yl)oxy)-3-chloropropan-2-yl butanoate (**2b**)

1-((1H-Indol-4-yl)oxy)-3-chloropropan-2-ol (**2a**) (0.08 g, 0.37 mmol) and butyric anhydride (0.075 mL, 0.46 mmol) were added to pyridine (0.05 mL, 0.62 mmol). The mixture was stirred for 24 h at rt. Extraction was performed with hexane and CH$_2$Cl$_2$ and washed with satd. NaCl solution. The organic phase was dried over anhydrous MgSO$_4$ and evaporated, yielding 0.08 g of a mixture of **2a** (84.6%, HPLC) and **2b** (11.0%, HPLC). Separation by flash chromatography using CH$_2$Cl$_2$ as eluent yielded 1% of **2b** (0.80 mg, 0.003 mmol). Spectroscopic data for **2b**: $^1$H NMR (400 MHz, CDCl$_3$): δ 8.22 (s, 1H, NH), 7.03-7.16 (m, 3H, Ar-H and CH), 6.64 (m, 1H, CH), 6.56 (dd, 1H, Ar-H), 4.19-4.37 (m, 3H, CH and CH$_2$), 3.75-3.90 (m, 2H, CH$_2$), 2.37 (td, 2H, CH$_2$, $J_1$ = 11.7 Hz, $J_2$ = 7.33, $J_3$ = 1.34 Hz), 1.69 (sext, 2H, CH$_2$, $J$ = 7.34 Hz), 0.97 (t, 3H, CH$_3$, $J$ = 7.64 Hz). $^{13}$C NMR (100 MHz, CDCl$_3$): δ 151.9, 137.4, 122.9, 122.8, 118.7, 105.2, 101.0, 99.7, 70.1, 68.6, 46.2, 36.4, 18.4, 13.6.

7-(3-Chloro-2-hydroxypropoxy)-3,4-dihydroquinolin-2(1H)-one (**3a**)

To a solution of **3** (1.6541 g, 10.1 mmol) in MeOH (30 mL), NaOH-solution (0.17 M, 30 mL, 0.5 eq) was added. Epichlorohydrin (1.565 mL, 20.0 mmol) was added dropwise to the reaction mixture which was then stirred at rt for 24 h. TLC (CHCl$_3$:CH$_2$Cl$_2$:EtOH, 10:9:1) showed full conversion of **3a** with two products: R$_f$ (**3e**) = 0.31, R$_f$ (**3a**) = 0.41. Insoluble by-products were filtered off. The filtrate was extracted with CH$_2$Cl$_2$ (3 × 20 mL). The organic phase was washed with satd. NaCl solution (2 × 10 mL), dried over anhydrous MgSO$_4$, and filtered before the solvent was removed under reduced pressure. This resulted in white crystals and a yellow, highly viscous oil in a mixture. The mixture was recrystallized from EtOH to yield **3a** as white crystals (1.1074 g, 5.05 mmol, 50% yield). $^1$H NMR (400 MHz, CD$_3$OD): δ 7.08 (m, 1H, Ar-H), 6.59 (m, 1H, Ar-H), 6.49 (m, 1H, Ar-H), 3.84–4.27 (m, 2 H,

CH$_2$), 2.88 (m, 2H, CH$_2$), 2.74–2.88 (m, 2H, CH$_2$), 2.55 (m, 2H, CH$_2$). $^{13}$C NMR (100 MHz, CD$_3$OD): δ 172.7, 158.2, 138.4, 128.3, 116.4, 108.5, 102.0, 68.9, 49.8, 43.5, 30.5, 24.0.

LiCl (1.0440g, 24.6 mmol) and AcOH (2.810 mL, 49.1 mmol) were added to a solution of **3a/3e** (1.0766 g, 4.91 mmol) in MeCN (10 mL). The reaction was stirred at rt and TLC showed full conversion of the starting material after 24 h. The reaction mixture was extracted with CH$_2$Cl$_2$ (3 × 20 mL), dried over anhydrous MgSO$_4$ and the solvent was removed under reduced pressure. **3a** was obtained as white crystals (0.9647 g, 3.77 mmol, 77% yield). $^1$H NMR (400 MHz, CD$_3$OD): δ 7.09 (m, 1H, Ar-H), 6.60 (m, 1H, Ar-H), 6.51 (m, 1H, Ar-H), 4.13 (m, 1H, CH), 4.06 (m, 2H, CH$_2$), 3.69–3.77 (m, 2H, CH$_2$), 2.87–2.90 (m, 2H, CH$_2$), 2.53–2.57 (m, 2H, CH$_2$). $^{13}$C NMR (400 MHz, CD$_3$OD) δ: 172.7, 158.2, 138.4, 128.3, 116.4, 108.5, 102.0, 69.6, 68.9, 45.4, 30.5, 24.0. HRMS (TOF-ASAP$^+$): [M+H]$^+$ = 256.0793 *m/z* (calc. mass: 256.0740, C$_{11}$H$_{15}$NO$_3$Cl).

5-(3-Chloro-2-hydroxypropoxy)-3,4-dihydroquinolin-2(1*H*)-one (**4a**) [26]

Epichlorohydrin (0.134 mL, 1.7 mmol) was added to a solution of **4** (0.1665g, 1.0 mmol) in H$_2$O (0.585 mL) and DMSO (0.375 mL). Aqueous solution of NaOH (9.5 M, 0.090 mL, 0.5 eq) was added to the reaction mixture. The solution was stirred at rt for 24 h where **4a** and **4e** slowly precipitated from the solution. TLC (hexane:*i*-PrOH, 4:1) showed full conversion after 24 h and the product mixture was filtered off, yielding **4a** and **4e** (0.1505 g) as white crystals in a 1:4 ratio, determined by $^1$H NMR. TLC (CHCl$_3$:acetone, 4:1), R$_f$ (**4a**) = 0.32, R$_f$ (**4e**) = 0.46. The mixture of **4e/4a** (1.0979 g, 5.01 mmol) was then dissolved in MeCN (5 mL). LiCl (1.0741 g, 25 mmol) and AcOH (2.865 mL, 50 mmol) were added, and the reaction mixture was stirred for 24 h until TLC showed full conversion of **4e**. Na$_2$CO$_3$-solution (pH 12) was added until the reaction mixture reached pH 7. The product was filtered off and **4a** was obtained as white crystals (0.7807 g, 3.1 mmol, 61% yield). $^1$H NMR (400 MHz, CD$_3$OD) δ 7.01 (m, 1H, Ar-H), 6.56 (m, 1H, Ar-H), 6.41 (m, 1H, Ar-H), 4.05 (m, 1H, CH), 3.98 (m, 2H, CH$_2$), 3.60–3.69 (m, 2H, CH$_2$), 2.86 (t, 2H, CH$_2$, *J* = 7.6 Hz), 2.43 (t, 2H, CH$_2$, *J* = 7.6 Hz). $^{13}$C NMR (100 MHz, CD$_3$OD) δ 172.5, 155.8, 138.6, 127.6, 111.9, 108.5, 106.3, 69.6, 69.0, 45.4, 29.6, 18.1. HRMS (TOF-ASAP$^+$): [M+H]$^+$ = 256.0743 *m/z* (calc. mass: 256.0740, C$_{11}$H$_{15}$NO$_3$Cl).

Practolol (*N*-(4-(2-hydroxy-3-(isopropylamino)propoxy)phenyl)acetamide) (**1c**)

Racemic *N*-(4-(3-chloro-2-hydroxypropoxy)phenyl)acetamide, **1a**, (0.220 g, 0.905 mmol) was dissolved in isopropylamine (0.519 mL, 6.33 mmol, 7.0 eq) and distilled H$_2$O (0.150 mL). The reaction mixture was stirred at rt for 48 h, and was concentrated under reduced pressure. The crude product was recrystallized from MeCN, and *N*-(4-(2-hydroxy-3-(isopropylamino)propoxy)phenyl)acetamide, **1c**, was collected as a white solid (0.115 g, 0.432 mmol, 48% yield). $^1$H-NMR (400 MHz, DMSO$_{d6}$). δ 9.74 (s, 1H, NH), 7.44 (d, 2H, Ar-H, *J* = 10.8 Hz), 6.84 (d, 2H, Ar-H, *J* = 10.8 Hz), 4.91 (d, 1H, OH, *J* = 5.3 Hz), 3.89–3.86 (m, 1H, CH), 3.80 (t, 2H, CH$_2$, *J* = 6.7 Hz), 2.73–2.62 (m, 2H, CH$_2$), 2.53–2.51 (m, 1H, CH), 1.99 (s, 3H, CH$_3$), 1.46 (s, 1H, NH), 0.95 (dd, 6H, 2 × CH$_3$, *J$_1$* = 2.2 Hz, *J$_2$* = 6.2 Hz). $^{13}$C-NMR (150 MHz, DMSO$_{d6}$): δ 167.7 (C = O), 154.4 (Ar-C), 132.4 (Ar-C), 120.4 (2 × Ar-C-H), 114.3 (2 × Ar-C-H), 70.9 (CH), 68.4 (CH$_2$), 50.0 (CH), 48.1 (CH$_2$), 23.7 (CH$_3$), 22.9 (2 × CH$_3$). IR (cm$^{-1}$, diluted): 3343 (m), 2969 (m), 1127 (m), 950 (s), 816 (m). HRMS (TOF-ASAP$^+$): [M+H]$^+$ = 267.1712 *m/z* (calc. mass: 267.1709, C$_{14}$H$_{23}$N$_2$O$_3$).

*3.8. Synthesis of Enantiomers*

Kinetic Resolution of **1a**

Racemic *N*-(4-(3-chloro-2-hydroxypropoxy)phenyl)acetamide, **1a**, (0.543 g, 2.24 mmol) was dissolved in dry MeCN (60 mL). Vinyl butanoate (2.56 g, 22.4 mmol) and molecular sieves (4Å) were added. The reaction was started by adding CALB (1.22 g) and stirred in an incubator shaker (30 °C, 200 rpm). Samples (150 µL) were collected regularly, concentrated, and dissolved in *i*-PrOH before HPLC analyses. After 26 h, the reaction was stopped by filtering off the enzyme and removing the solvent under reduced pressure. The crude product was purified twice by flash chromatography (CH$_2$Cl$_2$:EtOAc, 1:1). (*R*)-**1a** was collected as a white solid (0.208 g, 0.852 mmol, 38% yield, ee = 97%). $[\alpha]_D^{23}$ = −1.0 (c 1.0, i-PrOH). (S)-1-(4-Acetamidophenoxy)-3-chloropropan-2-yl butanoate, (S)-**1b**, was collected as a brown viscous liquid (0.33 g, 1.05 mmol, 47%yield, ee = 84%). $[\alpha]_D^{23}$ = +16.3 (c 1.1, MeCN). (S)-N-(4-(3-Chloro-2-hydroxypropoxy)phenyl)acetamide, (S)-**1a**, was obtained by hydrolysis of (S)-**1b** with CALB: The crude product was purified by flash chromatography (CH$_2$Cl$_2$:EtOAc, 1:1) and (S)-**1a** was collected as a light brown solid (0.0475 g, 0.195 mmol, 26% yield, ee = 81%) $[\alpha]_D^{20}$ = +11.99 (c 1.0, i-PrOH). The E-value was calculated by E&K calculator 2.1b0 PPC, E = 55. The NMR spectra for (S)-**1a** were in correspondence with the spectra for **1a**.

Kinetic Resolution of **2a**

Racemic 1-chloro-3-(1H-indol-4-yloxy)-propan-2-ol (**2a**) (18.5 mg, 0.08 mmol) was dissolved in dry CH$_2$Cl$_2$ (3 mL) and molecular sieves (4Å) were added. Vinyl butanoate (75 µL, 0.59 mmol) and immobilized CALB (36.8 mg) were added, and the reaction was stirred in the incubator shaker for 24 h (30 °C, 200 rpm) to reach 50% conversion. Samples (150 µL) were collected regularly for chiral HPLC analysis. (R)-**2a** was obtained in 92% ee and E = 66. NMR spectra were in accordance with the spectra for **2a**.

Kinetic Resolution of **3a**

Racemic 7-(3-chloro-2-hydroxypropoxy)-3,4-dihydroquinolin-2(1H)-one (**3a**) (0.7492 g, 2.93 mmol) was dissolved in dry MeCN (60 ml) and molecular sieves (4Å) were added. Vinyl butanoate (1.672 mg, 1.860 mL, 14.7 mmol) was added to the solution and the reaction was started by adding CALB (1.3724 g) and placing the container in an incubator shaker (30 °C, 200 rpm). The reaction was stopped after 24 h by filtering off CALB and the molecular sieves. The solvent was removed under reduced pressure yielding the mixture of ester (S)-**3a** and chlorohydrin (R)-**3a** as a brown oil, which were separated by flash chromatography (EtOAc:hexane:MeOH, 7:12:1). Chlorohydrin (R)-**3a** was isolated as white crystals (0.2573 g, 34% yield, ee = 96%), $[\alpha]_D^{20}$ = −9.9 (c 1.0, MeOH). Ester (S)-**3b** was obtained as a yellow oil (0.4436 g, 51% yield, 91% purity, 86% ee). $[\alpha]_D^{20}$ = +8.0 (c 1.0, MeOH). HPLC, eluent: hexane: i-PrOH:DEA, 90:10:0.4, $t_R$ (S)-**3b** = 56.30 min, $t_R$ (R)-**3b** = 63.47 min. E = 157. $^1$H NMR (R)-**3a** (600 MHz, CD$_3$OD): δ 7.06 (m, 1H, Ar-H), 6.56 (m, 1H, Ar-H), 6.48 (m, 1H, Ar-H), 5.33 (p, 1H, CH, J = 5.0 Hz), 4.15 (d, 2H, CH$_2$, J = 4.9 Hz), 3.83–3.86 (m, 2H, CH$_2$), 2.84–2.87 (m, 2H, CH$_2$), 2.52–2.54 (m, 2H, CH$_2$), 2.35 (td, 2H, CH$_2$, J = 7.4 Hz), 1.65 (sext, 2H, CH$_2$, $J_1$ = 7.4 Hz, $J_2$ = 15.0 Hz), 0.96 (td, 3H, CH$_3$, $J_1$ = 7.4 Hz, $J_2$ = 15.0 Hz). $^{13}$C NMR (150 MHz, CD$_3$OD); δ 175.8, 172.3, 157.6, 138.2, 128.1, 116.3, 108.2, 101.8, 70.8, 66.1, 48.2, 35.2, 30.1, 23.7, 16.1, 12.3. $^1$H NMR (S)-**3b** (600 MHz, CD$_3$OD). δ 7.06 (m, 1H, Ar-H), 6.56 (m, 1H, Ar-H) 6.48 (m, 1H, Ar-H), 5.33 (p, 1H, CH), 4.15 (d, 2H, CH$_2$, $J_1$ = 4.14 Hz), 3.83–3.86 (m, 2H, CH$_2$-Cl), 2.85 (t, 2H, CH$_2$, J = 7.10 Hz), 2.53 (t, 2H, CH$_2$, J = 7.60 Hz), 2.35 (t, 2H, CH$_2$, J = 7.10 Hz), 1.65 (sext, 2H, CH$_2$, J = 7.10 Hz), 0.96 (t, 3H, CH$_3$, J = 8.0 Hz).

Kinetic Resolution of **4a**

Racemic 5-(3-chloro-2-hydroxypropoxy)-3,4-dihydroquinolin-2(1H)-one (**4a**) (0.1724 g, 0.67 mmol) was dissolved in dry MeCN (20 mL) and molecular sieves (4Å) were added. Vinyl butanoate (0.428 mL, 3.75 mmol) was added to the solution and the reaction was started by adding CALB (0.6745 g) and placing the container in an incubator shaker (37 °C, 200 rpm). The reaction was stopped after 74 h by filtering off CALB and the molecular sieves. The solvent was removed under reduced pressure, yielding a mixture of ester (S)-**4b** and chlorohydrin (R)-**4a** as a brown oil, which were separated by flash column chromatography (EtOAc:hexane, 7:3). Chlorohydrin (R)-**4a** was isolated as white crystals (0.0652 g, 0.2549 mmol, 38% yield, ee = 96%), $[\alpha]_D^{20}$ = −9.9 (c 1.0, DMSO). HPLC eluent: hexane:EtOH:TFA (90:10:0.1), 0.4 mL/min. $t_R$(S)-**4a** = 64.10 min, $t_R$(R)-**4a** = 69.48 min, R$_s$ = 1.70. Ester (S)-**4b** was obtained as a colorless oil (0.0947 g, 0.029 mmol, 43% yield, 79% purity, 77% ee). $[\alpha]_D^{20}$ = +9.9 (c 1.0, DMSO). HPLC eluent: hexane:iPrOH:DEA, 95:5:0.4, 0.4 mL/min. $t_R$(S)-**4b** = 110.32 min, $t_R$(R)-**4b** = 118.43 min, R$_s$ = 1.65, E = 31. $^1$H NMR

(R)-**4a** (400 MHz, DMSO$_{d6}$). δ 10.02 (s, 1H, NH), 7.07 (t, 1H, Ar-H, J = 8.1 Hz), 6.59–6.61 (d, 1H, Ar-H, J = 8.2 Hz), 6.48–6.50 (d, 1H, Ar-H, J = 7.9 Hz), 5.57 (s, 1H, OH), 4.03 (m, 1H, CH), 3.96–3.97 (m, 2H, CH$_2$), 3.75–3.79 (dd, 1H, CH$_2$, J$_1$ = 4.6 Hz, J$_2$ = 11.1 Hz), 3.66–3.70 (dd, 1H, CH$_2$, J$_1$ = 5.5 Hz, J$_2$ = 11.1 Hz), 2.82 (t, 2H, CH$_2$, J = 7.7 Hz), 2.40 (t, 2H, CH$_2$, J = 7.7 Hz). $^{13}$C NMR (100 MHz, DMSO$_{d6}$): 170.5, 155.9, 139.8, 128.1, 111.7, 108.8, 106.2, 69.6, 69.1, 47.3, 30.3, 18.7. (0.0947 g, 0.29 mmol, 43% yield, ee = 77%). $[\alpha]_D^{20} = +9.99$ (c 1.0, DMSO) HPLC, eluent: hexane: i-PrOH:DEA, 95:5:0.4, 0.4 mL/min t$_R$(S)-**4b** = 110.32 min, t$_R$(R)-**4b** = 118.43 min, R$_s$ = 1.65. $^1$H NMR (S)-**4b** (400 MHz, DMSO$_{d6}$). δ 10.04 (s, 1H), 7.08 (t, 1H, Ar-H, J = 8.12 Hz), 6.60–6.62 (d, 1H, Ar-H, J = 8.01 Hz), 6.49–6.51 (d, 1H, Ar-H, J = 7.84 Hz), 5.35 (m, 1H, CH), 4.17–4.21 (dd, 1H, CH$_2$, J$_1$ = 4.14 Hz, J$_2$ = 10.62 Hz), 4.11–4.15 (dd, 1H, CH$_2$, J$_1$ = 6.16 Hz, J$_2$ = 10.60 Hz), 3.94–3.98 (dd, 1H, CH$_2$, J$_1$ = 4.16 Hz, J$_2$ = 11.84 Hz), 3.86–3.91(dd, 1H, CH$_2$, J$_1$ = 6.50 Hz, J$_2$ = 11.82 Hz), 2.77 (t, 2H, CH$_2$, J = 7.67 Hz), 2.39 (m, 2H, CH$_2$), 2.35–2.31 (m, 2H, CH$_2$), 1.61–1.51 (sext, 2H, CH$_2$, J = 7.32 Hz), 0.89 (t, 3H, CH$_3$, J = 7.39 Hz). $^{13}$C NMR (100 MHz, DMSO$_{d6}$): 172.6, 170.5, 155.5, 139.9, 128.2, 111.8, 109.1, 106.3, 71.1, 67.4, 43.8, 35.8, 30.2, 18.6, 18.4, 13.8.

### 3.9. Enantiopure Drug Derivatives

(S)-Practolol, (S)-N-(4-(2-hydroxy-3-(isopropylamino)propoxy)phenyl)acetamide, (S)-**1c**

(R)-N-(4-(3-chloro-2-hydroxypropoxy)phenyl)acetamide, (R)-**1a**, (0.175 g, 0.719 mmol) was dissolved in i-PrNH$_2$ (0.470 mL, 5.73 mmol, 8.0 equiv) and distilled H$_2$O (0.075 mL). The reaction mixture was stirred at rt for 96 h, and the solvent was removed under reduced pressure. The crude product was recrystallized from MeCN and (S)-N-(4-(2-hydroxy-3-(isopropylamino)propoxy)phenyl)acetamide ((S)-**1c**) was collected as a white solid (0.0313 g, 0.117 mmol, 16% yield, ee = 96%). Optical rotation of (S)-**1c**: $[\alpha]_D^{20} = -3.998°$ (c 1.0, EtOH). mp: 124.7–124.9 °C (lit. 130–131 °C) [25]. $^1$H NMR (600 MHz, DMSO$_{d6}$): δ 9.75 (s, 1H, NH), 7.45–7.44 (d, Ar-H, J = 8.4 Hz), 6.85–6.84 (d, Ar-H, J = 8.4 Hz), 4.95 (s, 1H, OH), 3.90–3.87 (m, 1H, CH), 3.81–3.80 (d, 2H, CH$_2$, J = 6.8 Hz), 2.72–2.65 (m, 2H, CH$_2$), 2.55–2.52 (m, 1H, CH), 2.00 (s, 3H, CH$_3$), 0.97 (dd, 6H, CH$_3$, J$_1$ = 2.0 Hz, J$_2$ = 6.3 Hz). $^{13}$C NMR (150 MHz, DMSO$_{d6}$): δ 167.7, 154.4, 132.5, 120.4, 114.4, 70.9, 68.3, 49.9, 48.2, 23.7, 22.7. IR (cm$^{-1}$, diluted): 3343 (m), 2969 (m), 1127 (m), 950 (s), 816 (m). HRMS (TOF-ASAP$^+$): [M+H]$^+$ = 267.1711 m/z (calc. mass [M+H]$^+$ = 267.1709, C$_{14}$H$_{23}$N$_2$O$_3$).

(S)-7-(3-(tert-Butylamino)-2-hydroxypropoxy)-3,4-dihydroquinolin-2(1H)-one (S)-**3c**

Chlorohydrin (R)-**3a** (0.2573, 1.0063 mmol) was dissolved in t-BuNH$_2$ (15.4 mL, 147 mmol, 147 equiv) and H$_2$O (4.6 mL) and stirred at rt for 8 h. t-BuNH$_2$ and H$_2$O were removed under reduced pressure, yielding the crude product as a mixture of a yellow oil and white crystals. The crude product was purified by flash chromatography (EtOAc:hexane:MeOH:TEA, 80:7:10:3), yielding (S)-**3c** as a yellow oil (0.2419 g, 0.9460 mmol, 82% yield, 89% purity (NMR), ee = 97%). $[\alpha]_D^{20} = -16.0$ (c = 1.0, MeOH). $^1$H NMR (600 MHz, CD$_3$OD) δ 7.08 (m, 1H, Ar-H), 6.59 (m, 1H, Ar-H), 6.50 (m, 1H, Ar-H), 4.02 (m, 1H, CH), 3.95 (m, 2H, CH$_2$), 2.88 (t, 2H, CH$_2$, J = 7.6) 2.73–2.81 (m, 2H, CH$_2$), 2.54 (t, 2H, CH$_2$, J = 7.6 Hz), 1.18 (s, 9H, C(CH$_3$)$_3$). $^{13}$C NMR (150 MHz, CD$_3$OD) δ 172.6, 158.3, 138.3, 128.1, 116.1, 108.4, 101.9, 70.8, 68.7, 50.4, 44.8, 30.7, 27.1, 24.2.

(S)-5-(3-(tert-Butylamino)-2-hydroxypropoxy)-3,4-dihydroquinolin-2(1H)-one (S)-**4c**

Chlorohydrin (R)-**4a** (0.0652 g, 0.2549 mmol) was dissolved in t-BuNH$_2$ (3.75 mL, 37.75 mmol) and H$_2$O (1.0 mL) and stirred at rt for 10 h. t-BuNH$_2$ and H$_2$O were removed under reduced pressure, yielding the crude product which was purified by flash chromatography (EtOAc:hexane:MeOH:TEA, 80:7:10:3), yielding (S)-**4c** as a yellow oil (0.0521 g, 0.1784 mmol, 70% yield, ee = 96%). $^1$H NMR (600 MHz, DMSO$_{d6}$) δ 10.02 (s, 1H, NH), 7.08 (m, 1H, Ar-H), 6.59 (m, 1H, Ar-H), 6.50 (m, 1H, Ar-H), 5.54 (s, 1H, NH), 5.37 (s, 1H, OH) 4.03 (m, 1H, CH), 3.95–4.05 (m, 2H, CH$_2$), 3.39 (t, 2H, CH$_2$, J = 7.6 Hz), 2.73–2.81 (m, 2H, CH$_2$), 2.54 (t, 2H, CH$_2$, J = 7.6 Hz), 1.21 (s, 9H, C(CH$_3$)$_3$). $^{13}$C NMR (150 MHz, DMSO$_{d6}$) δ 172.6, 158.3, 138.3, 128.1, 116.1, 108.4, 101.9, 70.8, 68.7, 58.2, 44.8, 30.7, 27.1 (3C), 20.4.

## 4. Conclusions

The *S*-enantiomers of practolol, carteolol and a carteolol derivative were produced with *ee*'s of > 96% from the enantiopure chlorohydrins from the CALB-catalyzed kinetic resolutions with preservation of the *ee*. The remaining enantiomer (in the hereby reported cases, the *S*-esters) may be converted to the wanted enantiomer by dynamic kinetic resolution, which we do not report here. The syntheses of the chlorohydrins **1a-4a** have been optimized with reduced amount of base, lowering of reaction temperature and lowering of reaction time compared to previously reported methods, giving moderate to high total yields. With the use of a catalytic amount of base and shorter reaction time, the formation of by-products was reduced. We propose a mechanism for the regeneration of the base in these reactions. We struggled to reproduce the synthesis of the enantiopure chlorohydrin as precursor for enantiopure carteolol, however, we have now obtained 96% *ee*, which we are in the process of improving. Absolute configurations of the produced enantiomers were determined based on both optical rotation values and comparison of CALB preference for one stereoisomer of similar secondary alcohols. Determination of absolute configuration is of utmost importance when reporting data of enantiopure compounds. We report here several enantiopure compounds which have not been reported previously.

**Supplementary Materials:** The following are available online at https://www.mdpi.com/article/10.3390/catal11040503/s1, $^1$H and $^{13}$C NMR spectra, MS and IR spectra and relevant chiral HPLC chromatograms.

**Author Contributions:** Investigation, writing, original draft preparation, E.E.J.; supervision and writing, review and editing, E.E.J.; investigation and partly writing of manuscript M.A.G., G.B.A., S.S.L., M.B.H., M.R. All authors have read and agreed to the published version of the manuscript.

**Funding:** This research received no external funding.

**Data Availability Statement:** The data presented in this study are available online.

**Acknowledgments:** EEA project 18-COP-0041 GreenCAM is thanked for support, SyncoZymes Co LTD, Shanghai, China is thanked for gift of CALB.

**Conflicts of Interest:** The authors declare no conflict of interest.

## References

1. Pfeiffer, C.C. Optical isomerism and pharmacological action, a generalization. *Science* **1956**, *124*, 29–31. [CrossRef] [PubMed]
2. Lehmann, P.A.; Rodriegues de Miranda, J.F.; Ariëns, E.J. Stereoselectivity and Affinity in Molecular Pharmacology. In *Drug Research/Fortschritte der Arzneimittelforschung/Progrés des recherches pharmaceutiques*; Birkhäuser: Basel, Switzerland, 1976; Volume 20, pp. 101–142. ISBN 978-3-0348-7094-8.
3. Causes of Death, Norway. 2019. Available online: http://statistikkbank.fhi.no/dar/ (accessed on 27 February 2021).
4. Ponikowski, P.; Anker, S.D.; AlHabib, K.F.; Cowie, M.R.; Force, T.L.; Hu, S.; Jaarsma, T.; Krum, H.; Rastogi, V.; Rohde, L.E.; et al. Heart failure: Preventing disease and death worldwide. *ESCHeart Fail.* **2014**, *1*, 4–25. [CrossRef] [PubMed]
5. Klouman, M.; Åsberg, A.; Widerøe, T.-E. The blood pressure level in a Norwegian population—The significance of inheritance and lifestyle. *Tidsskr. Nor. Laegeforen.* **2011**, *131*, 1185–1189. [CrossRef] [PubMed]
6. Patrick, G.L. *An Introduction to Medicinal Chemistry*, 5th ed.; Oxford University Press: Oxford, UK, 2013; p. 814.
7. Lund, I.T.; Bøckmann, P.L.; Jacobsen, E.E. Highly enantioselective CALB catalyzed kinetic resolution of building blocks for—Blocker atenolol. *Tetrahedron* **2016**, *72*, 7288–7292. [CrossRef]
8. Jacobsen, E.E.; Anthonsen., T. Single Enantiomers from Racemates. Lipase catalysed Kinetic Resolution of Secondary Alcohols. In situ Stereoinversion. *Trends Org. Chem.* **2017**, *18*, 71–83.
9. Blindheim, F.H.; Hansen, M.B.; Evjen, S.; Zhu, W.; Jacobsen, E.E. Chemo-Enzymatic Synthesis of Enantiopure Synthons as Precursors for (*R*)-Clenbuterol and other β2- agonists. *Catalysts* **2018**, *8*, 516. [CrossRef]
10. Andersen, S.; Refsum, H.; Tanum, L. Kirale legemidler. *Tidsskr. Nor. Laegeforen.* **2003**, *123*, 2055–2056. [PubMed]
11. Barrett, A.M.; Carter, J.; Fitzgerald, J.D.; Hull, R.; Le Count, D. A new type of cardioselective adrenoceptive blocking drug. *Br. J. Pharmacol.* **1973**, *48*, 340.
12. Pribble, A.H.; Conn, R.D. The use of practolol in supraventricular arrhythmias associated with acute illnesses. *Am. J. Cardiol. Heart.* **1975**, *35*, 645–650. [CrossRef]
13. Kahn., G. *Cardiac Drug Therapy*, 8th ed.; Humana Press: Heildeberg, Germany, 2015; p. 769.
14. Mulik, S.; Ghosh, S.; Bhaumik, J.; Banerjee, U.C. Biocatalytic synthesis of (*S*)-Practolol, a selective β-blocker. *Biocatalysis* **2016**, *1*, 130–140. [CrossRef]

15. Ader, U.; Schneider, M.P. Enzyme assisted preparation of enantiomerically pure β-adrenergic blockers III. Optically active chlorohydrin derivatives and their conversion. *Tetrahedron Asymmetry* **1992**, *3*, 521–524. [CrossRef]
16. Koch-Weser, J.; Frishman, W.H. Pindolol: A New β-Adrenoceptor Antagonist with Partial Agonist Activity. *N. Engl. J. Med.* **1983**, *308*, 940–944. [CrossRef]
17. Lima, G.V.; Da Silva, M.R.; de Sousa Fonseca, T.; de Lima, L.B.; de Oliveira, M.D.C.F.; de Lemos, T.L.G.; Zampieri, D.; Dos Santos, J.C.S.; Rios, N.S.; Gonçalves, L.R.B.; et al. Chemoenzymatic synthesis of (S)-Pindolol using lipases. *Appl. Catal. A Gen.* **2017**, *546*, 7–14. [CrossRef]
18. Liu, Y.; Zhou, X.; Zhu, D.; Chen, J.; Qin, B.; Zhang, Y.; Wang, X.; Yang, D.; Meng, H.; Luo, Q.; et al. Is pindolol augmentation effective in depressed patients resistant to selective serotonin reuptake inhibitors? A systematic review and meta-analysis. *Hum. Psychopharmacol.* **2015**, *30*, 132–142. [CrossRef] [PubMed]
19. Watson, P.G.; Barnett, M.F.; Parker, V.; Haybittle, J. A 7 year prospective comparative study of three topical β blockers in the management of primary open angle glaucoma. *Brit. J. Ophthalmol.* **2001**, *85*, 962. [CrossRef] [PubMed]
20. Schuster, A.K.; Erb, C.; Hoffmann, E.M.; Dietlein, T.; Pfeiffer, N. The Diagnosis and Treatment of Glaucoma. *Dtsch. Arztebl. Int.* **2020**, *117*, 225–234. [PubMed]
21. Norwegian Prescription Database. Available online: http://www.reseptregisteret.no/default.aspx (accessed on 27 February 2021).
22. Jacobsen, E.E.; Hoff, B.H.; Anthonsen, T. Enantiopure derivatives of 1,2-alkanediols. Substrate requirements for lipase B from Candida antarctica. *Chirality* **2000**, *12*, 654–659. [CrossRef]
23. Jacobsen, E.E.; Anthonsen, T.; el-Behairy, M.F.; Sundby, E.; Aboul-Enein, M.N.; Attia, M.I.; El-Azzouny, A.A.E.S.; Amin, K.M.; Abdel-Rehim, M. Lipase Catalysed Kinetic Resolution of Stiripentol. *Int. J. Chem.* **2012**, *4*, 7–13.
24. Anthonsen, H.W.; Hoff, B.H.; Anthonsen, T. Calculation of enantiomer ratio and equilibrium constants in biocatalytic pingpong bi-bi resolutions. *Tetrahedron Asymmetry* **1996**, *7*, 2633–2638. [CrossRef]
25. Danilewicz, J.C.; Kemp, J.E.G. Absolute Configuration by Asymmetric Synthesis of (+)-1-(4-Acetamidophenoxy)-3-(isopropylamino)-propan-2-ol (Practolol). *J. Med. Chem.* **1973**, *16*, 168–169. [CrossRef] [PubMed]
26. Manghisi, E.; Perego, B.; Salmibeni, A. A Process for the Preparation of 5-(3-Tert-Butylamino-2-Hydroxypropoxy)-3,4-Dihydrocarbostyryl. European Patent EP0579096 (A2), 19 January 1994.

MDPI
St. Alban-Anlage 66
4052 Basel
Switzerland
Tel. +41 61 683 77 34
Fax +41 61 302 89 18
www.mdpi.com

*Catalysts* Editorial Office
E-mail: catalysts@mdpi.com
www.mdpi.com/journal/catalysts

www.ingramcontent.com/pod-product-compliance
Lightning Source LLC
LaVergne TN
LVHW070637100526
838202LV00012B/828